ADVANCES IN
ORGAN BIOLOGY

Volume 10 • 2006

THE BIOLOGY OF THE EYE

ADVANCES IN ORGAN BIOLOGY
E. Edward Bittar, Series Editor

ADVANCES IN ORGAN BIOLOGY

THE BIOLOGY OF THE EYE

Edited by: JORGE FISCHBARG

Lazlo Z. Bito Professor of Physiology and Cellular Biophysics
Columbia University
New York, NY
USA

VOLUME 10 • 2006

2006

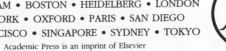

AMSTERDAM • BOSTON • HEIDELBERG • LONDON
NEW YORK • OXFORD • PARIS • SAN DIEGO
SAN FRANCISCO • SINGAPORE • SYDNEY • TOKYO
Academic Press is an imprint of Elsevier

ELSEVIER B.V.
Radarweg 29
P.O. Box 211, 1000 AE
Amsterdam, The Netherlands

ELSEVIER Inc.
525 B Street
Suite 1900, San Diego
CA 92101-4495, USA

ELSEVIER Ltd.
The Boulevard
Langford Lane, Kidlington
Oxford OX5 1GB, UK

ELSEVIER Ltd.
84 Theobalds Road
London WC1X 8RR
UK

First edition 2006

Library of Congress Cataloging in Publication Data
A catalog record is available from the Library of Congress.

British Library Cataloguing in Publication Data
A catalogue record is available from the British Library.

ISBN: 0-444-50925-9

∞ The paper used in this publication meets the requirements of ANSI/NISO Z39.48-1992 (Permanence of Paper).
Transferred to digital print 2007
Printed and bound by CPI Antony Rowe, Eastbourne

CONTENTS

LIST OF CONTRIBUTORS

Banmeet S. Anand

Missouri-Kansas City
Kansas City, Missouri

Darlene A. Dartt

Schepens Eye Research Institute
Boston, Massachusetts

Nicholas A. Delamere

KY Lions Eye Research Institute
University of Louisville School of
Medicine
Louisville, Kentucky

Sridhar Duvvuri

Missouri-Kansas City
Kansas City, Missouri

Bernd Ehinger

Department of Ophthalmology
University of Lund
Sweden

Niels Ehlers

Department of Ophthalmology
University of Arhus
Arhus, Denmark

Jorge Fischbarg

Laszlo Z. Bito Professor of Physiology
and Cellular Biophysics in
Ophthalmology
College of Physicians and Surgeons
Columbia University
New York, New York

Jesper Hjortdal

Department of Ophthalmology
University of Arhus
Arhus, Denmark

Robin R. Hodges   Schepens Eye Research Institute
                                   Boston, Massachusetts

Peter Koch Jensen   Department of Ophthalmology
                                   Rigshospitalet, Copenhagen
                                   Denmark

Jens Folke Kiilgaard   Department of Ophthalmology
                                   Rigshospitalet, Copenhagen
                                   Denmark

Morten la Cour   Department of Ophthalmology
                                   Herlev University Hospital
                                   Denmark

Larry S. Liebovitch   Professor and interim Director
                                   Center for Complex Systems and
                                   Brain Sciences
                                   Florida Atlantic University
                                   Boca Raton, Florida

Henrik Lund-Andersen   Department of Ophthalmology
                                   University of Copenhagen
                                   Herlev Hospital, Denmark

Ashim K. Mitra   Division of Pharmaceutical Sciences
                                   School of Pharmacy
                                   University of Missouri-Kansas city
                                   Kansas City, Missouri, USA

Mogens Holst Nissen   Institute of Medical Anatomy
                                   The Panum Institute
                                   University of Copenhagen
                                   Copenhagen, Denmark

Carsten Röpke   Institute of Medical Anatomy
                                   The Panum Institute
                                   University of Copenhagen
                                   Copenhagen, Denmark

Birgit Sander   Department of Ophthalmology
                                   University of Copenhagen
                                   Denmark

List of Contributors

J. Sebag	Professor of Clinical Ophthalmology Doheny Eye Institute University of Southern California and VMR Institute, California, USA
Tongalp Tezel	Assistant Professor of Ophthalmology and Visual Sciences University of Louisville, School of Medicine Kentucky Lions Eye Center Louisville, Kentucky USA
Klaus Trier	Trier Eye Clinic and Research Lab Hellerup, Denmark
Guido A. Zampighi	Departments of Neurobiology and Physiology David Geffen School of Medicine Los Angeles California
Driss Zoukhri	Schepens Eye Research Institute Boston, Massachusetts

PREFACE

The invitation to edit this book came from Dr. E. Edward Bittar, an excellent colleague and friend. Along with the invitation came exceedingly useful suggestions on the contents and format, for which Dr. Bittar is to be amply credited, as well as for his mentoring this author and advising on steps to move the book along during difficult periods.

This preface is the only place in which the Editors seemingly can thank the authors in public for the time and dedication they have spent for this volume. Seeing how much knowledge has been distilled into crisp text, one can feel the dedication and love for their field the contributors have. Partly as a consequence of the seas of acquaintances the editors navigate in, a large number of the chapters have been written by Scandinavians. That seems quite natural these days; that part of the World distinguishes itself in love for Academia and tradition in ophthalmic sciences. For them as well as the other authors, in this sampling the pleasure of imparting knowledge continues to be an important driving force in our world, which should be a sign of hope.

Editing a book of this sort presents a quandary: given two extremes, one can try to run a regimented production along narrow lanes, or can allow the authors latitude for them to write as they see fit. In this case, the second option carried the day hands down. The original instructions asked the authors to think of a potential audience of newcomers to the field trying to discover in relatively simple terms what is known about the eye, which are the areas receiving most of the attention, and where the excitement of crossing the border and venturing into the unknown lies in every case. I think the authors have responded brilliantly. Each one in his own style; some wrote short accounts, others wrote wonderful comprehensive reviews. The extension of each chapter in some way gives a measure of the width of the currents crossing that field, and of the complexities that the author felt compelled to communicate. The Editors have prudently opted for stepping aside and letting that be.

The subjects covered are an indication of the growth and diversification of areas of interest in the eye. Years ago, books in this area took great care in covering the anatomy, and justifiably so, as the future ocular surgeons needed to start their careers with the best of directions about the area they would be operating on. As the basic sciences progressed, books included growing sections on the functionality of the different organs. We have of course kept the anatomical separation of subjects and the functional descriptions. However, we have also chosen to add a chapter on the shape of the eye, in a way perhaps acknowledging that ophthalmology has begun modifying the corneal shape, and has begun asking what would it take to give the eye optimal shape for image formation. The chapters on drug delivery and immunology respond to the same activist approach, one in which as we learn the basics of the eye we learn as well about ocular characteristics that allow intervention or explain pathology. In addition, although this book addresses basic mechanisms, the chapters contain mentions to pathology and disease wherever these subjects arise naturally.

Speaking for the Editors, we have greatly enjoyed the reading of these materials. I hope the same intellectual fulfillment will be now felt by the readership.

New York, February 2005   Jorge Fischbarg

WHY THE EYE IS ROUND

Larry S. Liebovitch

Advances in Organ Biology
Volume 10, pages 1–19.
© 2006 Elsevier B.V. All rights reserved.
ISBN: 0-444-50925-9
DOI: 10.1016/S1569-2590(05)10001-9

ABSTRACT

An impressive characteristic about eyes is their round, spherical structure. This chapter explores the optical, mechanical, structural, phylogenic, and ontogenic reasons why eyes are round. This exploration is used as a starting point to describe how the different features of the eye are related to each other, and how the roundness is maintained by the inflow and outflow of fluid in the eye.

I. INTRODUCTION

If you look up into the night sky at the constellation of the Big Dipper and have 20/30 or better visual acuity and adequate night vision, you will see that the next-to-the-last-star in the handle of the dipper is actually two stars that are quite close together. One star is brighter than the other. The brighter star is called *Mizar* and the fainter *Alcor*. It is easy to fall into the trap described by the ancient Arabic proverb that, "He sees Alcor, but not the full moon." The lesson here is that the most outstanding fact about eyes is not something arcane, but the obvious fact that eyes are round (i.e., eyes are spheres). Therefore, this first chapter will focus on the fact that eyes are round. Why should eyes be round? What does it tell us about how eyes are constructed and how they work? Not only is this shape similar in different animals, but the variation in size of the vertebrate eye, from tree shrew to whale, is much smaller in proportion than the variation in size of these creatures. It will also be described how different features of the eye (Figure 1) are related to each other, and how the roundness is maintained and controlled by the formation, flow, and removal of fluid in the eye.

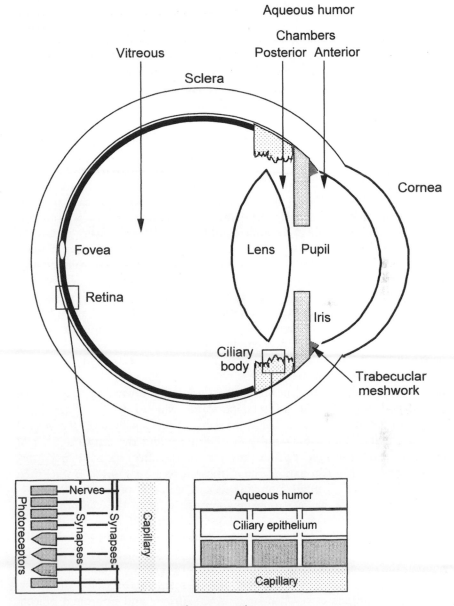

Figure 1. The eye.

II. WHY ARE THINGS ROUND?

When I first thought about the roundness of eyes, I realized I did not know why anything was round. So, I made a list of other round objects to help organize my thought process. My list consisted of the sun, the earth, the moon, oranges, frog urinary bladders, basketballs, and rocks. As you can see, the list consists of organic animate and inorganic inanimate objects (Volk, 1985).

A. Inanimate Objects

Before starting with the inanimate objects on the list, the understanding of the concept of *equilibrium* is needed. Consider your textbook, unopened, on a desk. Even though it is static, there are at least two forces at work, making it that way. It is actually in dynamic equilibrium, subject at every instant, to opposing forces, which balance it. Gravity is pulling the book down toward the center of the earth. The desk is pushing it up, preventing it from moving. All objects that appear static are actually in this balancing act of opposing forces. If one of the forces were stronger, it would change the object rapidly, until an opposing force balanced it, and then the object would again be at a new equilibrium. Objects change so rapidly when out of equilibrium that we are not likely to catch sight of them during that time.

What forces are balancing in these inanimate objects? How do those forces determine the shapes of these objects? In the sun, gravity pulls the gases of the sun together, pushing all its material toward its center. The inward pull of gravity raises the temperature, which raises the pressure of the gas in the sun until the outward pressure of the gas balances the inward pull of gravity. Both the inward pull of gravity and the outward push of gas pressure are *isotropic*. That is, they are equally effective in all directions. That is why the sun is round. If one of these forces were not isotropic, then the sun would not be round. Sometimes there are other pressures. If a star is rapidly rotating, or has a strong magnetic field, then the gas pressure is weaker along that axis. The gas collapses along that axis, and the star becomes a flattened disk. The weaker pressure along the axis balances the weaker gravitational force of the thin mass in the thickness of the disk, whereas the stronger pressure along the radius of the disk balances the larger gravitational force of the larger amount of mass in the radial direction. Thus, round objects exist when forces are *isotropic* and nonround objects when forces are not isotropic.

In the earth, the gravitational force pushing inward is balanced by the outward push of the strength of the rocks, a result of the push of electrons against each other in adjacent atoms. Both these forces are isotropic, and so the earth is round. In a basketball, the air pressure pushing outward is balanced by the tension on the fabric pushing inward. Again, both these forces are isotropic and so the basketball is round.

B. Animate Objects

In inanimate objects, a round configuration results from a balance of *isotropic* forces (i.e., forces experienced equally in all directions). But what determines the shapes of living things? The zoologist and classical scholar, D'Arcy Thompson attempted to answer this puzzling phenomena in his book, "*On Growth and Form*" first published in 1917 (Thompson, 1966). Although you may not be familiar with his publication, there is a good chance that you have seen reproductions of his drawings. His exquisite illustrations of forms of radiolaria, or how the shapes of animals change from one species to another have been prolifically copied. The seminal point of Thompson's book was that genes do not set the blueprint of the shape of an organism, but they set the rules of how the organism interacts with its environment. It is then this dynamic interaction between the organism and its environment that produces the structure.

For example, the *final* shape of the long bones in the arms and legs is dependent on forces between osseous cells and the forces of their environment. Since bone is alive, material is constantly being added and removed from biochemical reactions by cells within the bone. When a bone is bent, fluid flows inside the bone. The negative and positive ions in this fluid flow at different rates generating an electrical voltage. This voltage affects the cells in the bone, so that their enzymes add more calcium on the electrically negative side of the bend and remove more calcium on the electrically positive side of the bend. As a result, the bone is resculpted into a straighter shape. Bone is very strong at resisting compressive forces pushing inward on both ends. It is weak at resisting tensile forces pulling outward from both ends. The resculpting adds material where the bone is in compression and more material is needed. It removes material where the bone is in tension and excess material is wasted. Thus, the genes, through their complex programming of cells and their enzymes, have set the rule: add material where it is needed and remove material where it is not needed. The genes have set the rule of how the bone interacts with the environment. That rule and its interaction with the environment then generate the straight shape of the bone.

Such interactions also sculpt the eye and its surrounding tissues. In congenital glaucoma, the increased pressure in the eye stimulates the entire eye to develop to a larger size than normal. When an eye with retinal blastoma has to be enucleated at an early age to prevent cancer from spreading, the bones of that orbit do not grow as large as the other orbit, because the pressure of the eye is needed to stimulate their normal growth.

For the living things in my list, how much shape is determined solely by the genes and how much by the rules of interaction with the environment set by the genes? I have my own guesses about oranges and frog urinary bladders. What are your guesses? To answer these questions you must ask yourself, "What forces are balancing to determine the shape?" and "What is the mechanism of feedback between the world and the tissue?"

III.   WHY ARE EYES ROUND?

A. Optical Properties

My first guess was that since the most important function of the eye is to form our image of the world, there must be an optical reason why eyes are round.

The eye focuses light onto the retina. Most people think that this focusing is performed by the lens in the eye. However, light is bent most sharply when it passes through an interface of materials of different refractive indices. In the eye, the difference in refractive index is much larger at the air–tissue interface of the cornea (the clear front surface of the eye), than at the fluid–tissue–fluid interface of the lens. Thus, two-third of the focusing of light is done by the cornea and only one-third by the lens. The lens does the fine-tuning of the focusing of the image. The cornea controls the overall quality of the image. It is problems of the cornea that produce nearsightedness, farsightedness, or astigmatism that can be corrected by glasses or contact lenses.

Is the eye round to achieve the best optical image on the retina, where the light is detected and transformed into electrical signals? There are a number of different aberrations, ways in which the focus of images on the retina are not perfect. Important deviations include *spherical aberration* (where a light ray in the center of the cornea reaches a focus that is closer to the cornea than a ray at the periphery of the cornea), and *chromatic aberration* (where a ray of blue light reaches a focus that is closer to the cornea rather than a ray of red light). Another aberration is that the cornea focuses images onto a spherical surface rather than a flat surface. Moreover, this spherical surface has a different radius for vertical and horizontal images on the cornea. The retina of the eye is a spherical surface whose radius is a good compromise between those two different radii. This looks like a good reason why eyes might be round, but actually, it is only a very small effect.

In fact, the image of the world on the retina does not need to be in very good focus across the entire retina. Brown notes that "the optical characteristics of the eye ... are nicely matched to the receptors [photoreceptors] and neural components (Records, 1979)." Only a very small part of the retina, the fovea, requires light to be accurately focused. This is because the neural components of the retina that sense light only have high resolution in the fovea. There are about 100 million photoreceptors, rods, and cones in the eye that convert light into electrical signals. There are 1 million retinal ganglion nerve cells that carry the information out of the eye into the brain. This enormous number of nerve cells is about one-third of all the afferent nerve fibers bringing information into the brain. But even with this large number of nerve cells, there are still 100 photoreceptors for each nerve cell. Hence, the light from every photoreceptor does not individually reach the brain. Only in the fovea there is a 1:1 coupling between photoreceptors and nerve cells. Away from the fovea, the output from many photoreceptors is processed and blended together into far fewer nerve cells that reach the brain. Thus, throughout most of the retina the neural pixels (picture elements) are coarse. In most of the retina, the eye sacrifices spatial resolution for enhanced sensitivity at low-light levels, as well as enhanced resolution of how the light level is changing in time.

The spatial resolution is high only in the fovea, which senses an area that is about two degrees (2°) across, only four times the diameter of the full moon. Everything else in your image of the world is fuzzy. The look of the world, its sharp edges and beautiful colors, is an illusion generated high up in the neural pathway located in the visual cortex in the back of your head. The eye is not like a camera. It is more like an electronic information sampling system. The brain moves the high resolution, clear image fovea to sample interesting features such as an ornate edge or a flashing light. It samples phenomena that look interesting. What you see depends primarily on what you saw before and what you are thinking now. This information is combined into the fiction of a clear, stable world.

A sharp, clear image is not needed across most of the retina because the neural elements there that detect light do not have a high-spatial resolution. A coarse image is a nice match to the coarse neural elements. Most of the retina provides a wide angle, low-resolution detection system to spot potential predators. The spherical retina may provide a useful detector for such system. Sharp, clear images are only needed in the fovea, a region 3 mm in diameter. A spherical shape is not needed to produce a clear image over such a small target. For example, when light is dim, at the bottom of the ocean or late at night on the land, animals have developed long cylindrical eyes with large, fast (high f ratio) lenses that maintain focus and clear images to the central area of their flattened retinas.

Thus, it does not seem as if roundness is a necessity for optical efficiency.

B. Eye Movement

I used to ask scientists at eye research conferences why they thought eyes were round. Inevitably, the answer that I received was that eyes were round because this was the best shape for rapid and accurate eye movements. It is mechanically easy to rotate a round eye in a round socket to aim it at any direction. Spheres also have the lowest moment of inertia for their mass and thus require the least force to move.

Is this the reason why eyes are round? In his classic book on the vertebrate eye, Walls notes that the "primitive function of the eye muscles was not to aim the eye at objects at all [but] ... designed to give the eyeball the attributes of a gyroscopically stabilized ship, for the purpose of maintaining a constancy of the visual field despite chance buffetings and twistings of an animal's body by water currents and so on" (Walls, 1963).

Let's examine the evolutionary sequence (Lythgoe, 1979). Fishes lack the fovea needed for sharp vision. They do not need to aim their eyes accurately, so they do not follow objects with their eyes. Amphibians also have limited eye movement capabilities. Neckless frogs turn their entire bodies in order to change their direction of gaze. Reptiles show variation in their eye movement. Some, like the Gila monster, have eyes that are fixed in their head. Others, like the chameleon, can use one eye to look forward and the other to look backward at the same time. Birds, the descendants of dinosaurs, have better vision than humans. Some birds have extended high-resolution areas on their retinas that cover a huge field of view. Other birds have more pigments in their photoreceptors for enhanced color resolution or extra structures to deliver more oxygen to the retina. Yet, their eyes are fixed and immobile. It is only mammals that have rapid and accurate movements.

This idea of roundness to facilitate eye motion, which seems obvious to many scientists, when considered in more detail, seems less convincing. The evolutionary record is whispering to us that eyes were round before they moved rapidly or accurately. Thus, it does not seem as if the eye is round *primarily* for eye movement reasons.

C. Hollow

Perhaps it is the hollow inside, which is significant. A spherical shell, inflated with fluid, can provide a clear optical pathway to the retina unobstructed by bones and ligaments. The spherical shape also provides the shortest, therefore the quickest, pathways for oxygen and nutrients to reach the interior structures of the eye and for wastes to leave them. A convoluted interior space, with serpentine passageways, would reduce the efficiency of such diffusion.

But the eye has not taken full advantage of this unobstructed interior space. Except in the core of the fovea, one layer of blood vessels that nourish the retina and two layers of synapses of nerve cells, lie in front of the photoreceptors. Light passes through these cells to reach the photoreceptors. These obstructions affect the image on the retina. You have probably observed this blood flow. On a clear day, when you look at a bright blue sky (but not the sun, which can cause severe and permanent damage) you can see tiny white specks darting around. This image is called the *blue entoptic phenomenon*. The white specks are white blood cells moving in front of the photoreceptors. The photoreceptors become adapted to the more numerous red blood cells shadowed against the blue sky, but then detect and respond to the occasional white blood cell. Experimentally, the speed of the white dots on a computer screen has been matched with the speed of these white specks to measure relative retinal microcirculation. To calibrate the system, a few volunteers wore a neck cuff to reduce the circulation to head so that the speed of the dots on the computer screen could be related quantitatively to the blood flow in the retina.

The eye has taken some, but not complete advantage, of this hollow space. Thus, it does not appear that the eye is round *primarily* for structural reasons to create a hollow space.

D. Phylogeny and/or Ontogeny

Walls notes, "The great German anatomist Froriep once likened the 'sudden' appearance of the vertebrate eye in evolution to the birth of Atena, fully grown and fully armed, from the brow of Zeus." There are no intermediate anatomical adaptations. Animals either have eyes that form images or spots that detect the amount of light. Perhaps roundness is a consequence of evolutionary pressures that produced the vertebrate eye. This idea is supported by the anatomical evidence found in the eyes of the cephalopods, such as squid and octopus. Their eyes evolved separately from the vertebrate eye, yet except for some small differences, their anatomy is strikingly similar. One of the few differences is that the cephalopod eye has nerves, which travel from the back of the photoreceptors, rather than the front of them, so that they do not interfere with the light pathway to the photoreceptors. Tripathi notes, "The final resemblance between the two types of eye [cephalopod and vertebrate] ... makes this one of the most striking cases of convergence in evolutionary history (Davson and Graham, 1974)." Convergence means that similar adaptive pressures led to similar anatomical structures. Perhaps, those pressures also dictated the roundness of the eye.

Maybe the answer lies not in phylogeny, the evolutionary history of a species, but in ontogeny, the developmental history of each new individual.

The structures of the eye need to be axis-symmetric along the line of rotation which brings light through the eye into the retina. Perhaps, developmental processes that form spherical structures are the embryo's path of least resistance to form such axis-symmetric structures.

Although speculation on species-specific evolution or individual development is both interesting as well as attractive, the hard evidence in support of these ideas is lacking. Thus, it does not seem that the eye is round *primarily* for phylogenic or ontogenic reasons.

E. Conclusions

Neither optical, nor movement, nor structural, nor evolutionary, nor developmental reasons seem to be the primary reason why the eye is round.

IV. PRESSURE

Although we do not understand *why* the eye is round, we do understand *how* it is round. As explained earlier, the roundness of the eye reflects a balance of two opposing forces. The outward force exerted by the pressure of the fluid inside the eye is balanced by the inward tension in the shell of the eye.

A. Surface Tension

The tension in the outer layers of the eye is called surface tension. If we were to make a small cut on the eye, the surface tension would be the force pulling the two sides of the cut away from each other. For a given pressure inside, the sphere is the shape that has the lowest surface tension. Containers for gas under pressure of any shape other than spherical require stronger walls. In the inorganic world, it is harder to manufacture spheres than cylinders, thus, most gas containers are cylinders. However, the material of these cylinders must be made twice as strong as would be needed for a sphere to hold the same pressure of gas.

In a cylinder, the surface tension across a cut in a curved direction is equal to that for a sphere of the same radius under the same pressure, but the surface tension for a cut in the long direction has twice the surface tension. This is the reason that the skin of frankfurters always tears in the long direction when cooked. The surface tension is twice as great in the lengthwise direction. Since the frankfurter skin is equally strong in both directions, it always breaks along the long direction, where the force tearing at it is twice that of the force tearing at it in the curved direction.

B. Pressure in the Eye

The fluid that flows in the eye is called the *aqueous humor*. It flows out of the ciliary body, passes in front of the lens, moves through the pupil, and circulates in the space behind the cornea. As discussed earlier, the outward force from the fluid pressure of the aqueous humor inside the eye is *isotropic*, felt equally in all directions. The inward force of the surface tension in the outer shell of eye is also *isotropic*. The balance between these inward and outward forces determines the spherical shape of the eye.

Since the force of the fluid pressure inside the eye is isotropic, a pressure increase in one part of the eye causes a pressure increase everywhere throughout the eye. In glaucoma, the pressure increase in the aqueous humor in the front of the eye is transmitted to the back of the eye. Although the pressure increase is caused by events in the front of the eye, the damage to vision is due to the effects of this pressure in the back of the eye. The increased pressure crimps the retinal nerve and blood flow, killing retinal ganglion cells either by cutting off the transport of essential materials along the inside of their axons, or the blood supply that nourishes them from the outside. The loss of vision results from the death of these nerve cells.

The hardness of the eye to touch is not determined by the toughness of the fabric of the eye, but by the fluid pressure inside the eye. When the pressure is high, the eye is hard. When the pressure is low, the eye is soft.

However, this is not the whole story. There is an additional factor. I have always felt that when my bicycle tires are old, no matter how much I pump them up, they never feel quite as hard as new tires. In the eye too, when the fabric is compromised, the shape and hardness of the eye change. For example, the shape of the cornea changes in keratoconus where the collagen in the cornea is weakened. In pathological myopia, there is a slow mechanical yielding of the fabric, and the eye steadily enlarges in time.

V. AQUEOUS FLOW

A. Balance of Inflow and Outflow

The eye is round because it is inflated by the pressure from the fluid inside. Is that what is necessary to maintain its shape, that is, to fill it once with aqueous humor under pressure? Nothing lasts forever. For example, my bicycle tires lose about 20% of their air every week. In order to maintain the pressure in the eye, we need to push fluid in and have it leak out in a very precise system. At first thought, it seems unbelievably wasteful to push fluid into the eye just to let it leak out again, but it's actually the most basic biological trick to expend energy for the sake of control. Balancing the inflow and outflow of aqueous humor provides a way to maintain and control the pressure inside the eye.

Soon we will see in detail how the aqueous humor is produced, and how it leaks out of the eye. The important point to remember here is that there is a balance of inflow and outflow. If the inflow was greater than the outflow, the fluid inside the eye would continually increase, and the eye would burst. If the inflow were less than the outflow, the fluid inside the eye would continually decrease, and the eye would collapse.

The flow of aqueous humor out of the eye is driven by the pressure inside the eye. The resistance to the flow of aqueous out of the eye determines the intraocular pressure inside the eye. If it is hard for the aqueous humor to leave the eye, then more aqueous accumulates in the eye. This increases the pressure within the eye, which forces more aqueous out. The pressure continues to increase until the aqueous flow out of the eye equals the aqueous flow into the eye. The pressure at which this balance occurs is determined by the resistance to the outflow of aqueous humor leaving the eye.

Thus, there is always a balance in the amount of aqueous entering and leaving the eye.

B. Inflow

The aqueous humor is generated by the *ciliary body*, a wiggly layer of tissue, two cells thick, along the edge of the ciliary muscle in the inside angle of the eye, a little back from where the clear cornea merges into the white sclera. From the ciliary body, the aqueous humor flows into the posterior chamber behind the lens. Then it passes through the pupil into the anterior chamber in front of the lens.

C. Outflow

The aqueous humor in the anterior chamber leaves the eye by passing through a series of structures in the angle of the eye inside of where the cornea merges with the sclera. On its way out of the eye, the aqueous flows through a coarse filter and then a fine filter, called the trabecular meshwork. Then it flows through a layer of cells and into a tube called Schlemm's canal that circles the cornea. From the canal it flows through collecting channels that bring it to the veins. It is not known which of these structures offers the most resistance to the flow. Some recent evidence suggests that the cells that line Schlemm's canal offer the most resistance to the flow, and thus determine the intraocular pressure inside the eye.

VI. THE CILIARY BODY

A. Structure

Aqueous humor is produced by two layers of epithelial cells in the ciliary body. A layer of pigmented cells is attached to a membrane that borders the capillaries. Inside of this is a layer of unpigmented cells that borders the posterior chamber. Fluid flows from the plasma in the capillaries, through these two layers of cells, into the posterior chamber.

In science, we use numbers to get a feeling for places that we can never touch with our fingertips. We will now use *Reynolds, Peclet*, and *concentration* numbers to gain a better understanding of the nature of the production of aqueous humor from these cells of the ciliary body.

B. Numbers in Science

There seems to be quite a misunderstanding about how numbers are used in science. Numbers are used only for qualitative purposes. Numbers are never used for quantitative purposes. What an oxymoron!

Let me illustrate this with an example. What do you think is the average density of the sun? The average density, the sun's mass divided by its volume, is about 1.4 gm/cm^3. The importance of this number is that the density of coal is about 3 gm/cm^3, and the density of iron is about 8 gm/cm^3. Thus, knowing the average density of the sun immediately tells us that the sun cannot be a burning ball of coal or a red hot ball of iron. In fact, the light coming from the sun has strong spectral lines of carbon and iron, so it would not be unreasonable to think that the sun was made of coal or iron. Yet, the number of the density tells us that the sun must be made out of something else.

The importance of the number of the average density of the sun is not that we know that it is 1.414 gm/cm^3 rather than 1.415 gm/cm^3, but that in relationship to other facts, namely the density of other materials, this number tells us something. It gives us the qualitative information that the sun is not made out of coal or iron. This is how numbers are used in science, to reach qualitative conclusions about the nature of things.

Let us now use some numbers to get a feel of what it is like to be in the ciliary body where the aqueous humor is produced.

C. Reynolds Number

The wind flowing toward a beach ball is deflected by its curved surface. The air curves around the beach ball. It continues its curvy path after it flows past the beach ball. This tendency of objects in motion to remain in motion is called inertia. The beach ball changes the flow of the air most near its surface and has little effect on the air some distance away. There is friction as the nearer and more distant streams of air scrape against each other. This friction is called viscosity.

The Reynolds number is the ratio of inertia to viscosity (Purcell, 1977). When the Reynolds number is small, then a fluid flow is dominated by viscous friction. The flow is viscous, smooth, and regular. For example, a pearl in honey has a Reynolds number of about 0.01. When the Reynolds number is large, then a fluid flow is dominated by inertia. The flow is fast, disorderly, and turbulent. For example, a rowboat in a lake has a Reynolds number of about 10,000.

The Reynolds number for the flow of aqueous humor through the cells of the ciliary body is about 0.00001. This is the same Reynolds number that you would have if you were in a swimming pool filled with molasses and were told not to move any part of your body faster than 1 cm/min. Thus, the flow of aqueous through these cells is smooth, laminar, and stately. It is like the squishy motion of a macrophage slipping between endothelial cells that you may have seen in movies taken through microscopes. It is not like the flow of water out of your kitchen faucet, gurgling with disordered turbulence at Reynolds numbers of 10,000–100,000.

From our first number, the Reynolds number, we have learned to picture aqueous production as thick and regular, dominated by friction rather than by inertia.

D. Peclet Number

Place a drop of black ink into a clear mountain stream. The drop will spread in the water. This is called diffusion. The drop will also be carried downstream. This is called advection. The osmotic Peclet number is the ratio of diffusion to advection. When the Peclet number is small, the flow is dominated by advection. The ink drop will be swiftly carried a long way downstream before it has time to spread. When the Peclet number is large, the flow is dominated by diffusion. In this case, the ink drop will spread much faster than travel downstream.

The Peclet number inside the cells of the ciliary body is about 100. The motion of molecules inside the cell is dominated by diffusion. A molecule soon wanders from any one part of the cell to any other part of cell, precluding the necessity for a conveyer belt. It is not necessary to mechanically grab molecules and carry them from one part of the cell to another. Just wait a little while and the molecule will diffuse. This is true for all cells except for nerve cells with long axons where diffusion is not efficient. In such nerve cells, energy from ATP drags carrier molecules along complementary tracks. However, the microsituation inside the ciliary body differs from both the nerve cell situation and our macroenvironment, where all types of clever mechanical devices are needed to move things from one place to another. Sometimes membranes wall off compartments within a cell and reduce the efficiency of the diffusion of molecules. Molecules are then carried by proteins across those membranes from one compartment to another.

Thus, from this second number we have learned that diffusion, rather than the mechanical transport of our macroscopic world, is the mechanism that moves molecules around in the cells of the ciliary body.

E. Concentration Number

Another useful number is the concentration number of ions, such as sodium, potassium, and chloride in the intracellular and extracellular solutions. The concentration of a solution of sodium chloride adjusted to match the osmolarity of plasma, called isotonic saline, is about 300 millimoles/liter. Molecular weight units, such as moles per liter, are helpful for computing the amounts that must be weighed out on a scale in order to mix a solution with a given concentration. These units are not helpful to form a physical picture of these solutions.

A more useful unit of concentration is the number of water molecules for each ion. The number of ions in solution is similar in isotonic saline, the blood plasma, the aqueous humor, the solution in the cells in the ciliary body, and the solution in the spaces between those cells. In each of these solutions, there are about 150 molecules of water for each ion. You can now picture a few cubes of solution, each with about 1 ion and about a 150 water molecules. You can now understand why these ions are close enough to interact a little, but not too much. The interactions between the ions change the osmotic force by about 10% from that which would be generated if the ions did not interact with each other.

Thus, from our third number we have learned that both within and outside the cell, there are about 150 molecules of water for each ion of sodium, potassium, and chloride.

F. Fluid Transport

How do the cells in the ciliary body produce the aqueous humor? Proteins in the cell membranes of these cells could clutch water molecules and push them across the cell membrane. But these cells do something more clever, more subtle, and more efficient. If one ion is transported across the cell membrane, it will change the concentration on the other side. This change in concentration will osmotically induce the flow of water. How much water? In these solutions 1 ion is balanced by 150 molecules of water. Therefore, the movement of 1 ion will induce the movement of 150 molecules of water.

So, why pump 150 molecules of water, if you can induce the same fluid flow by pumping 1 ion? That is exactly how these cells generate the flow of aqueous humor. They transport ions across the cell membrane. The fluid that forms the aqueous humor is then osmotically driven by the changes in concentration caused by these ions.

For this elegant scheme to work, the water permeability of the cell membrane must be much larger than its ion permeability. This is needed so that the water easily flows across to join the ion that has been transported to the other side of the membrane, and so that the ion does not freely cross back across the membrane. In order to move 150 water molecules for each ion, the permeability of the cell membrane to water needs to be about 150 times that of its permeability to ions, which is indeed the case.

The location of the small spaces whose concentrations have been altered by the transport of ions is not known. They could be up against the front surfaces of the cells or in between them.

The flow of aqueous humor through the cells in the ciliary body is prodigious. In the usual units, the flow is about 1 $(\mu l/min)/cm^2$. A more meaningful question is to ask how long it takes for each cell to transport a volume of aqueous equal to its own cellular volume. Each cell transports its own volume in 2 min. The fluid floods through these cells.

G. Ion Transport

The transport of ions is complex. Outside cells, the movement of ions is driven by differences in electrical voltage and concentration. Across the cell membranes, ions are moved by proteins. Some of these proteins use energy from ATP to move the ions uphill against their electrical and concentration gradients. Some of these proteins bind a few different ions at a time, and then move them into and out of the cell. Some of these proteins are like big holes that allow many ions through at any one time.

As mentioned previously, cells use energy to pump ions out, and then let them leak back in again. This is not futile. By controlling the pump and leak, cells control the movement of ions and the voltage across the cell membrane.

A typical epithelial cell, like those in the ciliary body, has about 1,000,000 sodium–potassium ATPase protein molecules in its cell membrane. There are so many of these molecules that they are quite close together on the cell surface, about 10 nm apart. About 30 times a second, each of these molecules uses the energy from ATP to move a few sodium ions out of the cell and a few potassium ions into the cell.

However, the ions leak back across the cell membrane in a very different way. Each cell has about 100 ion channel protein molecules in the cell membrane. There are so few of these molecules that they are quite far apart on the cell surface, about 1000 nm apart. About 10 times a second, for about

1/200 of a second, each of these channels opens and 30,000 ions leak through.

Thus, the pump is very different from the leak of ions across the cell membrane. The pump works by using energy to move a few ions at a time, at many places, on the cell surface. The leak works passively to move many ions at a time, across a few places, on the cell surface.

The leak of ions through the ion channels is well separated in space and time on the cell surface. This makes it possible to measure the flow of ions through an individual channel. A piece of cell membrane, small enough to contain only one ion channel, is sealed in a micropipette called a patch clamp. The electronics are sensitive enough to resolve the picoampere current through an individual ion channel in such a patch when it opens for a brief time. The much smaller currents through the proteins that pump ions are too small to measure this way.

The proteins for the pump and leak are on different sides of the cells in the ciliary body. The ions pumped across one side of the cell leak across the other side of the cell thus, there is a net transport of ions. The pigmented and unpigmented layers of cells in the ciliary body transport ions in opposite directions. This is a complex machine, and how it works to produce the aqueous humor is not clear.

H. Active or Passive

The ciliary body of an animal can be removed and mounted in a chamber. The current measured due to the flow of the ions transported is about $8 \ \mu A/cm^2$. Since each ion will drag about 150 water molecules with it, we can use that current measurement to compute the rate of aqueous production. The result is that the ion transport accounts for only 2% of the aqueous production! It has been suggested that the failure of ciliary body to pump fluid in these experiments is due to the fact that the tissue has been damaged in the dissection, or that the capillaries are collapsed in the mounting so that fluid does not reach the cells for them to transport. Perhaps the ion flows are recirculated so that the net ionic current is not a complete measure of the amount of fluid transport.

However, the fact remains, that in most of the experiments of isolated ciliary tissue or isolated animal eyes, the ciliary body does not secrete aqueous humor. This has led some scientists to argue that the aqueous humor is produced by the filtration of fluid under pressure from the plasma across the ciliary body rather than by fluid transport driven by active ion transport.

It is hard to believe that the aqueous humor is generated by passive filtration from the plasma. Aqueous humor has different concentrations of ions and other substances than the plasma. This suggests that those ions and substances were actively pumped into the aqueous humor. The ciliary epithelium has features that are present in other epithelia that are known to move fluid by the active transport of ions. These features include ruffled edges to enlarge the area of the cell membrane and a large number of mitochondria to provide energy in the form of ATP for ion transport. It would be hard to understand why such specializations for ion transport would not be present in a tissue that was not transporting ions and fluid.

It is puzzling that the production of the aqueous humor by the ciliary body has been so difficult to demonstrate in these experiments.

VII.   LARGE SCALE AQUEOUS MOTIONS

The aqueous humor emerges from the posterior chamber, through the pupil, into the anterior chamber. The fluid in the anterior chamber can be stained with fluorescein dye. At this larger scale, the Peclet number is small, so the dye does not diffuse rapidly enough to stain the new, emerging aqueous. Thus, the flow of the new, clear aqueous can be followed from the posterior into the anterior chamber. Even at this scale, the Reynolds number is still low. Thus, the "jet" of aqueous emerging through the pupil has the form of an expanding ball. This is very different from the thin cone of water in the high Reynolds number jet that emerges from your kitchen faucet.

The interior of the eye is warmer than the air outside. Warm aqueous rises and cool aqueous falls. Thus, there is also a vertical convective circulation of aqueous humor. This circulation can sometimes be seen in the motion of small particles in the eye.

VIII.   CONTROL OF INTRAOCULAR PRESSURE

Intraocular pressure is determined by the balance of the aqueous inflow and outflow. How that balance is set and maintained? Many variables can act to alter that balance. Like other somatic phenomena, there is a circadian rhythm in the pressure: it is low at 7 a.m. and high at 7 p.m.

Diseases alter the pressure. In glaucoma the increased resistance to the outflow of aqueous humor increases the pressure within the eye.

Osmotic agents, such as alcohol, suck additional fluid from the eye. Increased outflow of fluid lowers the pressure.

Pharmacological agents that mimic or inhibit neurotransmitters can also change the pressure. The physiological implications of the effects of these agents are rarely discussed in the literature. Does it mean that synapses are involved in either the sensing of intraocular pressure or control of aqueous production or aqueous outflow? If there is a neural circuit, is it local, through the spinal cord, or all the way upstairs to the brain?

IX. SUMMARY

We do not really understand why eyes are round. The roundness might involve their optical properties or their ability to move rapidly and accurately. However, the evidence in support of these reasons seems weak. Still less evidence suggests that roundness arises from the need to maintain a clear interior space or is a result of the evolution of eyes in vertebrate species or their individual developmental history.

If we do not know *why* eyes are round, we know *how* they are round. They are round because they are inflated from within by pressure of the aqueous humor, the fluid in the eye.

The balance of the inflow and outflow of the aqueous humor determines the pressure within the eye. We are not sure how the aqueous is produced. We are not sure which structures determine the resistance to the outflow of the aqueous from the eye. We do not understand the factors that control the intraocular pressure during the day, or how they are altered in the course of certain diseases.

It is sobering to realize how little we know about the most salient features of the eye.

ACKNOWLEDGMENTS

This work was supported in part by NIH grant EY6234. I thank Lana Thompson for her suggestions on the manuscript.

REFERENCES

Davson, H. and Graham, L.T., Jr. (Eds.) (1974). The Eye: Comparative Physiology, Vol. 5. Academic Press, New York.

Lythgoe, J.N. (1979). The Ecology of Vision. Oxford University Press, New York.

Purcell, E.M. (1977). Life at Low Reynolds Number. Am. J. Phys. 45, 3–11.

Records, R. (Ed.) (1979). Physiology of the Eye and Visual System. Harper & Row, New York.

Thompson, W.D. (1966). On Growth and Form. (Bonner, J., Ed.). Cambridge University Press, New York.

Volk, T. (1985). Majesty of the Sphere: Why So Many Things in Nature Are Round. The Sciences, November/December 1985, pp. 46–50. New York Academy of Sciences, New York.

Walls, G.L. (1963). The Vertebrate Eye and Its Adaptive Radiation. Hafner Publishers, New York.

TEARS AND THEIR SECRETION

Darlene A. Dartt, Robin R. Hodges and
Driss Zoukhri

Advances in Organ Biology
Volume 10, pages 21–82.
© **2006 Elsevier B.V. All rights reserved.**
ISBN: 0-444-50925-9
DOI: 10.1016/S1569-2590(05)10002-0

ABSTRACT

The exposed surface of the eye is continuously covered by a thin film of fluid, the tear film, which covers the entire ocular surface, including the cornea (the clear "window" of the eye) and conjunctiva (the white part of the eye, which extends under the eyelid). The tear film is a complex fluid that is secreted by several different glands surrounding the eye. The epithelial cells of the ocular surface itself also secrete components of the tear film. The action of blinking spreads the film of tears over the whole surface of the eye and mixes the tears underneath the lids. The tear film serves as an interface between the external environment and the ocular surface and is the first layer of protection for the cornea and conjunctiva. It is constantly responding to stresses that include desiccation, bright light, cold, mechanical stimulation, physical injury, noxious chemicals, and bacterial, viral, and parasitic infection. The tear film also maintains the health of the cornea and conjunctiva by providing optimal electrolyte composition, pH, nutrient levels, and a complex mix of proteins, lipids, and mucin. To respond to these various external and internal requirements, exquisite control of the volume, composition, and structure of the tear film is required. This control arises from regulating secretion from the individual orbital glands and ocular surface epithelia. Regulation of tear secretion provides an extremely stable fluid that protects and maintains the cornea and conjunctiva and ensures that the transparent cornea provides the retina with its window to the world and ensures clear vision.

I. FUNCTIONS OF THE TEAR FILM

The tears provide a large number of diverse functions to protect and maintain the ocular surface. Tears are able to perform these varied functions because they are an exceedingly complex fluid made up of a diverse array of lipids, protein, electrolytes, water, and mucins all organized into a stable, specific structure. The tear film provides a smooth refracting surface that eliminates the many small irregularities of the cornea for clear refraction of light (Lamberts, 1994; Tiffany, 1995).

Tears also transport metabolites to, and remove waste products from the epithelial cells of the cornea (Lamberts, 1994). Since the corneal epithelium, stroma, and endothelium lack a blood supply, they must be provided with oxygen and nutrients (Tiffany, 1995). Carbon dioxide and metabolic waste products must also be removed. The nearest blood supply to the cornea is at the limbus, which connects the cornea to the adjacent sclera. The limbal and conjunctival blood vessels provide for these needs by supplying a small amount of O_2 and small nutrient molecules to the cornea and removing CO_2. When the eye is open, tears secreted from the orbital glands provide the bulk of O_2 and nutrients in their secretions and remove CO_2. When the eye is closed, the aqueous humor, which bathes the endothelial side of the cornea, supplies the cornea with its entire metabolic needs. The supply is not quite adequate, however, and the cornea swells slightly. It is interesting to note that differentiated (multiple-cell layered) corneas survive best in culture when they are placed at the air–fluid interface and not submerged (Zieske et al., 1994). This suggests that the unique ocular surface/air interaction is important to the structure and function of the cornea.

Another function of tears is to provide the entire ocular surface with a moist environment with the appropriate electrolyte composition. The ocular surface has a narrow range of pH, osmolarity, and ionic concentrations necessary for optimal function. Small changes in these variables, especially osmolarity and ion concentrations lead to ocular surface disease (Gilbard et al., 1988). Tear pH is maintained between 7.14 and 7.82 (Feher, 1993) by the buffers in the eye that are mainly bicarbonate (HCO_3^-), H^+, and proteins (H^+ acceptors). Tear osmolarity is normally 300–304 mOsm and is similar to that of plasma (Gilbard et al., 1978). An increase of 10 mOsm is enough to be deleterious to the ocular surface, especially the conjunctiva. Tear osmolarity is derived from the ionic composition of tears, which is unique when compared to plasma or other body fluids. Tears contain Na^+, K^+, Cl^-, HCO_3^-, Ca^{2+}, Mg^{2+}, and trace levels of other ions. Tears have a higher K^+ and Cl^- concentration and a similar Na^+ concentration compared to plasma (Feher, 1993). This implies that tears are not an ultrafiltrate of plasma, but are secreted by the orbital glands. Because this secretion is highly regulated, the ionic composition of tears (thus the osmolarity and pH) can be tightly controlled and the health of the ocular surface maintained.

Protein composition is also important for a healthy ocular surface. Tears contain a large number of proteins, which are secreted by the orbital glands. Tears do not normally contain serum proteins, although these can enter the tears from leaky conjunctival blood vessels under pathological conditions. One of the most important functions of tear proteins is preventing bacterial and viral infections. The major tear proteins, lysozyme, secretory immunoglobulin (IgA), lactoferrin, lipocalin, and peroxidase are antibacterial (Fullard, 1994). The high-molecular weight glycoproteins in tears known as mucins are also antibacterial and antiviral. Mucins are secreted onto the ocular surface and protect the underlying epithelial cells by binding to and entrapping bacteria and viruses (Corfield et al., 1997). The carbohydrate side chains of mucins, which are attached to the protein core, are able to bind to a wide variety of pathogens. Since each type of pathogen has a specific carbohydrate sequence to which it will bind, mucin carbohydrate side chains are heterogeneous and are able to bind a wide variety of microorganisms. When mucins bind to bacteria or viruses, they prevent them from attaching to the ocular surface and invading it (Corfield et al., 1997). Thus, mucins block microbial binding sites before the microorganism penetrates the ocular surface and prevent infection.

In addition to preventing bacterial and viral infection, the diverse array of tear proteins can also regulate many functions of the ocular surface. These functions include cellular migration and proliferation during wound repair, normal cellular differentiation, and secretion of electrolytes and water. Tears contain a wide variety of growth factors, cytokines and biologically active peptides (Sullivan and Sato, 1994). The known growth factors in tears include: epidermal growth factor (EGF); hepatocyte growth factor (HGF); transforming growth factor (TGF; α, $\beta1$, and, $\beta2$), basic fibroblast growth factor (bFGF), tumor necrosis factor (TNF)α, and granulocyte macrophage-colony stimulating factor (GM-CSF). The cytokines interleukin (IL)-1α and IL-1β and the neuropeptides substance P (Sub P) and endothelin 1 are also present.

Tears protect the eye from noxious stimuli, such as acids, bases, and other chemicals. Tears also remove particles and debris, such as eyelashes or makeup, from the ocular surface. Two components of tears are effective in this mechanism. External irritants of the ocular surface cause neurally mediated reflex secretion of water and electrolytes to neutralize and wash away the irritants. The same neural pathways also stimulate mucin secretion. Mucins physically entrap and remove irritants. As the cornea and conjunctiva are innervated with sensory nerves, reflex secretion of electrolytes, water, and mucins provide a rapid response to noxious stimuli. The blinking mechanism then washes irritants into the lacrimal drainage system effectively removing them from the ocular surface.

The blinking mechanism that helps remove tears from the surface of the eye occurs continuously, as does the horizontial movement of the globe (saccades). This means that the lids and ocular surface are in almost constant movement. The eyelids move vertically as well as horizontally during the blink, which occurs about 15 times per minute (Doane, 1980). During the blink the eyelids move 17–20 cm/s and generate enough force to push the ocular surface 1–6 mm into the orbit (Doane, 1980). The rapid continuous movement of the eyelids and ocular surface requires minimal frictional resistance to avoid mechanical damage to the surface of the eye (Corfield et al., 1997). The composition of tears, especially the mucin component, provides a fluid that is non-Newtonian in behavior. This means that when a shear force is applied, such as occurs during the blink, the viscosity of tears decreases. Thus, in spite of the force generated by the blink, the ocular surface is unscathed. The forces generated by blinking also illustrate the ability of tears to wash away foreign bodies from the ocular surface.

II.　ORGANIZATION OF THE TEAR FILM

It is well documented that the tear film consists of three layers: a lipid layer, an aqueous layer, and a mucous layer. The lipid layer is the outermost layer and is thought to be about 0.1 μm thick. The aqueous layer, the middle layer, is between 7 and 10 μm thick. The mucous layer is the innermost layer. Its thickness is somewhat controversial. The original measurement of the mucin layer was 0.2–1.0 μm thick. Prydal has proposed that the mucous layer is much thicker than previously thought, that is about 30 μm thick in humans (Prydal et al., 1992). Another possibility is that the mucous and aqueous layers are not distinct layers, but rather are a gradient of decreasing mucous and increasing aqueous concentration from the apical surface of the cornea and conjunctiva to the lipid layer. So, while it is well documented that the tear film contains the three major components, lipids, aqueous, and mucins, the exact relationship between the aqueous and mucin layers has not been determined.

　　The layered structure of the tear film is maintained by the blink, which redistributes the lipid layer over the other layers. Between blinks, the lipid layer slowly begins to breakup and dry spots or discontinuities begin to form (Holly, 1973). It is not known why these dry spots form. One hypothesis is that the mucins influence tear stability by lowering the surface tension (Holly and Lemp, 1971, 1977). Another hypothesis is that it is the breakup of the lipid layer causes tear instability (Doane, 1994). A third hypothesis is that corneal epithelial cells that have newly migrated to the epithelial surface have not yet developed their glycocalyx (mucin coat attached to the plasma membrane). These cells could initiate dry spot formation (Gipson et al., 1992; Sharma, 1993; Corfield et al., 1997). Regardless of its cause, the tear film breaks up between blinks and then reestablishes its normal complex structure.

III.　ORBITAL GLANDS AND OCULAR SURFACE EPITHELIA THAT SECRETE TEARS

Given the diverse components of tears and the complex structure of the tear film, it is not surprising that many different tissues contribute to the tear film. In fact, each layer of the tear film is secreted by specific orbital glands and ocular surface epithelia (Figure 1). The lipid layer is secreted primarily by the meibomian glands embedded in the tarsal plate of the upper and

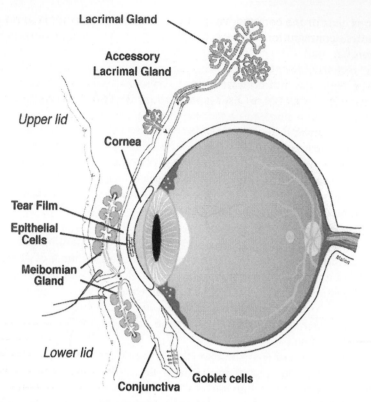

Figure 1. Schematic drawing showing the glands and epithelia of the eye and ocular surface that contribute to the tear film. Shown are the meibomian glands, which secrete the outer lipid layer; the main and accessory lacrimal glands, as well as the conjunctival and corneal epithelia, which secrete the middle aqueous layer; and the conjunctival goblet cells, which secrete the inner mucous layer. (From Dartt, D.A. and Sullivan D.A. (2000). Wetting of the ocular surface. In: Principles and Practice of Ophthalmology. (Albert, D. and Jakobiec, F. Eds.), 2nd edn., p. 960. W.B. Saunders Co., Philadelphia.)

lower eyelids and secondarily by the glands of Zeis and Moll. The aqueous layer is secreted largely by the main lacrimal gland and to a lesser extent by the accessory lacrimal glands (the glands of Kraus and Wolfring). The cornea also secretes electrolytes and water into the tear film. Aqueous portion of the tears could also originate from leakage across the conjunctival blood vessels or from electrolyte and water secretion by the conjunctival epithelial cells. The conjunctival epithelium may affect the composition of the tears by absorbing tear fluid or specific components of tears (e.g., glucose) similarly to the intestinal epithelium. The goblet and the stratified squamous cells of the conjunctival and corneal epithelia secrete the mucins. The relative contribution of these two types of cells to the mucous layer is unknown.

Until recently, secretion from the individual orbital glands and ocular surface epithelia had not been well characterized. Many of the earlier studies of secretion by these tissues used tear fluid, which can be collected easily. Because tears are a mixture of secretions from many different glands, collecting tears cannot draw an accurate picture of secretion from one gland. Even though the main lacrimal gland is primarily responsible for reflex tears, there is a small but significant contribution from the other orbital glands. To characterize secretion of each orbital gland, studies need to be performed on pure, uncontaminated secretion from individual glands or on *in vitro* preparations of individual glands or cells.

IV.   SECRETION OF THE LIPID LAYER OF THE TEAR FILM

A. Meibomian Glands

The lipid layer of the tear film is secreted primarily by the meibomian glands. Meibomian glands are sebaceous glands that lie in a parallel row in the upper and lower tarsal plates perpendicular to the lid margin. There are approximately 30–40 glands in the upper lid and 20–30 in the lower. Meibomian glands in the upper lid can be up to 10 mm long, while meibomian glands in the lower lid can be up to 6 mm long. The length of the gland depends on the position of the individual gland in the lid. The duct of each gland opens directly onto the inner margin of the eyelids. Tiffany has recently published an excellent review of meibomian gland structure and function (Tiffany, 1995).

B. Functional Anatomy

Sebaceous glands are oil-secreting glands that are found over most of the body except on the palms and soles (Thody and Shuster, 1989). Sebaceous glands are especially prevalent on the scalp, forehead, and face where they are associated with hairs and are known as pilosebaceous glands. Glands not associated with hairs are termed free sebaceous glands and are usually in transitional zones between skin and mucous membranes. These regions include the anogenital region, the periareolar skin, the border of the lips, and the eyelids. Thus, the meibomian glands are free sebaceous glands (not associated with hair follicles despite their proximity to the eyelashes) and are found at the border between the eyelid skin and the conjunctival mucous membrane (mucocutaneous junction).

All sebaceous glands have a similar basic structure. These glands consist of a single lobule (acinus) or a collection of lobules that coalesce into a system of ducts, which open directly onto the body surface (Thody and Shuster, 1989). The acinus or alveolus contains the lipid-secreting epithelial cells. The lipid is secreted into small ducts, which lead to a single, straight duct lined with four layers of ductal epithelium. Most sebaceous glands, the meibomian gland included, are surrounded by a connective tissue capsule, which separates individual acini and is rich in fibroblasts and capillaries. With the exception of the meibomian glands and the preputial glands of the rat, sebaceous glands are not innervated (Thody and Shuster, 1989). However, nerve fibers are found in the connective tissue stroma of the preputial and the meibomian glands.

There are two major cell types in meibomian glands and other sebaceous glands: the lipid-producing cells of the acinus and the stratified squamous cells of the ducts (Thody and Shuster, 1989). Unlike other sebaceous glands, the ductal cells in the meibomian gland do not secrete lipids. The lipid-producing cells can be subdivided according to their stage of differentiation into undifferentiated cells, maturing cells, and necrotic cells. The undifferentiated cells form a single layer and are in contact with the basement membrane (Figure 2). These cells are germinal basal cells with a small-flattened shape and do not synthesize and secrete lipids. The next several layers contain the maturing cells. These cells are rounded in appearance and larger due to the development of the smooth endoplasmic reticulum and golgi apparatus necessary for the synthesis of lipids. Lipids are stored in droplets surrounded by a membrane, which is known as a secretory granule. The closer the cell is to the duct, the more secretory granules the cell contains. Cells in the innermost zone are the necrotic cells. These cells are full of secretory granules and are 100–150 times larger than the peripheral cells. The nuclei have become pyknotic and there is little cytoplasm except for a few mitochondria. These cells are ready to degenerate. Just before this occurs, the secretory granules fuse to form large aggregates. Secretion occurs when the cells closest to the duct lyse releasing lipid droplets and cell debris. This mechanism of secretion is known as holocrine secretion.

The main duct of the meibomian gland consists of four layers of stratified squamous epithelium that is continuous with the epithelium of the acinus at one end and the mucocutaneous zone of the eyelid at the other end. The duct cells are a well-developed keratinized epithelium consisting of a basal cell layer, an intermediate cell layer, and a horny cell layer (Jester et al., 1981). These cells contain the markers characteristic of keratinization, including keratohyalin granules, lamellar granules, and bundles of keratin filaments (Thody and Shuster, 1989; Driver and Lemp, 1996). The stratum corneum (several layers of flat keratinized, nonnucleated cells) and stratum granulosum (also known as granular cell layer and is the start of keratinization)

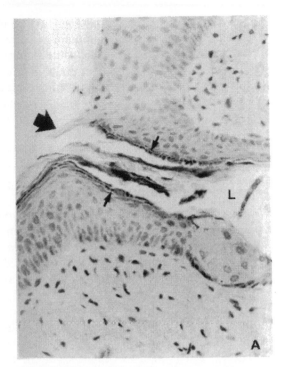

Figure 2. Light micrograph of an alveolus of a rabbit meibomian gland. Cells in the outer layer (large arrow) of the alveolus are undifferentiated and do not contain lipid secretory droplets. Clear lipid droplets are visible in the middle layers of cells. Cells in the inner layer (small arrow) disintegrate into the alveolar duct releasing lipid droplets to form the secretory product that is released into the lumen (L) of the central duct. H&E 350×. (From Jester, J.V., Nicolaides N., and Smith R.E. (1989). Meibomian gland dysfunction. I: Keratin protein expression in normal human and rabbit meibomian glands. Invest. Ophthalmol. Vis. Sci. 30, 927.)

usually found in keratinized epithelium are present in the distal part of the duct. The orifice of the meibomian glands consists of three concentric zones or cuffs (Jester et al., 1982). The inner cuff is the keratinized lining of the duct while the translucent cuff is the surrounding dermis while and the outer cuff is a subdermal structure (Tiffany, 1995).

Release of meibomian gland fluid onto the surface of the eye is related to the blink mechanism and is controlled by the coordinated relaxation and contraction of the pars marginalis and the pars palpebrae muscles. The secreted meibomian gland fluid is stored in the ducts and is kept there by contraction of the pars marginalis muscles that surround the orifice (Linton et al., 1961; Tiffany, 1995). At the same time the pars palpebrae of the

orbicularis oculi muscle is relaxed allowing the ducts to accumulate meibomian gland fluid. During the blink, the meibomian glands are milked by
contraction of the pars palpebrae and the stored fluid is released onto the
ocular surface by relaxation of the pars marginalis. The movement of
the eyelids during the blink then spreads the meibomian gland fluid over
the aqueous layer of the tear film.

C. Regulation of Secretion

While it is well known that blinking controls the release of meibomian gland
fluid from the ducts of the meibomian gland, it is not known what stimuli
induce secretion (the rupture of the necrotic acinar cells). Secretion can be
controlled by several mechanisms. One such mechanism is by regulating the
fusion of the secretory granule to the apical (duct side) plasma membranes
releasing of the contents of the cell. This could be initiated by the release of
hydrolytic enzymes from lysozymes into the cell cytoplasm. Another mechanism by which to regulate lipid secretion is regulation of the rate of lipid
synthesis by the endoplasmic reticulum and hence the maturation of the
acinar cells. Evidence suggests that androgen sex steroids regulate meibomian gland lipid synthesis and secretion. Indeed, human meibomian glands
have been shown to contain (Rocha et al., 2000; Wickham et al., 2000)
androgen receptor mRNA and protein in the nuclei of the acinar cells. In
addition, these glands also contain enzymes that either convert testosterone
and dihydroepiandrosterone into the potent androgen, 5α-dihydrotestosterone
(Rocha et al., 2000) or metabolize androgens into other androgenic forms
(Perra et al., 1990).

Another mechanism for the regulation of meibomian gland secretion
involves the nerves surrounding the acinar cells of the alveoli. Neural control
of the meibomian gland would be unique given that the other sebaceous
glands, with one exception (i.e., preputial gland of rat), are not innervated
(Thody and Shuster, 1989). Chung et al. have shown that meibomian glands
are richly innervated by several different nerve types (Chung et al., 1996).
Vasoactive intestinal peptide (VIP)-containing nerves, probably parasympathetic, are abundant and surround the acini (Figure 3) (Hartschuh et al.,
1983; Chung et al., 1996; Seifert and Spitznas, 1999). Sympathetic and
sensory nerves are also present, but are not as abundant as parasympathetic
nerves and are located mainly near blood vessels. Neuropeptide Y (NPY)-
containing nerves are also abundant with a similar distribution as VIP
(Chung et al., 1996). This is surprising as NPY is usually contained in
sympathetic nerves. However, the distribution of NPY and tyrosine hydroxylase, a marker for sympathetic nerves, is different. It is possible that
the NPY-containing nerve fibers in the meibomian gland are of parasympathetic origin. Thus, it seems likely that parasympathetic nerves play a role in

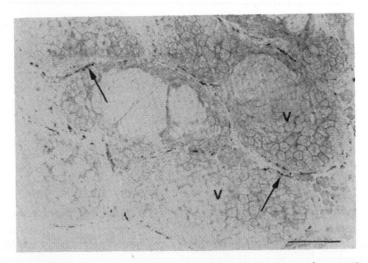

Figure 3. Fluorescence micrograph demonstrating the location of vasoactive intestinal peptide (VIP)-containing nerves around the alveoli (v) of the meibomian glands. Bar 0.1 mm. (From Seifert and Spitznas (1999) Vasoactive Intestinal Polypeptide (VIP) Innervation of the Human Eyelid Glands. Exp. Eye Res. 68: 685–692.)

the regulation of secretion from the meibomian gland. Indeed, Montagna and Ellis have shown that physostigmine, which prolongs the action of acetylcholine (a parasympathetic neurotransmitter), stimulates meibomian gland secretion in the rabbit (Montagna and Ellis, 1959). Nerves could also regulate the release of lipids from the secretory granules by stimulating fusion of the secretory granule and apical membranes. More research into the regulation of meibomian gland secretion is needed to determine the role of hormones and nerves in the control of secretion.

D. Secretory Product

Meibomian gland fluid is a complex lipid mixture that varies considerably between individuals (Nicolaides, 1986; Tiffany, 1987). The composition of meibomian gland fluid also differs from that of other sebaceous glands (Thody and Shuster, 1989). Meibomian gland fluid is 60–70% wax monoesters and sterol esters. Several types of diesters are also present. Minor constituents of meibomian gland fluid include hydrocarbons, triglycerides, diglycerides, free sterols (including cholesterol), free fatty acids, and polar lipids (including phospholipids). Meibomian gland fluid also contains the lipid components of the cellular membrane of the acinar cells themselves as a result of the holocrine mechanism of secretion. In addition, there are some unique diesters and triesters present as well as a large proportion of anteiso-branched chains in the acids and alcohols (Thody and Shuster, 1989). Their chain lengths are longer than those found in other sebum. Importantly, the mixture of lipids secreted by the meibomian gland melts at 35 °C and so is liquid on the surface of the eye Tiffany, 1995). The unique features of meibomian gland fluid are not surprising as it has a specialized role in the eye to preserve the transparent, fluid nature of the tears.

E. Function of Lipid Layer

The lipid layer is essential for the maintenance of the structural and refractive integrity of the ocular surface (Driver and Lemp, 1996). Patients with meibomian gland dysfunction have a 25–40% incidence of keratoconjunctivitis sicca, and meibomian gland dysfunction is one of the major causes of "evaporative" dry eye (Lemp, 1995; Driver and Lemp, 1996). Despite this, little is known about the function of the meibomian gland fluid. Experimental evidence has shown that meibomian gland fluid prevents spillover of the tears by containing tears within the palpebral opening; prevents damage of the lid margin skin by tears; and forms a seal over the exposed portion of the eye during sleep. It has been suggested, but not proven, that meibomian gland secretion also reduces evaporation in the open eye, prevents sebaceous lipids from entering the tears, and increases the stability of the tear film by interaction with the soluble components (Tiffany, 1987).

V. SECRETION OF THE AQUEOUS LAYER OF THE TEAR FILM

A. Main Lacrimal Gland

The main lacrimal gland is an almond-shaped gland located within the orbit of the eye and is the major contributor to the aqueous layer of the tear film. It is highly lobular with numerous ducts that open onto the superior portion of the surface of the eye (Lamberts, 1994).

Functional Anatomy

In humans, the main lacrimal gland is a multilobed, tubuloacinar exocrine gland (Figure 4). Each lobe contains many branched tubules of columnar secretory cells known as acinar cells (Walcott, 1994). In cross section, the tubules form an acinus or ring of secretory acinar cells, which converge to form interlobular ducts consisting of a single layer of cuboidal cells. The intralobular ducts empty into larger interlobular ducts consisting of

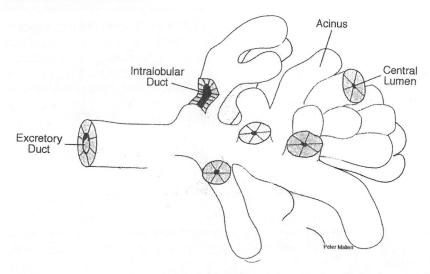

Figure 4. Schematic drawing of the lacrimal gland showing the tubules that contain acini. These acini surround central lumen of the ducts. These ducts coalesce to form intralobular ducts that join to form the main excretory duct that secretes lacrimal gland fluid onto the surface of the eye. (From Botelho, S.Y. (1964). Tears and the lacrimal gland. Sci. Am. 211, 78.)

two layers of epithelial cells. In humans, there are 6–12 ducts from which lacrimal gland fluid is secreted onto the ocular surface. In rats and rabbits, however, there is only one main excretory duct.

The main function of the acinar cells is to secrete protein, water, and electrolytes. Tight junctions surround each acinar cell on the luminal side creating a polarized cell (Figure 5). Thus, the plasma membrane can be separated into apical (luminal) and basolateral (serosal) components. The basal side of the cell contains the nucleus, endoplasmic reticulum, and Golgi apparatus while the apical side is filled with secretory granules that contain the protein secretory product. This polar structure of the acinar cell is the basis for the unidirectional secretion of proteins, electrolytes, and water.

The major function of the duct cells is to secrete electrolytes and water. Like acinar cells, the cuboidal duct cells are also polarized due to the presence of luminal tight junctions. Nuclei are located basolaterally, whereas the rough endoplasmic reticulum and mitochondria are more apical (Jakobiec and Iwamoto, 1979). While duct cells are known to secrete some proteins, such as EGF, they do not contain many secretory granules and their protein secretion is limited.

In addition to acinar and ductal cells, the lacrimal gland also contains myoepithelial cells. These flat cells contain multiple processes, which surround

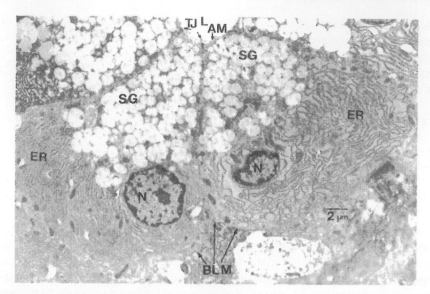

Figure 5. Transmission electron micrograph of the acinar cells of the main lacrimal gland. Acinar cells are pyramidal cells that are joined together by tight junctions TJ that separate the apical membrane (AM) from the lateral membrane (BLM). Secretory proteins are synthesized in the rough endoplasmic reticulum (ER), stored in the clear secretory granules (SG), and then secreted into the lumen (L). N, nucleus. (Courtesy of Dr. Kenneth R. Kenyon and Ms. Laila Hanninen.)

the basal area of the acinar and ductal cells. Myoepithelial cells contain smooth muscle actin in numerous long filaments and are thus thought to be contractile. In the salivary and mammary glands, there is evidence to suggest that myoepithelial cells contract the acinar cells to force fluid out of the secretory tubules. However, there is no evidence for this hypothesis in the lacrimal gland. Some investigators have proposed that myoepithelial cells serve as an exoskeleton for acini (Satoh et al., 1994). Receptors for the parasympathetic neurotransmitters, acetylcholine and VIP, have been found on myoepithelial cells (Lemullois et al., 1996; Hodges et al., 1997). This evidence supports a functional role for these cells, though the role has not been determined.

Other cell types present in the lacrimal gland include lymphocytes (containing IgA, IgG, IgM, IgE, and IgD), plasma cells, helper and suppressor T cells, B cells, dendritic cells, macrophages, bone marrow-derived monocytes, and mast cells. Lymphocytes are located between acinar and ductal cells, scattered throughout interstitial regions among the secretory tubules, or located within small, periductular, lymphoid aggregates (Sullivan, 1999). In humans, more than 50% of mononuclear cells in the lacrimal gland are plasma cells (Wieczorek et al., 1988). The vast majority of these plasma cells are IgA-positive. These cells synthesize and secrete IgA, which then is transported into acinar and ductal cells and secreted by these epithelial cells as secretory IgA (SIgA), an important component of the mucosal immune system of the eye.

Regulation of Secretion

The primary function of the lacrimal gland is to secrete proteins, electrolytes, and water onto the ocular surface. Since both the amount and the composition of lacrimal gland fluid are critical in maintaining a healthy ocular surface, it is critical that they be tightly regulated. Hormones and nerves play integral roles in the regulation of lacrimal gland secretion. Hormones are primarily responsible for stimulating constitutive protein secretion (secretion controlled by synthesis) while nerves are primarily responsible for stimulating regulated protein secretion (secretion controlled by release). Hormones from the hypothalamic–pituitary–gonadal axis, such as α-melanocyte stimulating hormone (α-MSH), adrenocorticotropic hormone (ACTH), prolactin, androgens, estrogens, and progestins have been shown to exert a significant influence on the lacrimal gland (Sullivan et al., 1998). Androgens are potent hormones that stimulate the secretion of SIgA and cystatin-related protein, and its action accounts for many of the gender-related differences seen in the lacrimal gland (Sullivan et al., 1998). If the hypothalamic–pituitary–gonadal axis is disrupted, (e.g., by hypophysectomy or anterior pituitary ablation) glandular atrophy results. It has also been shown that prolactin is synthesized in acinar cells and that vasopressin and ACTH may be produced by, or possibly accumulated within, lacrimal epithelial and myoepithelial cells, respectively (Frey et al., 1986; Mircheff, 1993; Djeridane, 1994; Markoff et al., 1995). In addition, glucocorticoids, retinoic acid, insulin, and glucagon are also known to affect various aspects of the lacrimal gland (Sullivan et al., 1998). These hormones interact with specific receptors through which their actions are mediated. The receptors for α-MSH, prolactin, androgens, estrogens, progestins, glucocorticoids, retinoic acid, and insulin are either transcribed and/or translated in lacrimal tissue.

Parasympathetic, sympathetic, and sensory nerves innervate the lacrimal gland. Parasympathetic innervation provides the primary input. A network of nerve fibers surrounds most acini (Ichikawa and Nakajima, 1962). While each individual acinar cell is not innervated, gap junctions electrically and chemically connect cells within an acinus so that even noninnervated cells can be neurally activated. Parasympathetic nerves are also located adjacent to duct cells and blood vessels. In the lacrimal gland, these nerves contain the neurotransmitters acetylcholine, a muscarinic, cholinergic agonist, and VIP. With stimulation, both neurotransmitters are released. Some VIP-containing nerves also hold another member of the VIP family of peptides, pituitary adenylate cyclase activating peptide (PACAP) (Elsas et al., 1996). While fewer than parasympathetic nerves, sympathetic nerves also innervate acinar, ductal, and vascular cells (blood vessels), but their importance is not known. Using antibodies to the enzymes that synthesize dopamine, norepinephrine, and epinephrine, it has been determined that the neurotransmitter released by sympathetic nerves is norepinephrine. These nerves also contain the peptide neurotransmitter NPY. Unlike the meibomian glands, NPY containing nerves have the same distribution as norepinephrine-containing nerves in the lacrimal gland, suggesting that sympathetic nerves contain both neurotransmitters. Sensory nerves are sparse in the lacrimal gland, but are adjacent to acinar cells. These nerves have been identified using antibodies to Sub P and calcitonin gene-related peptide (CGRP). Nerves containing galanin, usually found in sensory nerves, have also been detected (Adeghate and Singh, 1994). In addition, nerve fibers containing proenkephalin A-derived peptides (including met- and leu-enkephalin) have been demonstrated in the lacrimal gland (Lehtosalo et al., 1989). These fibers are numerous and surround acini and ducts, but not blood vessels. This distribution of nerve fibers along with the fact that sympathetic dennervation had no effect on these fibers suggests that they are of parasympathetic origin (Lehtosalo et al., 1989).

Neural reflexes are initiated by afferent sensory nerves in the cornea, conjunctiva, and nasal mucosa responding to mechanical, thermal, or chemical stimulation or by the optic nerve responding to light. These activate the efferent parasympathetic and sympathetic nerves of the lacrimal gland, which release their neurotransmitters. The neurotransmitters interact with specific receptors on the basolateral membranes of acinar and duct cells. This interaction activates a receptor specific to each neurotransmitter, which then initiates a cascade of intracellular events, known as signal transduction pathways. Activation of these signal transduction pathways induces fusion of the preformed secretory granules with the apical membrane to release secretory proteins into the lumen (regulated protein secretion). It also induces activation of ion channels and pumps in the apical and basolateral membranes to cause electrolyte and water secretion into the lumen. Peptide hormones released by activation of the hypothalamic–pituitary axis can also activate these same signal transduction pathways to cause regulated protein secretion. Steroid hormones, on the other hand, typically enter the acinar cells and interact with specific receptors in the nucleus. This interaction may increase the rate of synthesis of specific proteins. Once synthesized these proteins are packaged into secretory granules that, unlike granules containing regulated proteins, immediately fuse with the apical membrane and release secretory proteins into the lumen (constitutive protein secretion).

Signal Transduction Pathways Activated by Neurotransmitters and Peptide Hormones

In general, neurotransmitters and hormones are not cell permeant. Thus, the cells express protein receptors on their plasma membranes, which selectively interact with the extracellular stimuli (ligand). Signal transduction proceeds in three steps: initiation of the signal by interaction of the ligand with the receptor; amplification of the signal through the interaction of the receptor–G protein–effector enzyme, leading to the generation of second messenger molecules; and termination of the signal through the action of protein phosphatases and membrane pumps to bring the amount of phosphorylated proteins and ions, respectively, back to resting levels (Figure 6). The receptors can be subdivided into three classes based on their coupling system or mode of action: G protein-coupled receptors, tyrosine kinase coupled receptors, and intracellular receptors. G protein-coupled and tyrosine kinase coupled receptors will be described first. Intracellular receptors will be described in relation to steroid hormones.

G Protein-Coupled Receptors: G protein-coupled receptors are the largest family of receptors and include more than 1000 cloned receptors (Gutkind, 1998). In the lacrimal gland, as well as other tissues, the receptors for acetylcholine, VIP, norepinephrine, CGRP, α-MSH, and ACTH, are G protein-coupled receptors. These receptors are part of the seven transmembrane G protein-coupled receptor family (Morris and Malbon, 1999).

G Proteins: G proteins are made up of three polypeptides—an α subunit (molecular weight 39–46 kDa) that binds and hydrolyzes GTP, a β subunit (36 kDa), and a γ subunit (6–9 kDa). When GDP is bound, the α subunit associates with the βγ subunit to form an inactive heterotrimer that binds to the receptor. When an agonist binds to the receptor, the receptor undergoes a conformational change. This, in turn, causes a conformational change of the α subunit that decreases its affinity for GDP so that GDP comes off the active site. Because the cellular concentration of GTP is much higher than that of GDP, the GDP is replaced with GTP. This activates the α subunit, and it dissociates from both the receptor and the βγ subunit. The GTP is hydrolyzed to GDP by the intrinsic GTPase activity of the α subunit. Once GTP is hydrolyzed to GDP, the α and βγ subunits reassociate, become inactive, and reassociate with the receptor (Hamm and Gilchrist, 1996).

Molecular cloning has identified over 16 different G protein α subunits in mammalian cells. These can be divided into four major classes based on sequence similarity: α_s, α_i, α_{12}, and α_q (Gutkind, 1998). Their amino acid sequences are from 45 to 80% identical (Hamm and Gilchrist, 1996) with each group of the α subunits regulating a different set of target effectors. There are 5 known mammalian β subunits, and 11 γ subunits that are more heterogeneous than the α or the β subunits (Gutkind, 1998). It is

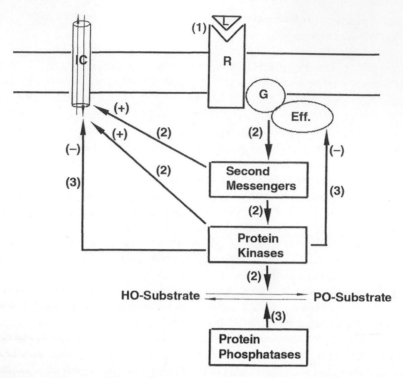

Figure 6. Schematic drawing showing the signal transduction pathways activated by neurotransmitters and peptide hormones. Typically, signal transduction proceeds in three steps: (1) initiation of the signal by interaction of the ligand (L) with the receptor (R); (2) amplification of the signal through interaction of the activated receptor with the G protein (G) and effector enzyme (Eff.) leading to the generation of second messengers which activate protein kinases and ion channels (IC) and; (3) termination of the signal through activation of protein phosphatases and inhibition of ion channels.

now well established that α and $\beta\gamma$ subunits can both positively and negatively regulate target effectors (Sunahara et al., 1996).

Target Effectors: These can be divided into three classes—phospholipases, adenylyl and guanylyl cyclases, and ion channels.

PHOSPHOLIPASES: Three types of phospholipases (phospholipase A_2, C, and D) have been described in mammals and are classified according to position on the phospholipid substrate they cleave (Figure 7). Phospholipase A_2 specifically hydrolyzes the fatty acid from the *sn*-2 position to generate a lysophospholipid and a free fatty acid. Phospholipase C (PLC) hydrolyzes the bond between phosphoric acid and glycerol to generate diacylglycerol (DAG). Phospholipase D (PLD) removes the X-group (alcohol component

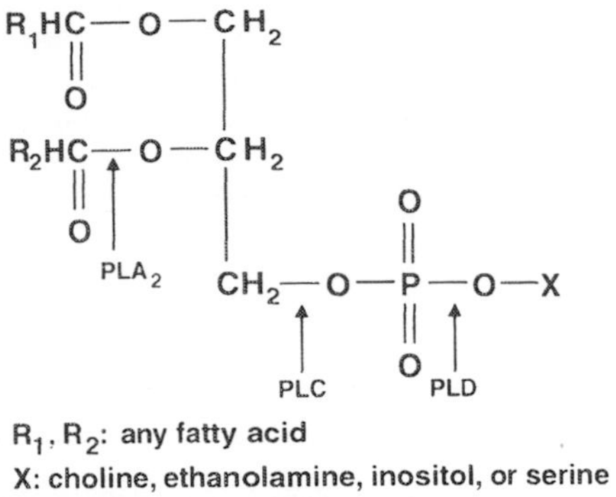

R_1, R_2: any fatty acid

X: choline, ethanolamine, inositol, or serine

Figure 7. Schematic drawing showing the location at which phospholipase A_2 (PLA$_2$), phospholipase C (PLC), and phospholipase D (PLD) cleave phospholipid substrates.

of the phospholipid) to generate phosphatidic acid (PA). As the role of phospholipase A_2 in the lacrimal gland has not yet been determined, we will focus on phospholipases C and D.

In mammals, there are two types of PLC: a phosphatidylcholine-specific PLC (PC-PLC) and a phosphoinositide-specific PLC (PI-PLC). PI-PLC is well characterized and will be referred to as PLC and will be discussed further. Molecular cloning has revealed that there are three families of PLCs: β, δ, and γ with molecular weights ranging from 85 to 154 kDa. Within each family, there are several cloned subtypes designated by adding Arabic numerals after the Greek letters (Cockroft, 1992). All three types of PLC catalyze the hydrolysis of the three common inositol-containing phospholipids: phosphatidylinositol (PI), phosphatidyl 4-monophosphate (PIP), and phosphatidyl 4,5-bisphosphate (PIP$_2$), with PIP and PIP$_2$ the preferred substrates. Members of the β PLC family are activated by G protein-coupled receptors whereas activation of the γ PLC family members occurs through tyrosine phosphorylation by tyrosine kinase coupled receptors, such as those activated by growth factors. Neither the receptors nor the transducers that are coupled to any of the δ PLC members are known (Berridge, 1993; Tsunoda, 1993). Hydrolysis of PIP$_2$ generates two-second messenger molecules: inositol 1,4,5-trisphosphate (IP$_3$) and DAG. IP$_3$ acts by raising the intracellular concentration of Ca^{2+} and DAG by activating protein kinase C (PKC) (Berridge, 1993).

IP$_3$, a water-soluble molecule, diffuses to the endoplasmic reticulum where Ca^{2+} is stored in an inactive, bound form. IP$_3$ interacts with specific receptors on the endoplasmic reticulum to release Ca^{2+} into the cytosol (Berridge, 1993). The receptors, to which IP$_3$ binds, initially identified by structure–activity and radioligand-binding studies, have been purified and functionally reconstituted into lipid vesicles from which single channel events have been resolved. The IP_3 receptor is a homotetramer (four identical subunits) of ≈ 310 kDa each and constitutes one of the largest of all known ion channels. Binding sites for IP_3 are located within the N-terminal domain, whereas the C-terminal regions form the intrinsic Ca^{2+} channel. Multiple isoforms of IP_3 receptor have been cloned. They share significant similarity to each other and are encoded by at least four genes. Regulation of the IP_3 receptor is complex in that it binds multiple IP_3 molecules, is desensitized by IP_3 itself, is phosphorylated by protein kinase A, and has biphasic sensitivity to cytoplasmic Ca^{2+} levels (Berridge, 1993).

Nishizuka et al. first reported the purification of PKC and showed that it is the target of DAG, establishing DAG as a second messenger molecule (Kishimoto et al., 1977; Nishizuka, 1984). Protein kinase C was also shown to be the cellular receptor for phorbol esters, a class of tumor promoters (Nishizuka, 1984). Molecular cloning and biochemical techniques have shown that PKC is a family of closely related enzymes consisting of at least 11 different isozymes. The PKC family has been divided into three categories based on structural and functional criteria (Figure 8) (Jaken, 1996). A first group, termed classical or conventional PKCs (cPKC), including PKCα, -βI, -βII and -γ isoforms, have a Ca^{2+} and DAG-dependent kinase activity.

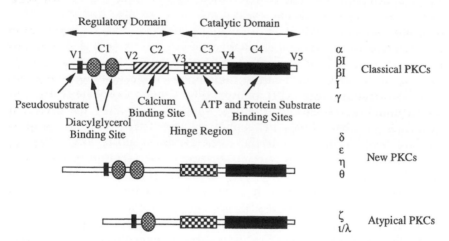

Figure 8. Schematic drawing showing the different structural domains of protein kinase C (PKC) and the individual PKC isoforms that comprise the three families of PKC isoforms. C1–4, constant regions, V1–5, variable regions. Reprinted from Advanced Experimental Biology, Vol. 438, Dartt, D.A., Hodges, R.R., Zoukhri, Signal transduction pathways activated by cholinergic and α_1-adrenergic agonists in the lacrinal gland, pp.113–121., 1998, with permission from Springer-Verlag.

A second group, termed new or novel PKCs (nPKC), including PKC-δ, -ϵ, -θ, and -η, isoforms, are Ca^{2+}-independent and DAG-stimulated kinases. A third group termed atypical PKCs (aPKC), including PKCζ, and -ι/λ isoforms, are Ca^{2+} and DAG-independent kinases. Protein kinase C isoforms have a molecular weight ranging from 74 to 115 kDa. The brain is the only known tissue to contain all isoforms of PKC, whereas peripheral tissues usually express between three and eight isoforms.

Structurally, PKC can be divided into two domains: a regulatory domain in the amino terminal half and a catalytic domain in the carboxyl terminal half (Figure 8). The regulatory domain of cPKC contains two conserved regions C1 and C2. The C1 region contains two tandem repeats of a cysteine-rich motif responsible for DAG and phorbol ester binding. The aPKC isoforms, unlike cPKC and nPKC isoforms, contain only one cysteine-rich motif and hence do not bind DAG/phorbol esters. The C2 region, called CalB, interacts with phospholipids in a Ca^{2+}-dependent manner and is implicated in translocation to membranes and activation of cPKC isoforms. The C2 region in nPKC lacks critical aspartate residues necessary for Ca^{2+} binding and thus is Ca^{2+} independent. Atypical PKC isoforms lack the C2 region entirely and also do not need Ca^{2+} for their enzymatic activity (Newton, 1997). In addition, all PKC isoforms have a pseudosubstrate sequence in their N-terminal part, which interacts with the catalytic domain to keep the enzyme inactive in resting cells. The catalytic domain of PKC contains two additional conserved regions, C3 and C4 responsible for ATP and protein substrate binding, respectively.

Activation of PKC modulates a plethora of cellular processes including secretion and exocytosis, cell proliferation and differentiation, and modulation of ion conductance. PKC isoforms do not show selectivity towards protein substrates *in vitro*. There is evidence, however, indicates that selectivity of action of PKC isoforms is dictated, in part, by their cellular localization. Intracellular proteins, such as RACKs (receptors for activated C-kinase) or proteins interacting with C-kinase (PICKs) target selective isoforms of PKC to different loci (Aderem, 1992; Mochly-Rosen et al., 1998). Such compartmentalization will limit the accessibility/availability of substrates to a given isoform of PKC.

Phospholipase D has been partially purified from many sources and has been cloned from yeast, bacteria, plant, and mammalian sources. The first reported mammalian PLD (PLD1) has 1072 amino acids and a molecular mass of 124 kDa (Meier et al., 1999). A second PLD (PLD2) which has 932 amino acids and 51% sequence identity to PLD1 has been cloned from a mouse embryonic library (Colley et al., 1997). The subcellular distribution and the mechanism of activation of PLD1 and PLD2 are different. PLD2 localizes predominantly in the plasma membrane, whereas PLD1 is perinuclear (i.e., in endoplasmic reticulum, Golgi, and late endosomes) (Colley et al., 1997). Protein kinase C, ARF, RhoA, and PIP2 regulate PLD1, whereas PLD2 is stimulated by PIP_2, but not PKC, ARF, or RhoA (Houle and Bourgoin, 1999).

Phospholipase D catalyzes the hydrolysis of phospholipids producing PA and the free polar head group (Billah and Anthes, 1990). Phosphatidic acid by itself or after its conversion to DAG by a PA phosphohydrolase is an important second messenger molecule. In the presence of a primary alcohol, PLD possesses the unique ability to catalyze a transphosphatidylation reaction in which the phosphatidyl moiety of the phospholipid substrate is transferred to the primary alcohol producing the corresponding phosphatidylalcohol (Dawson, 1967). Accumulation of such unique transphosphatidylation products has been used to detect PLD activity unambiguously in diverse cell types. Receptor-mediated activation of PLD has been demonstrated in many cell types. The mechanism by which receptors couple to PLD is poorly understood but evidence suggests that diverse mechanisms might be involved. Receptor activation of PLD appears to occur through mechanisms involving PKC activity, Ca^{2+}, G proteins, or receptor-linked tyrosine kinases. Phospholipase D is activated by phorbol esters, suggesting a major role for PKC in regulating PLD (Nishizuka, 1992). Agents mobilizing Ca^{2+}, such as Ca^{2+} ionophores, also activate of PLD (Exton, 1988). Since PKC activation and Ca^{2+} mobilization are downstream to PLC stimulation, it has been suggested that PLD activation may be secondary to receptor activation of PLC. However, a direct coupling of PLD to receptor activation, via a G-protein, has also been shown (Cockroft, 1992). Phospholipase D can also be activated by growth-factors binding to receptor tyrosine kinases, such as EGF receptor, or by pharmacological agents that enhance the accumulation of tyrosine-phosphorylated proteins, suggesting that tyrosine phosphorylation may represent an alternative pathway to PLD activation (Bourgoin and Grinstein, 1992).

Adenylyl and guanylyl cyclases: Adenylyl cyclase cleaves ATP to generate cyclic AMP (cAMP), whereas guanylyl cyclase generates cyclic GMP (cGMP) from GTP. cAMP and cGMP interact with and activate protein kinase A and G, respectively, leading to the cellular response. Adenylyl cyclase is by far the best-characterized mammalian cyclase and will be discussed further. Readers can find a review on guanylyl cyclase by Foster et al. (1999).

Molecular cloning has identified several isoforms of mammalian adenylyl cyclases forming a family of at least 10 enzymes (I through X) (Sunahara et al., 1996). Adenylyl cyclases have a molecular mass of $\approx$120 kDa, ranging from 1064 to 1248 amino acid residues with a complex topology within the membrane. A short cytoplasmic amino terminus is followed by six transmembrane spanning regions and a large cytoplasmic domain ($\approx$40 kDa). The motif is then repeated (Taussig and Gilman, 1995). The overall amino acid sequence similarity among the different isoforms of adenylyl cyclase is

Table 1. Regulation of Adenylyl Cyclases

Regulator	AC Subtype	Effect
Forskolin	All	Stimulation
Gsα	All	Stimulation
Giα	ACI, V, VI	Inhibition
Gzα	ACI, V	Inhibition
Goα	ACI	Inhibition
Gβγ	ACI	Inhibition
	ACII, IV	Stimulation
Ca^{2+}/Calmodulin	ACI, III, VIII	Stimulation
Ca^{2+}	ACV, VI	Inhibition
PKC	ACII, V	Stimulation
PKA	ACV, VI	Inhibition

Sunahara, R., Dessauer, C. and Gilman, A. (1996). Complexity and diversity of mammalian adenylyl cyclases. Ann. Rev. Pharmacol. Toxicol. 36, 461–480.

roughly 60% (Sunahara et al., 1996). All isoforms of adenylyl cyclase appear to be expressed in the brain, apparently in region-specific patterns, whereas there are substantial differences in patterns of expression in peripheral tissues.

Regulation of adenylyl cyclase enzymatic activity is complex and isoform specific. All adenylyl cyclase isoforms are activated by Gsα (Sunahara et al., 1996). The βγ subunit complex of Gs has been shown to exert an inhibitory effect on type I adenylyl cyclase and stimulatory effects on types II and IV. The effects of α subunits of the G proteins Gi, Go, and Gz, as well as that of Ca^{2+}, PKC, and protein kinase A (PKA), on adenylyl cyclase isoforms are summarized in Table 1 (Sunahara et al., 1996).

Most of the biological effects of cAMP are mediated through PKA. Protein kinase A is a ubiquitous serine and threonine protein kinase. When inactive, PKA consists of a complex of two catalytic (C) subunits and two regulatory (R) subunits. Binding of cAMP to the R subunit alleviates an autoinhibitory contact that releases the active C subunit. The active kinase is then free to phosphorylate-specific protein substrates. Mammalian cells contain three C subunit isoforms, α, β, and γ, with Cα and Cβ being predominant, and two R subunits, RI and RII (Dell'Acqua and Scott, 1997).

As discussed for PKC, PKA can phosphorylate a plethora of protein substrates. To confer specificity to the external signal, both the levels of cAMP and the activation states of PKA must be tightly controlled. Total cAMP levels are determined by a balance of cellular adenylyl cyclase to produce cAMP and phosphodiesterase activities to breakdown cAMP. There are also signal-terminating mechanisms, such as desensitization of adenylyl cyclase. In addition, mammalian cells contain a protein kinase inhibitor (PKI) which sequesters free C subunits and mediates the movement of the subunit to the nucleus, which is free of R subunits. Evidence showed that differences in subcellular targeting of PKA isoforms are additional factors contributing to specificity in cellular responses. Several proteins ranging in size from 15 to 300 kDa have been detected in a variety of tissues and shown to associate with PKA. These proteins have been named A-kinase anchoring proteins (AKAP) and shown to target different PKA isoforms to different loci in the cell (Dell'Acqua and Scott, 1997).

Tyrosine Kinase Coupled Receptors: All known receptor tyrosine kinases phosphorylate themselves on tyrosine residues (autophosphorylation) in response to agonist binding. This autophosphorylation plays a critical role in mediating the actions of growth factors (Sternberg and Alberola-Ila, 1998). The phosphotyrosine residues appear to serve as highly selective binding sites to which cytoplasmic signaling molecules bind. These signaling molecules mediate the pleiotropic responses of cells to growth factors. The best-studied receptors are the platelet-derived growth factor receptor (PDGFR) and the EGF receptor (EGFR). These studies showed that upon ligand binding, the intracellular region of the PDGFR or the tail region of the EGFR binds several signaling molecules, including phosphatidylinositol 3-kinase (PI3-K), PLCγ, GTPase-activating factor (GAP), and c-Src with each molecule binding to specific phosphotyrosine residues on the receptor (Malarkey et al., 1995).

Activation of the growth factor receptor also induces the tyrosine phosphorylation of the adaptor protein, Shc and the consequent formation of a Shc-GRB2 complex (Figure 9). This interaction causes the recruitment of SOS, which is a guanosine nucleotide exchange factor for Ras. SOS facilitates an exchange of GTP for GDP on Ras, which activates Ras. Ras, in turn, activates a cascade of kinases including MAPKKK, also known as cRaf1, and MAPKK, also known as MEK1 and MEK2 (Davis, 1993; Sternberg and Alberola-Ila, 1998). Ultimately mitogen-activated protein kinase (MAPK), also known as extracellularly regulated kinase (ERK), is activated. There are two types of MAPK–p44 (ERK1) and p42 (ERK2). When phosphorylated on both serine and threonine residues, MAPK becomes active. Activated MAPK phosphorylates key enzymes and also translocates to the cell nucleus where it regulates the expression of genes that cause proliferation, differentiation, and other cell functions (Treisman, 1996; Robinson and Cobb, 1997).

Recent investigations have found that G protein-linked receptors can activate the growth factor receptor stimulated pathway and hence activate growth, cell proliferation, and other cell functions controlled by the nucleus (Gutkind, 1998).

Signal Transduction Pathways in the Lacrimal Gland

In the lacrimal gland, acetylcholine and VIP, the parasympathetic neurotransmitters, are potent stimuli of regulated protein and electrolyte/water secretion. Norepinephrine, the sympathetic neurotransmitter, is also

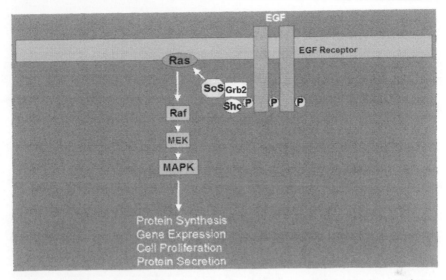

Figure 9. Schematic drawing of a growth receptor-activated signaling pathway. Epidermal growth factor (EGF) is used as the example. Epidermal growth factor binds to its receptor, a tyrosine kinase (TK), which dimerizes and autophosphorylates (P). The phosphorylation attracts the adapter proteins Shc, Grb2, and SOS. SOS is a guanine nucleotide exchange factor that activates Ras. Ras activates a cascade of kinases Raf, MEK, and mitogen-activated protein kinase (MAPK). MAPK translocates to the nucleus where it activates genes to cause protein synthesis, gene expression, proliferation, and protein secretion.

a potent stimulus of regulated protein secretion, but a weak stimulus of electrolyte/water secretion. Neuropeptide Y, the sympathetic neurotransmitter, and CGRP, the sensory neurotransmitter, are weak stimuli of protein secretion. Substance P, the sensory neurotransmitter, does not appear to stimulate either protein or electrolyte/water secretion while α-MSH and ACTH are potent, effective stimuli of protein secretion, but their effect on electrolyte and water secretion is unknown (Dartt, 1994). The signaling pathways activated by these agonists in the lacrimal gland will be discussed.

Cholinergic Agonist-Activated Signal Transduction Pathway

Acetylcholine, released from parasympathetic nerves, activates muscarinic receptors on the basolateral membrane of lacrimal gland cells (Figure 10). Of the five receptor subtypes (M_{1-5}) identified, only the M_3 or glandular subtype is present in the lacrimal gland (Mauduit et al., 1993). These receptors are coupled to the $G_{\alpha q/11}$ subtype of G protein, which is, in turn, coupled to PLC (Mauduit et al., 1993; Meneray et al., 1997). As previously discussed

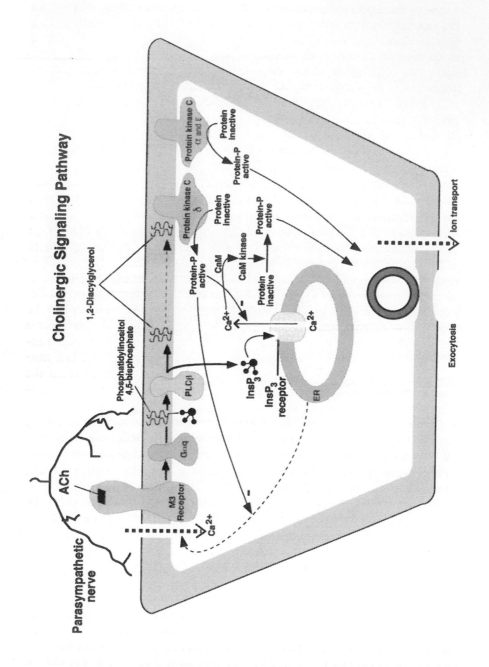

Cholinergic Signaling Pathway

1,4,5-IP$_3$ and DAG are produced from the hydrolysis of PIP$_2$ by PLC. 1,4,5-IP$_3$ interacts with its receptors on the endoplasmic reticulum to open channels to release Ca^{2+} into the cytoplasm. Depletion of these Ca^{2+} stores leads to an increase in the influx of extracellular Ca^{2+} across the plasma membrane. Ca^{2+}, either alone or through Ca^{2+}/calmodulin-dependent protein kinases to phosphorylate-specific substrates, stimulates secretion.

The DAG formed from the hydrolysis of PIP$_2$ activates PKC (Figure 10). Five isoforms of PKC are expressed in the rat lacrimal gland: one cPKC, PKCα; two nPKCs, PKCδ and -ϵ; and one aPKC, PKC-ι/λ. Using immunofluorescence techniques, these isoforms were shown to have a differential localization (Zoukhri et al., 1997a). This is of great interest because part of the selectivity of action of PKC isoforms is dictated by their localization (i.e., accessibility to their substrates). Cell fractionation techniques combined with western blotting showed that cholinergic agonists caused a differential translocation of PKC isoforms. PKCϵ translocated from the cytosol to the plasma membrane 30 s after cholinergic stimulation and stayed associated with the membrane at 10 min. In contrast, PKCδ transiently translocated to the membrane reaching a maximum by 1 min and returning to the cytosol by 5 min, whereas we could not detect a significant translocation of PKCα. PKC-ι/λ, as expected for aPKC isoforms, did not translocate in response to cholinergic agonists (Zoukhri et al., 1997a). In an attempt to define the role that individual PKC isoforms might play in regulating lacrimal gland functions in response to cholinergic stimulation, we have synthesized isoform-specific peptide inhibitors of PKC. These peptides were derived from the unique pseudosubstrate sequences of PKCα, -δ, and -ϵ and were myristoylated at their N-terminus to make them cell permeant. Using this strategy we showed that cholinergic agonists activate PKCα and -ϵ to a larger extent than PKCδ to induce protein secretion (Zoukhri et al., 1997b). In contrast, Ca^{2+} elevation stimulated by cholinergic agonists was negatively modulated by PKCδ and to a lesser extent by PKCϵ, but not by PKCα (Zoukhri et al., 2000).

Activation of Phospholipase D: Taking advantage of the transphosphatidyl reaction catalyzed by PLD to produce a phosphatidylalcohol we showed

---

Figure 10. Schematic drawing of signal transduction pathway activated by cholinergic agonists in lacrimal gland acinar cells to stimulate protein and electrolyte/water secretion. M$_3$, subtype 3 of muscarinic receptors; G$_{\alpha q/11}$, q/11 subtype of α-subunit of a guanine nucleotide-binding protein; InsP$_3$, inositol trisphosphate; protein-P, phosphorylated protein; ER, endoplasmic reticulum. Reprinted from Advanced Experimental Biology, Vol. 438, Dartt, D.A., Hodges, R.R., Zoukhri, Signal transduction pathways activated by cholinergic and α_1-adrenergic agonists in the lacrinal gland, pp.113–121., 1998, with permission from Springer-Verlag.

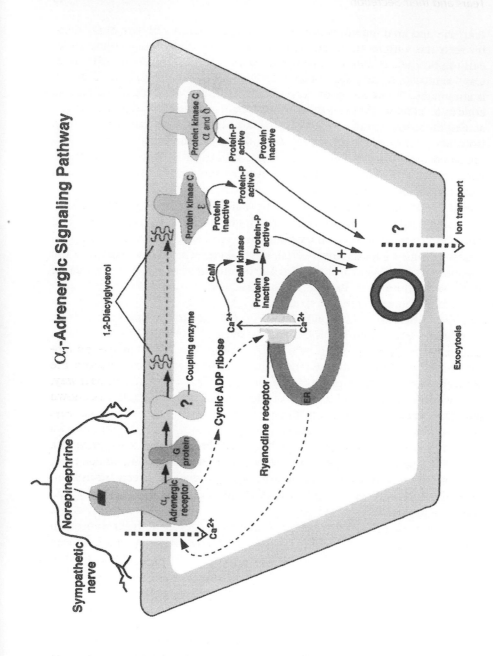

α₁-Adrenergic Signaling Pathway

that the lacrimal gland contains a PLD activity. Cholinergic agonists, through the muscarinic receptor, stimulate both the hydrolytic activity of PLD to produce PA, and the transphosphatidyl reaction. However, if either Ca^{2+} is mobilized or PKC is activated only the transphosphatidyl reaction is stimulated. This finding has two fundamental implications. First, cholinergic agonist activation of PLD in the lacrimal gland is not secondary to the activation of PLC by these agonists. Second, lacrimal gland may contain two isoforms of PLD, one regulated through receptor activation, and the second regulated through Ca^{2+} and PKC. Another finding, unique to the lacrimal gland, is that α_1-adrenergic agonists failed to activate both properties of PLD (Zoukhri and Dartt, 1995).

Ca^{2+}, Ca^{2+}/calmodulin-dependent protein kinases, and PKC phosphorylate-specific substrates that cause lacrimal gland protein and electrolyte and water secretion. The phosphorylation targets are not known. The second messengers also stimulate water and electrolyte secretion by activating ion channels and ion pumps in the basolateral and apical membranes to cause ions, followed by water, to move into the lumen.

α_1-Adrenergic Agonist-Activated Signal Transduction Pathway

Norepinephrine, released from sympathetic nerves, binds to α_1- and β-adrenergic receptors on lacrimal gland cells. β-Adrenergic receptors are coupled to AC to activate a cAMP-dependent signal transduction pathway, as is VIP and will be discussed subsequently. Molecular cloning has shown that there are three subtypes of α_1-adrenergic receptors named α-$_{1A/D}$, α-$_{1B}$, and α-$_{1C}$ and the subtype of α_1-adrenergic receptor present in the lacrimal gland has been identified as α-$_{1D}$ (unpublished observations) (Figure 11). Neither the subtype of G protein nor the effector enzyme is known. In most exocrine tissues, α_1-adrenergic agonists activate the same pathway as cholinergic agonists (i.e., activation of PLC and PLD). However, in the lacrimal gland α_1-adrenergic agonists do not activate PLC or PLD (Hodges et al., 1992; Zoukhri and Dartt, 1995). Activation of the α_1-adrenergic receptors in the lacrimal gland does lead to a slight increase in [Ca^{2+}], which has been proposed to occur through generation of cyclic ADP ribose and activation of ryanodine receptors that release Ca^{2+} into the cytosol (Gromada et al.,

Figure 11. Schematic drawing of signal transduction pathway activated by α_1-adrenergic agonists in lacrimal gland acinar cell to stimulate protein and electrolyte/water secretion. G, guanine nucleotide-binding protein; ADP, adenosine diphosphate; protein-P, phosphorylated protein; ER, endoplasmic reticulum; Calmodulin. Reprinted from Advanced Experimental Biology, Vol. 438, Dartt, D.A., Hodges, R.R., and Zoukhri, Signal transduction pathways activated by cholinergic and α_1-adrenergic agonists in the lacrinal gland, pp.113–121., 1998, with permission from Springer-Verlag.

1995). However, it appears that the major mechanism by which α_1-adrenergic agonists stimulate secretion is by activating specific isozymes of PKC. Using the myristoylated pseudosubstrate-derived peptides discussed previously, we found that α_1-adrenergic agonists activate PKCε to stimulate protein secretion. Surprisingly, inhibition of PKCα and -δ isoforms led to increased protein secretion suggesting that these two isoforms inhibit protein secretion triggered by α_1-adrenergic agonists (Zoukhri et al., 1997b). This is in contrast to the stimulatory effect that PKCα and -δ isoforms have on protein secretion when activated by cholinergic agonists or phorbol esters. This implies that the effect (inhibitory or stimulatory) of a given isoform of PKC is stimulus-dependent and might in part depend on the cellular localization of this isoform. Interestingly, α_1-adrenergic agonists did not promote the translocation of PKC again contrasting with the effect of cholinergic agonists further supporting the role of PKC location in dictating its action (Hodges et al., 1992).

As with cholinergic agonists, the PKC activated by α_1-adrenergic agonists stimulates secretion by phosphorylating and activating a specific set of protein substrates. This leads to activation of ion channels and pumps to induce protein and electrolyte/water secretion.

VIP-Activated Signal Transduction Pathway

Vasoactive intestinal peptides released from parasympathetic nerves, norepinephrine released from sympathetic nerves using β-adrenergic receptors, and the peptide hormones α-MSH and ACTH from the bloodstream each activate cAMP-dependent signal transduction pathways (Dartt, 1994) (Figure 12). The VIP-dependent pathway will be described. Vasoactive intestinal peptides interact with specific VIP receptors located on the basolateral membranes of lacrimal gland cells. Two types of VIP receptors have been identified, VIPRI and VIPRII, both of which are present in the lacrimal gland (Hodges et al., 1997). VIPRI is the predominate receptor. VIPRI is located in acinar and duct cells, whereas VIPRII is located in the myoepithelial cells that surround the acini. Activation of the VIP receptors stimulates the $G_{\alpha s}$ subtype of G proteins (Meneray et al., 1997). The effector enzyme activated by VIP is adenylyl cyclase, which produces cAMP from ATP. There are at least three isoforms of AC present in the lacrimal gland: ACII was present only on myoepithelial cells; ACIII was present in occasional ducts, blood vessels, and myoepithelial cells; and ACIV was present on all acinar cells in a subcellular structure that appeared to be Golgi apparatus or perhaps endoplasmic reticulum (Hodges et al., 1997). The subcellular location of ACIV appeared to be in the Golgi or ER, which was unusual, as one would expect to find it on the basolateral membrane in the same location as the VIP receptors. It is possible that ACIV might be recruited from the

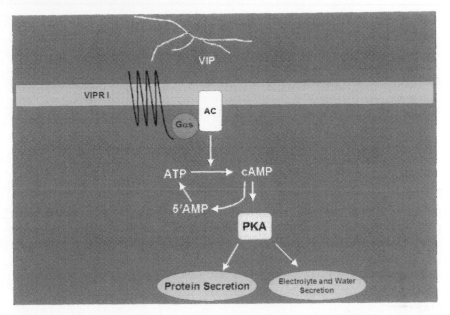

Figure 12. Schematic drawing of vasoactive intestinal peptide (VIP)- or norepinephrine-activated signal transduction pathway used by lacrimal gland acinar cells to stimulate protein and electrolyte/water secretion. $G_{\beta\gamma}$, β, and γ subunits of a guanine nucleotide binding protein; GDP, guanine diphosphate; ATP, adenosine triphosphate; cAMP, cyclic adenosine monophosphate; 5'-AMP, 5''-analog of adenosine monophosphate; protein-P, phosphorylated (activated protein). (Modified from Dartt, D.A. (1994). Regulation of tear secretion. Adv. Exp. Med. Biol. 350, 5.)

intracellular membranes upon stimulation with VIP. Alternatively, other isoforms of adenylyl cyclase might be present on acinar basolateral membranes and used for VIP stimulation. It is notable that myoepithelial cells contain VIP receptors and adenylyl cyclase, suggesting that VIP also activates them, and they could play a role in secretion or in another cellular process activated by VIP.

The increase in the cellular level of cAMP by AC activates PKA, which phosphorylates a specific set of protein substrates. This in turn activates of ion channels and pumps to induce protein and electrolyte/water secretion. The cAMP signal is terminated by the enzyme cAMP phosphodiesterase, which converts cAMP to the inactive 5'-AMP. cAMP levels can be increased experimentally by forskolin, which directly activates adenylyl cyclase, and by cAMP phosphodiesterase inhibitors, such as isobutylmethylxanthine (IBMX), pentoxifylline, theophylline, and papaverine, that prevent breakdown of cAMP.

In addition to VIP, β-adrenergic agonists, α-MSH, and ACTH each activate the cAMP pathway. There are differences in secretion stimulated by each agonist. VIP is a potent stimulus of protein and electrolyte/water secretion whereas β-adrenergic agonists are weak stimuli of protein and electrolyte/water secretion. α-MSH and ACTH are potent stimuli of protein secretion, but their effects on electrolyte/water secretion are unknown. This suggests that there is differential coupling of cAMP-dependent receptors to $G_{\alpha s}$ and adenylyl cyclase.

In the lacrimal gland three different signal transduction pathways exist for stimulating lacrimal gland secretion: cholinergic agonist-, α_1-adrenergic agonist-, and VIP-activated. Since the nerves innervating the lacrimal gland contain at least two different neurotransmitters, stimulation of one type of nerve can release two or more agonists and, using separate signal transduction pathways activate secretion. Thus, the lacrimal gland is designed to secrete with several backup mechanisms to ensure that scretion can occur. In spite of this design, lacrimal gland secretion can be compromised, resulting in a spectrum of diseases classified as dry eye syndromes or keratoconjunctivitis sicca. In particular aqueous deficient, as opposed to evaporative, dry eye does occur.

EGF-Activated Signal Transduction Pathway

The lacrimal gland contains EGF receptors, which can activate the MAPK pathway in the lacrimal gland (Marechal et al., 1996) (Figure 9). Epidermal growth factor binds to its receptor in the lacrimal gland and dimerizes causing tyrosine phosphorylation of each subunit. This attracts Shc, GRB2, and SOS. SOS activates Ras and ultimately the MAPK cascade to phosphorylate p44 and p42 MAPK (Ota et al., 2003).

The activation of G protein-coupled receptors by cholinergic and α_1-adrenergic agonists activates MAPK through different mechanisms (Ota et al., 2003). α_1-Adrenergic agonists transactivate the EGF receptor either directly or by activating nonreceptor tyrosine kinases. This causes activation of Shc, GRB2, and SOS. SOS activates Ras, which then activates the MAPK cascade to activate p44 and p42 MAPK. In contrast, cholinergic agonists do not activate the EGF receptor or activate Shc, GRB2, or SOS. They do activate p44 and p42 MAPK probably by activating PKC. The step at which cholinergic agonists act is probably distal in the pathway to Ras and is perhaps at MAPKKK. Activation of MAPK by EGF or by cholinergic and α_1-adrenergic agonists negatively modulates lacrimal gland protein secretion (Ota et al., 2003).

Inhibitor-Activated Signal Transduction Pathway: Inhibitors of secretion are also present in the lacrimal gland. To date one family of inhibitors, the enkephalins, has been identified. Four different enkephalin sequences contained in preproenkephalin A, including met- and leu-enkephalin, have been identified in nerves surrounding acinar and ductal cells (Walcott, 1990). Enkephalins bind to δ opioid receptors, which activate the $G_{\alpha i}$ subtype of G proteins. This interaction prevents the activation of AC by $G_{\alpha s}$ by VIP, β-adrenergic agonists, α-MSH, or ACTH, preventing an increase in electrolyte, water, and protein secretion by these agonists (Cripps and Bennett, 1994).

Signal Transduction Pathways Activated by Steroid Hormones

The steroid hormones, androgens, are a major regulator of constitutively secreted proteins from the lacrimal gland. However, unlike neural and peptide hormone stimulation in which secretion occurs in seconds or minutes, androgens typically provide long-term regulation taking hours or days to be effective. Thus, steroids use a very different signal transduction pathway (Sullivan, 1999a).

One constitutively secreted protein whose secretion is controlled by androgens is secretory IgA (SIgA), the predominant immunoglobulin in tears and the primary mediator of the secretory immune system of the eye. It consists of polymeric IgA (pIgA) coupled to J chain and secretory component. IgA and J chain are produced by plasma cells, whereas the acinar and ductal cells synthesize secretory component (SC). Secretory component is synthesized as a precursor molecule called the polymeric immunoglobulin (pIg) receptor. This receptor is incorporated into the basolateral membrane of acinar cells and functions as an IgA receptor. The pIg receptor is composed of extracellular, membrane-spanning, and intracellular (cytoplasmic) domains. The pIg receptor binds to pIgA, which is coupled to J chain, and the receptor–ligand complex is endocytosed into distinct vesicles within the acinar cell (Figure 13). These vesicles are sorted directly to the apical membrane bypassing the lysosomal compartment. Within the vesicular component, the membrane spanning and cytoplasmic domains are cleaved from the extracellular domain of the pIg receptor to form SIgA. Once the vesicles reach the apical membrane, they immediately fuse with the apical membrane releasing SIgA into the glandular lumen to mix with the other secreted proteins, electrolytes, and water.

The classical pathway for androgens to control secretion is by controlling protein synthesis. Androgens diffuse into the nucleus and bind to specific, high-affinity receptors (Figure 13). These receptors are members of the steroid/thyroid hormone/retinoic acid family of ligand-activated transcription factors. The androgen–receptor complex then associates with a response element in the regulatory region of the SC target gene and dimerizes with another sex steroid-bound complex. In combination with appropriate coactivators and promoter elements, this increases SC gene transcription and

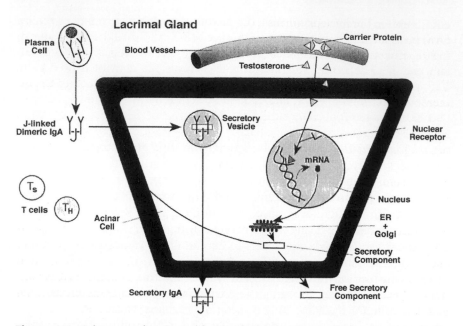

Figure 13. Schematic drawing of lacrimal gland acinar cell showing the classic mechanism for androgen (indicated here by testosterone) regulation of secretory immunoglobulin A (secretory IgA) secretion. ER, endoplasmic reticulum, T_S, suppressor T cell; T_H, helper T cell. (From Dartt, D.A. and Sullivan, D.A. (2000). Wetting of the ocular surface. In: Principles and Practice of Ophthalmology. (Albert, D. and Jakobiec, F., Eds.), 2nd edn., p. 970. W.B. Saunders Co., Philadelphia.)

eventually SC synthesis. This pathway is the most likely mechanism by which androgens control SC in the lacrimal gland as high-affinity androgen-specific receptors, which adhere to DNA, have been identified in acinar epithelial cells. Also, androgens enhance SC mRNA levels and androgen receptor, transcription or translation antagonists inhibit androgen-induced SC production (Sullivan, 1999a).

It is possible that androgens may also affect the secretion of other proteins or electrolytes/water through nonclassic pathways. These pathways are fast, occurring in seconds or minutes, and involve alteration of membrane fluidity, control of neurotransmitter receptors, and interaction with specific plasma membrane receptors (Brann et al., 1995). Androgens may also work by another recently discovered, nonclassic pathway. Lewin et al. hypothesized that androgens bind to the sex-hormone binding globulin, which interacts with its receptor located on the plasma membrane (Lewin, 1996). This activates AC to produce cAMP and activate PKA. Protein kinase A in turn activates cAMP response element binding protein (CREB).

CREB enters the nucleus, forms a dimer with itself, binds to, and activates a cAMP response element (CRE) on a gene. This stimulates transcription of the gene, which leads ultimately to translation. In addition, neurotransmitters and peptide hormones that stimulate the production of cAMP (e.g., VIP, β-adrenergic agonists, αMSH, and ACTH) could also stimulate CREB thereby affecting long-term lacrimal gland function.

The androgen-induced synthesis and secretion of SC by lacrimal gland acinar cells may also be influenced by neurotransmitters, cytokines, and secretagogues. VIP, β-adrenergic agonists, cAMP analogs (e.g., 8-bromoadenosine 3′: 5′-cyclic monophosphate), and other agents which stimulate the cAMP signaling pathway enhance androgen-induced SC secretion. This effect on SC is also observed with the cytokines IL-1α, IL-1β, TNF α, and prostaglandin E_2 (PGE$_2$). In contrast, androgen-induced secretion of SC is decreased by cholinergic agonists. In addition to these agents, insulin, extracellular calcium, high-density lipoprotein, and factors from the thyroid and adrenal glands may also modify androgen action on SC synthesis and secretion by lacrimal tissue (Sullivan, 1999a). Thus, the interactions between the neural, hormonal, and immune pathways for regulation of lacrimal gland secretion are complex and need to be investigated further.

Secretory Product

Water and Electrolyte Secretion: Electrolyte and water secretion occurs by very different mechanisms than protein secretion and has been studied in detail for the main lacrimal gland. It is important to note that it is not valid to sample tears when studying lacrimal gland electrolyte and water secretion, but rather it is necessary to analyze pure lacrimal gland fluid collected directly from the excretory duct. The electrolyte composition of the lacrimal gland fluid is different from interstitial fluid. The Na^+ and K^+ concentrations are greater in lacrimal gland fluid than in interstitial fluid and the Cl^- concentration is lower. However, the electrolyte composition of the lacrimal gland fluid changes with flow rate, and fluid secretion can vary 130-fold from basal to maximal stimulation (Gilbard and Dartt, 1982). In rat exorbital lacrimal gland fluid, the Na^+ concentration increases while the K^+ concentration decreases with increasing secretory rate (Alexander et al., 1972). In rabbit lacrimal gland fluid, the concentrations of Na^+, Cl^-, and Ca^{2+} decrease as the rate of secretion increases (Botelho et al., 1976). Thus, lacrimal gland fluid osmolarity in the rabbit also changes with secretory rate decreasing as flow rate increases (Gilbard and Dartt, 1982).

Lacrimal gland fluid is produced in two stages accounting for the variation of electrolyte composition with secretory rate. The first stage is the production of primary fluid by the acinar cells and secretion of the fluid into the acinar lumina. This fluid is predominantly NaCl and has plasma-like

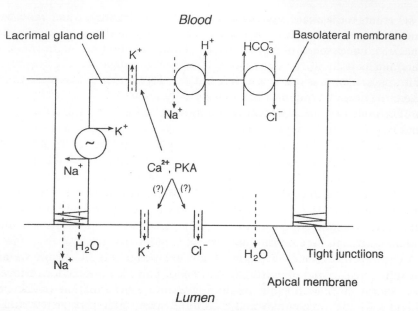

Figure 14. Schematic drawing of lacrimal gland acinar cell showing the mechanism of electrolyte and water secretion. Possible roles of Ca^{2+} and PKA (protein kinase A) in activating ion movements are also indicated. $\sim$, Na^+, K^+-ATPase; dashed arrows, passive ionic movements; solid arrows, active ionic movements. (From Dartt, D.A. (1992). Physiology of tear production. In: The Dry Eye. (Lemp, M.A. and Marquardt, R., Eds.), p. 65. Springer-Verlag, Berlin.)

electrolyte concentrations. The second stage is the modification of primary fluid by duct cell secretion. The duct cells secrete a KCl rich solution so that the final secreted fluid is rich in K^+. It has been estimated that as much as 30% of the volume of the final lacrimal gland fluid is secreted by the duct cells (Mircheff, 1994).

Acinar cell NaCl secretion is driven by the sodium pump or Na^+, K^+-ATPase located on the basolateral membranes that use ATP to transport Na^+ out and K^+ into the cells (Figure 14). Na^+ enters the cell from the interstitium across the basolateral membrane using a favorable electrochemical gradient maintained by the Na^+, K^+-ATPase. Cl^- also enters the cell using the same transporters as Na^+ against its electrochemical gradient. K^+ leaves the cells through a passive permeability channel in the basolateral membrane, which completes the cycle necessary for the activity of the Na^+, K^+-ATPase. Cl^-, following a favorable electrochemical gradient, leaves the cells through passive Cl^- permeability channels in the apical membrane and into the lumen.

Na^+ enters the secreted fluid using a paracellular pathway, which favors the movement of Na^+ into the lumen. There are two possible sets of transporters that could be responsible for the transport of Na^+ and Cl^- across the basolateral membranes (Mircheff, 1994). The first set is the NaCl and Na–K–2Cl cotransporters, and the second set is the Na/H and Cl/HCO₃ exchangers. Experimental evidence indicates that the Na/H and Cl/HCO₃ exchangers are activated in lacrimal gland acinar cell secretion though a role for the NaCl and Na–K–2Cl cotransporters cannot be completely ruled out.

Secretion of KCl from duct cells has not been well characterized but probably uses mechanisms similar to those used by acinar cells. It has been hypothesized that Na^+, K^+-ATPase in the basolateral membrane pumps K^+ into the cell and Na^+ out. The K^+ exits the cell through the apical membrane via K^+ permeability channels, down a favorable electrochemical gradient. Secretion of Cl^- could occur using the same passive permeability mechanisms as in the acinar cells (Mircheff, 1994).

The last component of lacrimal gland fluid is water. Movement of water is driven by hydrostatic and osmotic gradients with water passively following electrolyte movement. Thus, water follows Na^+ and moves into the lumen. Passive water permeability channels, aquaporin 4 and 5 have been identified in the lacrimal gland (Raina et al., 1995; Ishida et al., 1996). Lacrimal gland fluid is isotonic at high secretory rates, but becomes hypertonic at low secretory rates suggesting that water is reabsorbed by the duct system if the fluid remains in it long enough. This could occur if a hormone, similar to antidiuretic hormone, which regulates water permeability in the kidney, could regulate the water permeability of the ducts. There is no evidence to support or refute this hypothesis.

The ion transporters (the pumps and permeability channels) are activated through the signal transduction pathways and result in electrolyte and water secretion. Ca^{2+} and PKA are known to activate the passive K^+ and Cl^- permeability channels (ion channels), which are among the first transport processes altered to induce secretion. In other epithelia these protein kinases are also known to phosphorylate and thus activate the Na^+/HCO_3^- and Na/H exchanger and to alter the activity of the Na^+, K^+-ATPase. Thus, protein kinases stimulate electrolyte and water secretion by phosphorylating and activating ion channels and pumps. The details of this activation and the sequence of events initiated by stimulation are not understood in the lacrimal gland.

Mechanism of Protein Secretion: Protein secretion occurs when the secretory granule membranes fuse with the apical membrane. For constitutive protein secretion, such as SIgA, the granules are not stored and fusion occurs readily. Due to the transient nature of these granules, the mechanism of this fusion has not been studied. For regulated protein secretion, used by most proteins secreted by the lacrimal gland, granules are stored and are visible in light or electron micrographs. Fusion is controlled and does not occur until appropriately stimulated. While the mechanism for controlling regulated secretory granule fusion in the lacrimal gland has not been studied, granule fusion for intracellular vesicle transport and neurotransmitter secretion from nerves has been well studied.

In nerves, a network of proteins on both the secretory granule and apical membranes necessary for granule fusion to occur. On its cytoplasmic surface the secretory granules contain a RAB-GTP complex and v-SNARE (Rothman and Wieland, 1996; Rothman and Sollner, 1997). RAB proteins are a family of guanine triphosphatase proteins while SNAREs are a family of proteins crucial for vesicle docking at the intended area. v-SNAREs are the "pilot" proteins, which direct the granules to the correct destination while t-SNAREs are the receptors on the target membranes, which capture the granules. v- and t-SNAREs interact with one another through α-helical coils on their cytoplasmic sides.

After a secretory granule buds from the trans-Golgi network, the secretory granule moves to the receptor membrane (Rothman and Wieland, 1996; Rothman and Sollner, 1997). RAB-GTP displaces Sec-1, which allows v-SNARE to interact with t-SNARE. Membrane fusion occurs when N-ethylmaleimide-sensitive fusion protein (NSF) and soluble NSF attachment protein (SNAP) hydrolyze ATP to ADP. The released energy disrupts the SNARE complex allowing membrane fusion to occur releasing granule contents into the acinar lumen. While little is known about the roles of PKC and PKA in membrane fusion, it is known that Ca^{2+} regulates fusion. To date, SNAP and SNARE proteins have not been identified in the lacrimal gland, though they are present in other exocrine tissues.

Proteins Secreted by the Lacrimal Gland: The main lacrimal gland secretes only some of the proteins present in tears. Tear proteins also originate from several sources, including the main and accessory lacrimal glands, the ocular surface epithelium, and the conjunctival blood vessels. To identify which proteins the main lacrimal gland secretes, it is again necessary to analyze pure lacrimal gland fluid. Thus far, this has only been accomplished in animal models, so not all the proteins secreted by the human lacrimal gland have been identified. Recently another approach has been employed that involves analysis of the lacrimal gland itself for the mRNA of secreted proteins. Table 2 is a list of proteins identified as being synthesized and secreted by the lacrimal gland. In addition, proteins that are transcribed and/or translated in the lacrimal gland, which may be secreted by this tissue, are also listed in Table 2. Basic fibroblast growth factor and plasminogen activator are present in tears and may originate in the lacrimal gland (Sullivan, 1999). Many more proteins are likely to be identified using molecular techniques.

Table 2. Proteins Secreted by the Lacrimal Gland

Apolipoprotein D	Basic fibroblast growth factor	β-Amyloid protein precursor
Convertase decay-accelerating factor	Cystatin-related protein	Cystatin
Endothelin-1	Granulocyte-monocyte colony-stimulating factor	Epidermal growth factor
Lacritin		
Group II phospholipase A_2	Immunoglobulin G	Hepatocyte growth factor
Immunoglobulin M	Interlukin-1β	Interlukin-1α
Lactoferrin	Monomeric immunoglobulin A	Lysozyme
Peroxidase	Polymeric immunoglobulin A	Plasminogen activator
Prolactin	Secretory component	Retinoic acid
Secretory immunoglobulin A	Transforming growth factor-α	Tear lipocalins
Transforming growth factor-β1	Tumor necrosis factor-α	Transforming growth factor-β2

B. Accessory Lacrimal Glands

The accessory lacrimal glands are mini-lacrimal glands weighing about 10% of the main lacrimal gland and located within the conjunctival epithelium (Seifert et al., 1994). The number of accessory lacrimal glands varies from 4 to 42 in the upper conjunctiva of humans, with 6 or fewer in the lower conjunctiva (Allansmith et al., 1976). Despite their size, they contribute significantly to the aqueous layer of the tear film by secreting proteins, electrolytes, and water.

Functional Anatomy

Each accessory lacrimal gland consists of a single excretory duct composed of one to two cell layers. The excretory duct divides into smaller intralobular ducts which either terminate blindly or form secretory tubules. The secretory tubules branch off laterally in large numbers, terminating in secretory end pieces. Unlike the main lacrimal gland, the accessory lacrimal gland does not contain acini (Seifert et al., 1994). Instead, it is the epithelial cells of the secretory tubules that are similar to the acini of the main lacrimal gland. These are polar cells with a basally located nucleus and bordering a centrally located endoplasmic reticulum and Golgi apparatus. Moreover, secretory granules are present in these cells–sparsely in some and abundantly in others. Secretory granules have varying diameters and degrees of electron translucence.

Regulation of Secretion

The accessory lacrimal glands' secretion was originally believed to contribute to basal, but not stimulated, tear secretion. This implied that their secretion was not regulated. However, Seifert and Spitznas showed that nerves are present in accessory lacrimal glands surrounding glandular epithelial, myoepithelial, vascular endothelial, plasma cells, and fibroblasts (Seifert and Spitznas, 1993). Like the main lacrimal gland, the accessory lacrimal gland has a dense neural plexus as identified by antibodies to two different neuronal marker proteins, PGP and S-100 (Seifert et al., 1997). Because many of the nerve endings contain small clear vesicles and a few large, dense core vesicles, it has been suggested that these nerves are parasympathetic and thus would contain the neurotransmitters acetylcholine and VIP. Only a single nerve fiber was found that was structurally consistent with sympathetic nerves. Sensory nerves were also identified in the accessory gland tissue because a few nerves containing the sensory neurotransmitters Sub P and CGRP were identified by immunohistochemistry in association with acini, blood vessels, and ducts (Seifert et al., 1997). While definitive identification of the major types of nerves present requires additional research, the conclusion may be reached that accessory lacrimal glands are innervated.

Neural innervation of the accessory lacrimal glands suggests that their protein, electrolyte, and water secretion can be regulated. In a dry eye rabbit model in which the orbital glands, the Harderian and nictitans glands, have been removed and the lacrimal gland excretory duct sealed off, the accessory lacrimal glands are believed to be the major contributors to the aqueous layer of the tear film. Using this model, Gilbard et al. showed that accessory lacrimal glands are functionally innervated by topical application of pilocarpine, a cholinergic, muscarinic agonist, to the ocular surface (Gilbard et al., 1990). However, this stimulated fluid secretion was *not* blocked by the muscarinic antagonist, atropine. This indicates that the secretion was *not* mediated by the cholinergic neurotransmitter (Gilbard et al., 1990). The pilocarpine-induced secretion was inhibited by the administration of a local anesthetic, suggesting that a neural reflex, probably via afferent sensory nerves in the cornea and efferent parasympathetic or sympathetic nerves in the conjunctiva, mediated fluid secretion. In contrast, topical application of VIP also stimulated secretion, but this effect was *not* blocked by local anesthetic. This is consistent with acetylcholine and VIP release by parasympathetic nerves; however, VIP alone stimulates accessory lacrimal gland secretion. This model did not test sympathetic neurotransmitters.

The dry eye rabbit model was also used to determine the effect of the peptide hormones, α-, β-, γ-MSH, and glucagon. Each stimulated fluid secretion, suggesting that regulation of accessory lacrimal gland secretion may have a hormonal component.

In the main lacrimal gland, VIP, α-, β-, and γ-MSH, and glucagon stimulate secretion through the cAMP-dependent signal transduction pathway. These compounds were also tested in the dry eye rabbit model. Topical application of compounds known to increase cellular cAMP levels also stimulated fluid secretion. For example, forskolin, an activator of adenylyl cyclase, 8-bromo cAMP, a membrane-permeable cAMP analog, and IBMX, an inhibitor of cAMP-dependent phosphodiesterase, all stimulated fluid secretion. It is not clear if VIP, α-, β-, and γ-MSH, and glucagon use the cAMP-dependent signal transduction pathway in accessory lacrimal glands, although it is reasonable to suggest that they do. Experiments have not been undertaken to determine whether activation of other signal transduction pathways also stimulates accessory lacrimal gland secretion. Based on evidence to date, it may be suggested that parasympathetic nerve stimulation to release VIP activates a cAMP-dependent pathway to stimulate accessory lacrimal gland fluid secretion, and that peptide hormones in the blood-stream may also activate this pathway.

Secretory Product

Several of the proteins secreted by the accessory lacrimal glands have been identified by histochemical and immunohistochemical techniques. Proteins, such as lysozyme, lactoferrin, SC, and IgA are secreted by both the accessory lacrimal and main lacrimal glands (Gillete et al., 1980, 1981).

Despite the small volume of electrolytes, water, and protein secreted by the accessory lacrimal glands, they appear to be able to provide a stable tear film under nondiseased conditions. Maitchouk et al. (1998) showed that, in monkeys in which the main lacrimal glands were surgically removed, tear production by accessory lacrimal glands was sufficient to maintain a stable tear layer. In addition, the protein profile in tears from the experimental animals was essentially identical to that of normal tears.

C. Corneal Epithelium

Corneal epithelia can also secrete water and electrolytes into tears and are a possible source for these components in tears, although the magnitude of this secretion is probably minor compared with that of the main lacrimal gland. Since the cornea is the primary refractive element of the eye, maintenance of corneal transparency is essential to vision. To stay transparent, the cornea must be relatively dehydrated. Excess water can enter the cornea when the metabolism of the cornea changes. This can occur during contact lens wear as carbon dioxide accumulates behind the lens, acidifying the cornea and causing it to swell and lose transparency. In corneal endothelia, Na^+, Cl^-, and water are transported primarily by Na^+, K^+-ATPase. However, the corneal epithelia also contains Na^+, K^+-ATPase and is thus able to transport Na^+, Cl^-, and water as well. In addition, the cornea acts as a protective barrier between the external environment and the remainder of the eye and as such prevents pathogen entrance and fluid loss. Since corneal epithelium "is tight" and relatively impermeable, the cornea must be able to transport electrolytes and water while maintaining the integrity (tightness) of its intercellular junctions (Rich et al., 1997).

Functional Anatomy

The cornea consists of an outer nonkeratinized, multilayered epithelium, a middle connective tissue stroma, and an inner layer of cells–the endothelium–that border the anterior chamber. The cornea is transparent because it does not contain blood vessels, and the cells within its layers are very regular and precisely arranged.

The corneal epithelium is five to seven cell layers thick (Gipson, 1994). The apical layers contain three to four layers of flattened stratified squamous cells. Under these cells are one to three layers of midepithelial cells. The bottom layer consists of one layer of columnar basal cells. The epithelium is attached to the underlying basement membrane and stroma via well-developed adhesion complexes known as hemidesmosomes.

Regulation of Secretion

The cornea, one of the most densely innervated tissues in the body, has extensive sensory innervation and modest sympathetic innervation (Marfurt, 1999). Sensory nerves contain the peptide neurotransmitters Sub P, CGRP, and gallanin. Substance P and CGRP are not only present in the same nerves but also in the same secretory vesicles. Sympathetic nerves contain the neurotransmitters norepinephrine, serotonin, and NPY. In a few species, sparse numbers of parasympathetic nerves containing the neurotransmitter VIP are present. While there is a high concentration of the parasympathetic neurotransmitter acetylcholine in the cornea, most of it is associated with epithelial cells, not with parasympathetic nerves.

Corneal nerves function by modulating epithelial cell proliferation and mitosis, modulating cell migration during wound healing, and interacting with sensory fibers to exert trophic influences on the cornea. Loss of neural function is particularly devastating to the corneal epithelium, since a failure of reflex tearing and blinking results in exposure keratopathy and can progress to frank corneal ulcers.

By releasing norepinephrine, which activates β-adrenergic receptors, sympathetic nerves are the primary regulators of electrolyte and water secretion from the cornea into tears (Figure 15) (Klyce and Crosson, 1985). Stimulation of electrolyte and water secretion also occurs through norepinephrine activation of α_1-adrenergic receptors, although this is a minor pathway (Akhtar, 1987). Serotonin and dopamine released from sympathetic nerves also stimulate corneal tear secretion, although they act

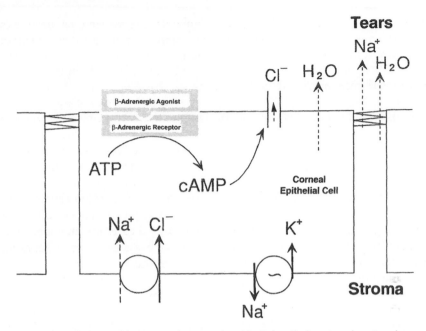

Figure 15. Schematic drawing of corneal epithelial cell showing the signal transduction pathway activated by β-adrenergic agonists to stimulate electrolyte and water secretion into tears. $\sim$, Na^+, K^+-ATPase; dashed arrows, passive ionic movements; solid arrows, active ionic movements. (Modified from Dartt, D.A. (1992). Physiology of tear production. In: The Dry Eye. (Lemp, M.A. and Marquardt, R., Eds.), p. 65. Springer-Verlag, Berlin.)

through a presynaptic mechanism, facilitating the release of norepinephrine from sympathetic nerves. Finally, there are muscarinic cholinergic receptors on corneal epithelial cells, but their function is unknown, as corneal acetylcholine is not of neural origin. The function of acetylcholine is also not known (Edelhauser et al. (1994). The Cornea; Smolin and Thoft).

Stimulation of Electrolyte and Water Secretion into Tears through Signal Transduction Pathways

β-Adrenergic agonists regulate Na^+, Cl^-, and water secretion by the corneal epithelium. In contrast to lacrimal gland and conjunctival stratified squamous cell receptors, which are located on the basolateral membranes, corneal epithelial cell receptors are found on both the basolateral and apical membranes. Binding of β-adrenergic agonists to their receptors activates the cAMP-dependent pathway (as previously described in the lacrimal gland) to activate adenylyl cyclase (Figure 15), increasing cAMP levels that activate PKA. Activated PKA then phosphorylates-specific protein substrates that directly or indirectly mediate fluid secretion. Forskolin, permeable cAMP analogs, and cAMP phosphodiesterase inhibitors can also stimulate corneal epithelial fluid secretion.

The mechanism of Na^+, Cl^-, and water secretion in the cornea is similar to that in the acinar cells of the lacrimal gland. Cl^- enters the corneal epithelial cells across the basolateral membrane through a $Na^+/K^+/2Cl^-$ cotransporter and leaves the cell by Cl^- channels in the apical membrane. Maintenance of the Na^+ gradient by the Na^+, K^+-ATPase in the basolateral membrane provides the energy for Cl^- to enter the cell across the basolateral membrane—against its electrochemical equilibrium. This raises the intracellular Cl^- to a high enough level that it can passively exit the cell across the apical membrane into the tears. Na^+ and water then enter the tears by a paracellular pathway (Rich et al., 1997). K^+ channels in the basolateral membrane provide an exit for K^+ to balance the Na^+ entering the cell (Rae and Farrugia, 1992).

Elevating the intracellular cAMP level has been shown to increase Na^+ and Cl^- currents, while muscarinic agonists (which are Ca^{2+} and protein kinase C dependent) and cGMP activate the K^+ current (Farrugia, and Rae, 1992). Crosson et al. (1986) found that activation of PKC by addition of phorbol esters stimulated Cl^- secretion. In addition, immunohistochemical and western blotting techniques have shown that all of the 11 PKC isoforms are present in the corneal epithelium (unpublished observations). One or more of these isoforms could play a role in muscarinic agonist-stimulated fluid secretion. Additional experiments are necessary to clarify the role of muscarinic agonists in stimulating corneal epithelial fluid secretion and to identify the specific components of the signaling pathway used.

β-Adrenergic agonists stimulate fluid secretion by activating PKA to phosphorylate-specific protein substrates. The protein substrates in the corneal epithelium, as in the lacrimal gland, are unknown. It is likely that similar ion permeability channels and transport proteins in the corneal epithelium are activated as in the lacrimal gland. For example, the same water transporter, aquaporin 5 (AQP5), has been identified in the corneal epithelium, as in the lacrimal gland (King and Agre, 1996). Aquaporin 5 is present only in the corneal epithelium, and not in the corneal stroma or endothelium. This suggests a specific role for water transport, probably as a component of fluid secretion, in both corneal epithelium and the lacrimal gland. Interestingly, AQP5 has a consensus site for phosphorylation by PKA. Thus, activation of this kinase by β-adrenergic agonists in the cornea could activate AQP5. Other possible substrates for PKA are the Na^+, K^+-ATPase and the $Na^+/K^+/2Cl^-$ cotransporter, both of which also contain consensus sequences for phosphorylation by PKA.

Thus, the corneal epithelium, while a minor source, is an important one for the electrolytes and water secreted into tears. Since the secretion of electrolytes and water is not the main purpose of the cornea, it must be accomplished without alteration of the transparency and barrier function, the cornea's major functions. In aqueous deficiency dry eye diseases in which the major sources of the tears have been affected, the cornea could secrete enough fluid to prevent the worst sequelae of this disease, namely corneal ulcers.

D. Conjunctival Epithelium

The conjunctiva covers the inner surface of the upper and lower eyelids and extends to the edge of the cornea (the limbus). Unlike the cornea, the conjunctiva is not transparent, so its hydration does not need to be precisely regulated. Moreover, it is not as regularly organized at the cellular or tissue level as the cornea and it contains blood vessels, whereas, the cornea does not. The conjunctiva performs several important functions. First, it protects the ocular surface from bacterial and viral infection by synthesizing and secreting antibacterial proteins and mucus and by recruiting the cellular components of the immune system. Second, the conjunctiva is another source of electrolytes and water for the aqueous layer of the tear film. Since the human conjunctiva occupies 17 times more surface area than the cornea, it probably contributes a larger proportion of electrolytes and water to the tear film, and is potentially a significant source of these compounds (Watsky et al., 1988).

There are two possible ways by which the conjunctiva could contribute to the aqueous layer of the tear film. (1) Its stratified squamous cells could secrete electrolytes and water. (2) Conjunctival blood vessel permeability could increase allowing plasma to leak; this would probably only occur under pathological conditions (e.g., inflammation) or with topical application of compounds that increase conjunctival blood vessel permeability. (Although this latter mechanism can produce a large volume of fluid, it will not be discussed here.) The conjunctiva also absorbs electrolytes and water, thereby, modifying the tear film through absorption as well as secretion.

Functional Anatomy

The conjunctiva is composed of a surface epithelium and an underlying vascularized stroma that contains nerves and lymphoid tissue. Stratified squamous cells are the major cell type of the conjunctival epithelium and, according to various reports, are organized into 2–3, 5–7, or 10–12 cell layers (Gipson, 1994). The bottom cell layer is attached to a basement membrane and underlying stroma by the same type of adhesion complexes that are found in the cornea. All layers of the stratified squamous cells have numerous small clear vesicles in their cytoplasm. These vesicles appear to contain mucus and could function as the second mucus system of the eye, which will be discussed in a subsequent section. Unlike those in other stratified squamous epithelia, conjunctival goblet cells are intercalated between stratified squamous cells. The goblet cells are major contributors to the mucous layer of the tear film and will be discussed subsequently.

Regulation of Secretion

Sensory, sympathetic, and parasympathetic nerves innervate the conjunctiva, although the innervation is less dense than that in the cornea (Ruskell, 1985; Elsas et al., 1994). These nerves are unmyelinated and branch frequently as they enter the stroma and basal epithelial cell layer. Both a subepithelial plexus, which often terminates on blood vessels, and an intraepithelial plexus, which abuts the base of epithelial cells, are formed by the nerve endings. Sensory nerves, which contain the neurotransmitters Sub P and CGRP, are plentiful (Luhtala and Uusitalo, 1991). Sympathetic nerves contain the neurotransmitters norepinephrine and NPY, while parasympathetic nerves contain acetylcholine and VIP (Dartt et al., 1995). Both sympathetic and parasympathetic nerves innervate conjunctival stromal blood vessels and epithelial cells, with sympathetic innervation predominating.

Stimulation of Electrolyte and Water Secretion by the Conjunctival Epithelium through Signal Transduction Pathways

Stimulation that induces conjunctival electrolyte and water secretion has only recently been investigated; thus, information about it is limited. Because the application of epinephrine to the basolateral side of the tissue stimulates Cl^- secretion into tears, it appears that sympathetic nerves are one conjunctival stimulus (Shi and Candia, 1995). This secretion is probably mediated by β-adrenergic receptors activating the cAMP-dependent signal transduction pathway. Forskolin, an activator of adenylyl cyclase, was found to mimic the effect of epinephrine; whereas, isoproternol, a selective β-adrenergic agonist, increased cellular cAMP levels in cultured conjunctival epithelial cells (Shi and Candia, 1995). Unlike the β-adrenergic receptors in the corneal epithelium, which are located on the apical side of the epithelium, those in the conjunctiva appear to be on the basolateral side, similar to their location in the lacrimal gland.

Several different receptors are present in cultured conjunctival epithelial cells and could function in secretion. In these cells, the parasympathetic neurotransmitter VIP and the sympathetic agonists dopamine and serotonin increased the cellular cAMP level in cultured conjunctival epithelial cells (Sharif et al., 1997). No neurotransmitters tested (Substance P or NPY) increased cellular inositol phosphate production, although bradykinin, platelet activating factor, and histamine did. Thus, a role for cAMP-dependent agonists in stimulating conjunctival electrolyte and water secretion is likely. The role of inositol phosphate/Ca^{2+}-dependent agonists needs further investigation.

Electrolyte and Water Secretion

A model for conjunctival Na^+, Cl^-, and water secretion similar to that in lacrimal gland acinar cells (Figure 12) and corneal epithelial cells has been proposed by Kompella et al. (1993) and Shi and Candia (1995). Cl^- is secreted across the apical membrane following its electrochemical gradient. Na^+ enters tears via a paracellular pathway between conjunctival cells, and water follows passively. Coupled $Na^+/K^+/2Cl^-$ or Na^+Cl^- cotransporters appear to be the Cl^- entry mechanism on the basolateral side. Furthermore, it appears that a Na^+/H^+ exchanger is not operative in these cells. The Na^+/K^+-ATPase located on the basolateral side of the cell provides the energy for the cotransporters. Approximately 60–75% of the conjunctiva's active ion transport is Cl^- secretion into tears. Na^+-glucose absorption from the tears accounts for the remaining percentage. Thus, the conjunctival epithelium is able both to secrete Na^+, Cl^-, and water into the tears and to reabsorb Na^+ and glucose from the tears, although their relative contributions to the final composition and amount of tears is unknown. However, the amount of ions secreted by the conjunctiva was fourfold greater than that of the cornea (Kompella et al., 1993). The higher transport rate and larger surface area of the conjunctiva than those of the cornea suggest that the conjunctiva may play a more prominent role than the cornea in electrolyte and water secretion into the tear film.

E. Function of the Aqueous Layer of the Tear Film

Tear proteins and electrolytes/water serve various functions for the ocular surface. The main and accessory lacrimal glands secrete proteins, such as SIgA, and to some extent IgG and IgM, which are a part of the secretory immune system of the eye and protect the ocular surface from microbial agents and toxic compounds. Furthermore, these proteins inhibit bacterial adhesion and colonization, interfere with viral attachment and internalization, decrease parasitic invasion, and prevent toxin-induced damage in mucosal epithelial cells (Childers et al., 1989; Sullivan, 1999). The eye's susceptibility to infectious and allergic diseases may be significantly limited by the ocular secretory immune system, which may also help maintain both corneal and conjunctival integrity and visual function (Sullivan, 1999).

Other lacrimal proteins that are known or thought to be secreted from the main and accessory lacrimal glands have diverse effects on the ocular surface. (1) The major tear components lysozyme and lactoferrin possess antibacterial activity. Furthermore, lactoferrin may suppress activation of the classical complement pathway through inhibition of C3 convertase, modulate ocular inflammatory reactions, and lessen ultraviolet B radiation-induced peroxide formation in corneal epithelial cells. (2) Secretory PLA_2, a group II phospholipase A_2, is a lipolytic enzyme that catalyzes the hydrolysis of the acyl ester bond of phosphoglycerides. This enzyme may synergize with lysozyme to degrade bacteria and act independently as an antibacterial agent. (3) Peroxidase may have bactericidal, virucidal, and fungicidal actions. (4) Tear lipocalins (formerly known as specific tear prealbumin) may have antibacterial effects. (5) Apolipoprotein D, a glycoprotein that has the ability to bind phospholipids and cholesterol, may contribute to the surface spreading of meibomian gland lipids, or function as a clearance factor, thereby protecting the cornea from harmful lipophilic molecules. (6) Convertase decay-accelerating factor protects against autologous complement activation. (7) Plasminogen activator, a serine protease, is chemotactic for leukocytes and also catalyzes the conversion of plasminogen to the active proteolytic enzyme, plasmin. (8) β-Amyloid protein precursor and cystatin, which inhibit serine and cysteine proteases, may aid in corneal wound healing. (9) HGF, TGF-α, TGF-β1, TGF-β2, IL-1α, IL-1β, TNF-α, GM-CSF, endothelin-1, bFGF, and EGF are growth factors and cytokines and may play an important role in the proliferation, motility and/or differentiation of corneal and conjunctival epithelial cells, and/or in the wound healing of the ocular surface. Some of these cytokines also appear to have receptors on corneal, limbal and/or conjunctival epithelial cells, stromal keratocytes (fibroblasts) and may control the expression of IL-6, IL-8, HGF, keratocyte growth factor (KGF), EGF receptor, platelet-derived growth factor receptor-beta, bFGF, TGF-β1, monocyte-CSF, and GM-CSF in these cells (Sullivan, 1999).

In response to trauma or irritation of the ocular surface, various lacrimal gland proteins, such as cytokines and growth factors, may be synthesized and/or secreted. These proteins, in turn, may act within the tissue or serve to heal and/or protect the ocular surface. For example, a stimulus, such as wounding that activates sensory nerves in the cornea enhances the production of TNF-α mRNA in acinar epithelial cells of the lacrimal gland (Thompson et al., 1994). Similarly, cholinergic agonists appear to stimulate human lacrimal gland secretion of EGF and TGF-β1 (Yoshino et al., 1996a,b).

The ocular secretory immune system is also modulated by parasympathetic nerves. In birds, carbachol increases IgG output from the Harderian gland by binding to muscarinic acetylcholine receptors on IgG plasma cells (Brink et al., 1994; Cameron, 1995). In contrast, carbamyl choline acutely (i.e., within hours) enhances, but chronically (i.e., within days) decreases, basal-, cholera toxin- and androgen-induced SC production by rat lacrimal gland acinar epithelial cells (Sullivan, 1999). The mechanisms behind the long-term inhibitory action of carbachol, possibly preventable by atropine, are unknown, but may involve suppression of cAMP or alteration of gene activity (Sullivan, 1999). The parasympathetic neurotransmitter VIP and the β-adrenergic agent isoproterenol also increase basal- and androgen-stimulated SC production by rat acinar epithelial cells, and cAMP appears to increase SIgA output by human main and accessory lacrimal glands *in vitro* (Hunt et al., 1996; Sullivan, 1999).

Experimental animal data suggest that the secretory immune system of the eye may also be regulated by cytokines. For example, IL-1α, IL-1β, and TNF-α have been shown to stimulate the acinar cell synthesis and secretion of SC (Kelleher et al., 1991); both IL-6 and IL-5 have been shown to stimulate IgA synthesis in lacrimal tissue explants (Pockley and Montgomery, 1990–1991), and TGF-β has been shown to enhance IgA output from rat lacrimal tissue, whether alone or in combination with IL-2, IL-5, IL-6, or IL-5 plus IL-6 (Rafferty and Montgomery, 1993). In summary, complex interaction between the ocular surface and the lacrimal gland regulates protein synthesis and secretion that in turn affects the ocular surface.

Tears have a unique electrolyte composition that is different from plasma. Maintenance of the specific electrolyte concentration and osmolarity of tears is important to the health of the ocular surface; relatively small changes that occur in aqueous deficient dry eye can cause serious changes in the ocular surface.

VI. SECRETION OF THE TEAR FILM MUCOUS LAYER

A. Goblet Cells

Goblet cells are interspersed among the conjunctival stratified squamous cells and are one of the most important sources of mucins in the mucous layer of the tear film. Corneal and conjunctival stratified squamous cells also secrete mucins into tears and will be discussed after the goblet cells.

Functional Anatomy

Goblet cells are present throughout the conjunctiva epithelium either as single cells or in clusters of cells (Figure 16). In humans, goblet cell clusters have been identified as mucous crypts (Kessing, 1968). Goblet cells are

DARLENE A. DARTT et al.

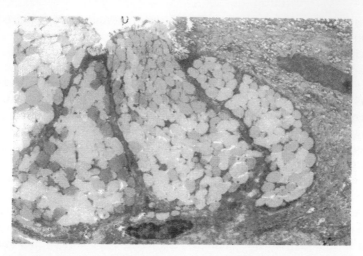

Figure 16. Electron micrograph of a cluster of goblet cells located between strati-fied squamous cells in the rat conjuntiva. Goblet cells extend from the basement membrane to the tear film and contain numerous electro-lucent secretory granules on the tear-film side.

highly polarized cells that extend to the conjunctival basement membrane in rats and mice, but not in humans. At the apical surface of the conjunctiva, tight junctions separate the goblet cell basolateral membrane from its apical membrane that abuts the tear film. Furthermore, goblet cells contain the synthetic enzymes for the synthesis and secretion of mucins. Their basal regions contain the nucleus, rough endoplasmic reticulum, and an especially well-developed Golgi apparatus. The apical portions of the goblet cells contain a large volume of secretory granules, which store the synthesized mucins and are each enclosed with a membrane. Mucin secretion occurs when secretory granule membranes fuse with each other and then with the apical membrane. Once a goblet cell has been stimulated to secrete, its entire granule content is released; this is known as apocrine secretion. It is hypothesized that the goblet cell body remains intact to resynthesize mucins and secrete again.

Because of the importance of goblet cells and the fact that various ocular surface diseases present with an increase or decrease in mucin levels, attempts have been made to correlate changes in goblet cell number with the degree of ocular pathology (Norn, 1992), a correlation that might have significant diagnostic potential. However, the analysis of alterations in goblet cell density has proven difficult. For example, histochemical techniques, such as Alcian blue and periodic acid/Schiff's reagent (AB/PAS) have typically identified conjunctival goblet cells. These procedures stain mucins in the secretory granules, but not in the rest of the cell. Furthermore, these techniques have not identified goblet cells that have secreted, because they secrete their entire content of secretory granules. Consequently, the use of AB/PAS or similar methods to determine goblet cell numbers present in different diseases only identify goblet cells that have *not* secreted, not those that have recently secreted and are in the process of resynthesizing mucins. Techniques, such as the use of an antibody to keratin 7, which bind to the goblet cell body, will allow identification of goblet cells despite the absence of secretory product (Krenzer and Freddo, 1997). Another complication in the analysis of goblet cells is the finding that in all species their number per unit area varies over the surface of the conjunctiva. It is, therefore, crucial to take samples from the same area of the conjunctiva and to include several controls, including the contralateral eye, if possible, or a large number of concurrent normal samples. Thus, the changes in goblet cell number and their role in ocular surface diseases need further clarification.

Regulation of Secretion

Early studies suggested that unlike the conjunctiva, which is innervated by sensory, sympathetic, and parasympathetic nerves, goblet cells were not innervated (Figure 17) (Ruskell, 1985). However, studies using antibodies to specific neurotransmitters demonstrated that parasympathetic and sympathetic nerves surround the basolateral membranes of a large population of goblet cells in the rat (Dartt et al., 1995). Sensory nerves, on the other hand, do not surround the goblet cells, but are found around neighboring stratified squamous cells. Nerves also appear to stimulate goblet cell mucin secretion. A sensory (neural) stimulus to the cornea induced goblet cell secretion in the conjunctiva (Kessler et al., 1995). Since this secretion was blocked by local anesthetic, nerves were implicated in the process. This secretion could occur either by via efferent parasympathetic and sympathetic nerves in the conjunctiva or via antidromic stimulation of sensory nerves in the conjunctiva (Figure 17). *In vitro* use of the parasympathomimetic agonist carbachol or the parasympathetic neurotransmitter VIP also stimulated goblet cell secretion in the rat (Dartt et al., 1996; Rios-Garcia et al., 1997). Goblet cell secretion was not stimulated by topical application of sympathetic agonists. The sensory neurotransmitter Sub P did not stimulate goblet cell secretion; however, the other sensory neurotransmitters CGRP and gallanin have not been tested. Thus, parasympathetic nerve stimulation appears to induce goblet cell mucin secretion.

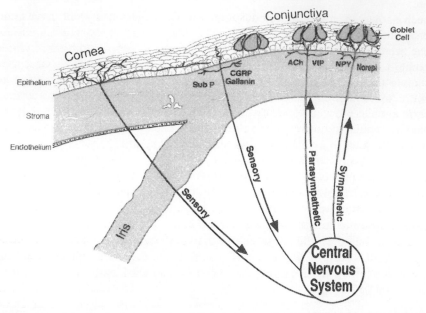

Figure 17. Schematic drawing of corneal and conjunctival epithelium showing the location of peripheral nerves in relation to conjunctival goblet cells, as well as possible mechanisms of neural stimulation. Stimuli to the cornea or conjunctiva activate sensory nerves that by a reflex mechanism activate parasympathetic or sympathetic nerves surrounding the goblet cells. Stimuli may also cause antidromic activation of conjunctival sensory nerves or of corneal sensory nerves that have collaterals to the conjunctiva. Although sensory nerves do not surround goblet cells, antidromic stimulation of sensory nerves causes the release of sensory neuropeptides that could diffuse to goblet cells. (© 1995 from Localization of nerves adjacent to goblet cells iun rat conjunctiva by Dartt, D.A., McCarthy D.M., Mercer, H.J., et al. (1995). Curr. Eye Res. 14, 993. Reproduced by permission of Taylor & Francis, Inc., http://www.taylor and francis.com.)

Signal Transduction Pathways that Stimulate Goblet Cell Mucous Secretion

The signal transduction pathways activated by parasympathetic and sympathetic agonists have recently begun to be investigated and characterized. Parasympathetic nerves release the neurotransmitters acetylcholine and VIP, both of which stimulate goblet cell mucin secretion. Acetylcholine interacts with muscarinic receptors located on goblet cells. Two subtypes of muscarinic receptors, M_2 and M_3, are present on goblet cell membranes. Activation of either of these receptors stimulates secretion. The next step presumably is activation of $G_{q/11}$, G protein, and PLC, although this has not been shown in goblet cells. Phospholipase C activation usually causes the production of IP_3 and DAG. IP_3 causes a release of intracellular Ca^{2+}. Increasing the intracellular $[Ca^{2+}]$ using the Ca^{2+} ionophore ionomycin stimulates goblet cell mucin secretion (Rios et al., 1999). Either Ca^{2+} alone or $Ca^{2+}/$ calmodulin-dependent protein kinases can stimulate secretion. Inhibitors of $Ca^{2+}/$calmodulin-dependent protein kinases did not specifically block cholinergic agonist-induced goblet cell secretion. Thus, cholinergic agonists do not use $Ca^{2+}/$calmodulin-dependent protein kinases to induce goblet cell secretion but do use Ca^{2+}.

Diacylglycerol may be involved since phorbol esters (which activate PKC), stimulate goblet cell secretion. There are 7 of the known PKC isoforms present in goblet cells. It is not clear, however, if cholinergic agonists activate PKC to stimulate secretion in goblet cells, because PKC inhibitors alone appear to stimulate mucin secretion.

The second parasympathetic neurotransmitter, VIP, also stimulates goblet cell mucin secretion. VIP type II, but not type I, receptors are located on the basolateral membrane of goblet cells (Dartt et al., 1996). The remainder of the signaling pathway activated by VIP is not known. VIP can either elevate intracellular Ca^{2+}or increase cellular cAMP levels. Either of these two signaling pathways could be present in goblet cells.

Sympathetic nerves in the conjunctiva contain norepinephrine and NPY. Norepinephrine can interact with both α- and β-adrenergic receptors. β-Adrenergic receptors, in particular β_1-adrenergic receptors, have been detected on goblet cells. α_1-Adrenergic receptors have not yet been investigated. It is not known if norepinephrine or NPY stimulates goblet cell mucin secretion.

An additional pathway, a purinergic pathway, has been identified in goblet cells. Activation of the P2Y2 subtype of purinergic receptor stimulates goblet cell secretion (Jumblatt and Jumblatt, 1998). The nucleotides ATP and UTP mediated this secretion. ATP and UTP can be released from several sources including nerves, goblet cells, other epithelial cells, and many other cell types. This could be a neurally mediated pathway, functioning with parasympathetic and sympathetic neurotransmitters, or a paracrine or autocrine pathway.

The accumulated evidence suggests a major role for nerves, especially parasympathetic nerves, in regulating goblet cell secretion. The role of inhibitors, as well as peptide and steroid hormones, in controlling goblet cell secretion has not yet been investigated.

Goblet Cell Mucin Secretion

Mucins are a heterogeneous collection of high-molecular weight glycoproteins in which carbohydrates represent a substantial proportion of the mass (Gipson and Inatomi, 1997). Mucins consist of a polypeptide backbone rich in serines and threonines and containing a variable number of tandem repeats. The protein backbone is highly glycosylated with carbohydrate chains of varying length and composition that are linked to the amino acids serine and threonine by O-glycosidic bonds. Human genes that synthesize the peptide backbone of mucins have recently been cloned and divided into nine types, designated MUCs 1–9 (Gipson and Inatomi, 1997). MUC5AC has been found in conjunctival goblet cells but not in corneal or conjunctival stratified squamous cells. MUC4 has been found in the conjunctiva in both goblet cells and stratified squamous cells (Inatomi et al., 1996). It is difficult to characterize the mucins synthesized and secreted by the goblet cells, because of the heterogeneous glycosylation of the peptide backbone. However, it is known that the mucin's protein core is synthesized in the endoplasmic reticulum, and the carbohydrate side chains are added in the Golgi apparatus. The synthesized mucins are stored in secretory granules until appropriate stimulation occurs. The processes by which secretory granules fuse with each other and with the apical cell membrane are unknown. It is likely, however, that they are similar to those in other exocytotic processes (e.g., neurotransmitter release and lacrimal gland protein secretion). In contrast to nerves and lacrimal glands where only a small percentage of secretory granules are released, in goblet cells all the secretory granules are released simultaneously.

B. Corneal and Conjunctival Stratified Squamous Cells

A second source of mucus in tears is corneal and conjunctival stratified squamous cells. The outer cell layer of the corneal epithelium and conjunctival epithelium contains numerous small, electron-lucent secretory vesicles. Initially these vesicles were shown to contain carbohydrates that were subsequently identified as mucins. Corneal epithelium expresses the MUC1 gene, a transmembrane mucin (Gipson and Inatomi, 1997). Conjunctival stratified squamous cells express MUC1 and MUC4 (Inatomi et al., 1996). However, MUC5AC expression is found only in goblet cells.

Despite the fact that both the cornea and conjunctiva are innervated, the role of nerves in regulating mucin secretion by their stratified squamous cells has not been investigated. In fact, no regulation, neural or hormonal, has been determined to date.

The goblet cell mucins' contribution to tear film mucous layer relative to that of the stratified squamous cells is not known. Possibly goblet cell and stratified squamous cell mucin secretions are differentially regulated and could respond to different stimuli, thus ensuring protection of the ocular surface in response to a wide variety of stimuli. Another possibility is that the different mucins have distinct roles in the ocular surface. Gipson and Inatomi (1997) have suggested that the membrane-spanning mucin MUC1 produced by stratified squamous cells extends from the apical membrane of these cells into the glycocalyx (Gipson and Inatomi, 1997). The gel-forming mucin MUC5AC, on the other hand, is secreted by goblet cells and forms the mucous layer of the tears. MUC4's role has not yet been identified. Further research is needed to determine the regulation of stratified squamous cell mucin secretion and the relative roles of stratified squamous cell and goblet cell mucin secretions.

C. Function of the Tear Film Mucous Layer

The mucous layer of the tear film has two major functions. First, it stabilizes the tear film (prevents tear breakup) and aids in tear spread. Second, it plays a major role in protecting the ocular surface from the environment (i.e., preventing microbial invasion, physical and chemical trauma, and desiccation of the ocular surface).

VII. ROLE OF TEAR SECRETION IN MAINTAINING THE OCULAR SURFACE

A healthy ocular surface is dependent upon the amount and composition of the tear film. Our knowledge to date, although incomplete, suggests that the amount and composition of the tear film is tightly controlled by regulation of the orbital gland and ocular surface epithelial secretions that produce the tear film. Regulation of secretion from these glands and epithelia is complex, requiring coordination of secretion from several sources to produce the multilayered tear film. An alteration in the tear film caused by change in the amount or type of secretion from any of the glands and epithelia can lead to ocular surface disruption and eventually to surface disease. Given that numerous tissues secrete tears and that their various secretions are regulated in diverse ways, it is not surprising that ocular surface diseases encompass a variety of tear-deficient disorders known as dry eye syndromes.

ACKNOWLEDGMENTS

This research review was supported by NIH grants EY06177 and EY09057.

REFERENCES

Adeghate, E. and Singh, J. (1994). Immunohistochemical identification of galanin and leucin-enkephalin in the porcine lacrimal gland. Neuropeptides 27, 285–289.

Aderem, A. (1992). The MARCKS Brothers: A family of protein kinase C substrates. Cell 71, 713–716.

Akhtar, R. (1987). Effects of norepinephrine and 5-hydroxytryptamine on phosphoinositide-PO_4 turnover in rabbit cornea. Exp. Eye Res. 44, 849–862.

Alexander, J., van Lennep, E. and Young, J. (1972). Water and electrolyte secretion by the exorbital lacrimal gland of the rat studied by micropuncture and catheterization techniques. Pflugers Arch. 337, 299–309.

Allansmith, M., Kajiyama, G., Abelson, M. and Simon, M. (1976). Plasma cell count of main and accessory lacrimal glands and conjunctiva. Am. J. Ophthalmol. 82, 819–826.

Berridge, M. (1993). Inositol trisphosphate and calcium signalling. Nature 361, 315–325.

Billah, M. and Anthes, J. (1990). The regulation and cellular functions of phosphatidylcholine hydrolysis. Biochem. J. 269, 281–291.

Botelho, S., Martinez, E., Pholpramool, C., Prooyen, H., Janssen, J. and De Palau, A. (1976). Modification of stimulated lacrimal gland flow by sympathetic nerve impluses in rabbit. Am. J. Physiol. 230, 80–84.

Bourgoin, S. and Grinstein, S. (1992). Peroxides of vanadate induce activation of phospholipase D in HL-60 cells. Role of tyrosine phosphorylation. J. Biol. Chem. 267, 11908–11916.

Brann, D., Hendry, L. and Mahesh, V. (1995). Emerging diversities in the mechanism of action of steroid hormones. J. Steroid Biochem. Mol. Biol. 52, 113–133.

Brink, P., Walcott, B., Roemer, E., Grine, E., Pastor, M., Christ, G. and Cameron, R. (1994). Cholinergic modulation of immunoglobulin secretion from avain plasma cells: The role of calcium. J. Neuroimmunol. 51, 113–121.

Cameron, R., Walcott, B., Claros, N., Mendel, K. and Brink, P. (1995). Cholinergic modulation of immunoglobulin secretion from avian plasma cells: The role of cyclic mononucleotides. J. Neuroimmunol. 61, 223–230.

Childers, N., Bruce, M. and McGhee, J. (1989). Molecular mechanisms of immunoglobulin A defense. Annu. Rev. Microbiol. 43, 503–536.

Chung, C., Tigges, M. and Stone, R. (1996). Peptidergic innervation of the primate meibomian gland. Invest. Ophthalmol. Vis. Sci. 37, 233–245.

Cockroft, S. (1992). G-protein-regulated phospholipases C, D and A_2-mediated signalling in neutrophils. Biochim. Biophys. Acta 1113, 135–160.

Colley, W., Sung, T., Roll, R., Jenco, J., Hammond, S., Altshuller, Y., Bar-Sagi, D., et al. (1997). Phospholipase D2, a distinct phospholipase D isoform with novel regulatory properties that provides cytoskeletal reorganization. Curr. Biol. 7, 191–201.

Corfield, A.P., Carrington, S.D., Hicks, S.J., Berry, M. and Ellingham, R. (1997). Ocular mucins: Purification, metabolism and functions. Prog. Retin. Eye Res. 16, 627–656.

Cripps, M. and Bennett, D. (1994). Inhibition of lacrimal function by selective opiate agonists. Adv. Exp. Med. Biol. 350, 127–132.

Crosson, C., Klyce, S., Bazan, H. and Bazan, N. (1986). The effect of phorbol esters on the chloride secreting epithelium of the rabbit cornea. Curr. Eye Res. 5, 535–541.

Dartt, D. (1994). Regulation of tear secretion. Adv. Exp. Med. Biol. 350, 1–9.

Dartt, D. (1994). Signal transduction and activation of the lacrimal gland. In: Principles and Practices of Ophthalmology: Basic Sciences. (Albert, D., Jakobiec, F. and Robinson, N., Eds.), pp. 458–465. W.B. Saunders, Philadelphia.

Dartt, D., McCarthy, D., Mercer, H., Kessler, T., Chung, E. and Zieske, J. (1995). Localization of nerves adjacent to goblet cells in rat conjunctiva. Curr. Eye Res. 14, 993–1000.

Dartt, D., Kessler, T., Chung, E. and Zieske, J. (1996). Vasoactive intestinal peptide-stimulated glycoconjugate secretion from conjunctival goblet cells. Exp. Eye Res. 63, 27–34.

Davis, R. (1993). The mitogen-activated protein kinase signal transduction pathway. J. Biol. Chem. 268, 14553–14556.

Dawson, R. (1967). The formation of phosphatidylglycerol and other phospholipids by transferase activity of phospholipase. D. Biochem. J. 102, 205–210.

Dell'Acqua, M. and Scott, J. (1997). Protein kinase A anchoring. J. Biol. Chem. 272, 12881–12884.

Djeridane, Y. (1994). Immunohistochemical evidence for the presence of vasopressin in the rat harderian gland, retina and lacrimal gland. Exp. Eye Res. 59, 117–120.

Doane, M. (1980). Interactions of eyelids and tears in corneal wetting and the dynamics of the normal human eyeblink. Am. J. Ophthalmol. 89, 507–512.

Doane, M. (1994). Abnormalities of the structure of the superficial lipid layer on the *in vivo* dry-eye tear film. Adv. Exp. Med. Biol. 350, 489–493.

Driver, P. and Lemp, M. (1996). Meibomian gland dysfunction. Surv. Ophthalmol. 40, 343–367.

Edelhauser, H., Geroski, D. and Ubels, L. (1994). In: The Cornea. (Smolin, G. and Thoft, R.A., Eds.). Little Brown and Company, Boston, MA.

Elsas, T., Edvinsson, L., Sundler, F. and Uddman, R. (1994). Neuronal pathways to the rat conjunctiva revealed by retrograde tracing and immunocytochemistry. Exp. Eye Res. 58, 117–126.

Elsas, T., Uddman, R. and Sundler, F. (1996). Pituitary adenylate cyclase-activating peptide-immunreactive nerve fibers in the cat eye. Graefes Arch. Clin. Exp. Ophthalmol. 234, 573–580.

Exton, J. (1988). Mechanisms of action of calcium-mobilizing agonists: Some variations on a young theme. FASEB J. 2, 2670–2676.

Farrugia, G. and Rae, J. (1992). Regulation of a potassium-selective current in rabbit corneal epithelium by cyclic GMP, carbachol, and diltiazem. J. Membr. Biol. 129, 99–107.

Feher, J. (1993). Pathophysiology of the Eye. Akademiai Kiado, Budapest.

Foster, D., Wedel, B., Robinson, S. and Garbers, D. (1999). Mechanisms of regulation and function of guanylyl cyclases. Rev. Physiol. Biochem. Pharmacol. 135, 1–39.

Frey, W., Nelson, J., Frick, M. and Elde, R. (1986). Prolactin immunoreactivity in human tears and lacrimal gland. In: The Preocular Tear Film: Health, Disease and Contact Lens wear. (Holly, F., Ed.), pp. 798–807. Dry Eye Institute, Lubbock, TX.

Fullard, R. (1994). Tear proteins arising from lacrimal tissue. In: Principles and Practices of Ophthalmology: Basic Sciences. (Albert, D., Jakobiec, F. and Robinson, N., Eds.), pp. 473–479. W.B. Saunders, Philadelphia.

Gilbard, J. and Dartt, D. (1982). Changes in rabbit lacrimal gland fluid osmolarity with flow rate. Invest. Opthalmol. Vis. Sci. 23, 804–806.

Gilbard, J., Farris, R. and Santanaria, J. (1978). Osmolarity of tear microvolumes in keratoconjunctivitis sicca. Arch. Ophthalmol. 96, 677–681.

Gilbard, J., Rossi, S., Gray, K., Hanninen, L. and Kenyon, K. (1988). Tear film osmolarity and ocular surface disease in two rabbit models for keratoconjunctivitis sicca. Invest. Ophthalmol. Vis. Sci. 29, 374–378.

Gilbard, J., Rossi, S., Heyda, K. and Dartt, D. (1990). Stimulation of tear secretion by topical agents that increase cyclic nucleotide levels. Invest. Ophthalmol. Vis. Sci. 31, 1381–1388.

Gipson, I. (1994). Anatomy of the conjunctiva, cornea, and limbus. In: The Cornea. (Smolin, G. and Thoft, R., Eds.), pp. 3–24. Brown and Company, Little, Boston.

Gipson, I. and Inatomi, T. (1997). Mucin genes expressed by the ocular surface epithelium. Prog. Retin. Eye Res. 16, 81–98.

Gipson, I., Yankauckas, M., Spurr-Michaud, S., Tisdale, A. and Rinehart, W. (1992). Characteristics of a glycoprotein in the ocular surface glycocalyx. Invest. Ophthalmol. 33, 218–227.

Gromada, J., Jorgensen, T. and Dissing, S. (1995). The release of intracellular Ca^{2+} in lacrimal acinar cells by alpha-, beta-adrenergic and muscarinic cholinergic stimulation: The roles of inositol triphosphate and cyclic ADP-ribose. Pflugers Arch. 429, 751–761.

Gutkind, J. (1998). The pathways connecting G protein-coupled receptors to the nucleus through divergent mitogen-activated protein kinase cascades. J. Biol. Chem. 273, 1839–1842.

Hamm, H. and Gilchrist, A. (1996). Hetertrimeric G proteins. Curr. Opin. Cell Biol. 8, 189–196.

Hartschuh, W., Weihe, E. and Reinecke, M. (1983). Peptidergic (neurotensin, VIP, substance P) nerve fibers in the skin. Immunohistochemical evidence of an involvement of neuropeptides in nociception, pruritis and inflammation. Br. J. Dermatol. 109, 14–17.

Hodges, R., Dicker, D. and Dartt, D. (1992). a_1-Adernergic and cholinergic agonists use separate signal transduction pathways in lacrimal gland. Am. J. Physiol. 262, G1087–G1096.

Hodges, R., Zoukhri, D., Sergheraert, C., Zieske, J. and Dartt, D. (1997). Identification of vasoactive intestinal peptide receptor subtypes in the lacrimal gland and their signal transducing components. Invest. Ophthalmol. Vis. Sci. 38, 610–619.

Holly, F. (1973). Formation and rupture of the tear film. Exp. Eye Res. 15, 515–525.

Holly, F. and Lemp, M. (1971). Wettability and wetting of corneal epithelium. Exp. Eye Res. 11, 239–250.

Holly, F. and Lemp, M. (1977). Tear physiology and dry eyes. Surv. Ophthalmol. 22, 69–87.

Houle, M. and Bourgoin, S. (1999). Regulation of phospholipase D by phosphorylation-dependent mechanisms. Biochim. Biophys. Acta 1439, 135–149.

Hunt, S., Spitznas, M., Seifert, P. and Rauwolf, M. (1996). Organ culture of human main and accessory lacrimal glands and their secretory behaviour. Exp. Eye Res. 62, 541–554.

Ichikawa, A. and Nakajima, Y. (1962). Electron microscope study on the lacrimal gland of the rat. Tohoku J. 77, 136–149.

Inatomi, T., Spurr-Michaud, S., Tisdale, A.S., Zhan, Q., Feldman, S.T. and Gipson, I.K. (1996). Expression of secretory mucin genes by human conjunctival epithelia. Invest. Ophthalmol. Vis. Sci. 37, 1684–1692.

Ishida, N., Maruo, J. and Mita, S. (1996). Expression and characterization of lacrimal gland water channels in Xenopus oocytes. Biochem. Biophys. Res. Commun. 224, 1–4.

Jaken, S. (1996). Protein kinase C isozymes and substrates. Curr. Opin. Cell Biol. 8, 168–173.

Jakobiec, F. and Iwamoto, T. (1979). The ocluar adnexa: Lids, conjunctiva, and orbit. In: Ocular Histology. (Fine, B. and Yanoff, M., Eds.), pp. 290–342. Harper and Row, Philadelphia.

Jester, J.V., Nicolaides, N. and Smith, R.E. (1981). Meibomian gland studies: Histologic and ultrastructural investigations. Invest. Ophthalmol. Vis. Sci. 20, 537–547.

Jester, J., Rife, L., Nii, D., Luttrull, J., Wilson, L. and Smith, R. (1982). In vivo biomicroscopy and photography of meibomian glands in a rabbit model of meibomian gland dysfunction. Invest. Ophthalmol. Vis. Sci. 22, 660–667.

Jumblatt, J. and Jumblatt, M. (1998). Regulation of ocular mucin secretion by P2Y2 nucleotide receptors in rabbit and human conjunctiva. Exp. Eye Res. 67, 341–346.

Kelleher, R., Hann, L., Edwards, J. and Sullivan, D. (1991). Endocrine, neural, and immune control of secretory component output by lacrimal gland acinar cells. J. Immunol. 146, 3405–3412.

Kessing, S. (1968). Mucous gland system of the conjunctiva. A quantitative normal anatomical study. Acta Ophthalmol. 95 (suppl.), 1.

Kessler, T., Mercer, H., Zieske, J., McCarthy, D. and Dartt, D. (1995). Stimulation of goblet cell mucous secretion by activation of nerves in rat conjunctiva. Curr. Eye Res. 14, 985–992.

King, L.S. and Agre, P. (1996). Pathophysiology of the aquaporin water channels. Annu. Rev. Physiol. 58, 619–648.

Kishimoto, A., Takai, Y. and Nishizuka, Y. (1977). Activation of glycogen phosphorylase kinase by a calcium-activated, cyclic nucleotide-independent protein kinase system. J. Biol. Chem. 252, 7449–7452.

Klyce, S. and Crosson, C. (1985). Transport processes across the rabbit corneal epithelium: A review. Curr. Eye Res. 4, 323–331.

Kompella, U.B., Kim, K.-J. and Lee, V.H.L. (1993). Active chloride transport in the pigmented rabbit conjunctiva. Curr. Eye Res. 12, 1041–1048.

Krenzer, K. and Freddo, T. (1997). Cytokeratin expression in normal human bulbar conjunctiva obtained by impression cytology. Invest. Ophthalmol. Vis. Sci. 38, 142–152.

Lamberts, D. (1994). Physiology of the tear film. In: The Cornea. (Smolin, G. and Thoft, R., Eds.), pp. 439–455. Brown and Company, Little, Boston.

Lehtosalo, J., Uusitalo, H., Mahrberg, T., Panula, P. and Palkama, A. (1989). Nerve fibers showing immunoreactivities for proenkephalin A-derived peptides in the lacrimal glands of the guinea pig. Graefes Arch. Clin. Exp. Ophthalmol. 227, 455–458.

Lemp, M. (1995). Report of the National Eye Institute/Industry workshop on clinical trials in dry eyes. CLAO J. 21, 221–232.

Lemullois, M., Rossignol, B. and Mauduit, P. (1996). Immunolocalization of myoepithelial cells in isolated acini of rat exorbital lacrimal gland: Cellular distribution of muscarinic receptors. Biol. Cell 86, 175–181.

Lewin, D. (1996). From outside or in, sex hormones tweak prostate cells. J. NIH Res. 8, 29–30.

Linton, R.G., Curnow, D.H. and Riley, W.J. (1961). The meibomian glands: An investigation into the secretion and some aspects of the physiology. Br. J. Ophthal. 45, 718–723.

Luhtala, J. and Uusitalo, H. (1991). The distribution and origin of substance P immunoreactive nerve fibres in the rat conjunctiva. Exp. Eye Res. 53, 641–646.

Maitchouk, D., Beuerman, R., Varnell, R. and Pedroza-Schmidt, L. (1998). Effects of lacrimal gland removal on squirrel monkey cornea. Adv. Exp. Med. Biol. 438, 619–624.

Malarkey, K., Belham, C., Paul, A., Graham, A., McLees, A., Scott, P. and Plevin, R. (1995). The regulation of tyrosine kinase signalling pathways by growth factor and G-protein receptors. Biochem. J. 309, 361–375.

Marechal, H., Jammes, H., Rossignol, B. and Mauduit, P. (1996). EGF receptor mRNA and protein in rat lacrimal acinar cells: Evidence of its EGF-dependent phosphotyrosilation. Am. J. Physiol. 270, C1164–C1174.

Marfurt, C. (1999). Nervous control of the cornea. In: Nervous Control of the Eye. (Burnstock, G. and Sillito, A., Eds.), pp. 41–92. Harwood Academic Publishers, Amsterdam.

Markoff, E., Lee, D., Fellows, J., et al. (1995). Human lacrimal glands synthesize and release prolactin. Endocrinology 152 (suppl.), 440A.

Mauduit, P., Jammes, H. and Rossignol, B. (1993). M_3 muscarinic acetylcholine receptor coupling to PLC in rat exorbital lacrimal acinar cells. Am. J. Physiol. 264, C1550–C1560.

Meier, K., Gibbs, T., Knoepp, S. and Ella, K. (1999). Expression of phospholipase D isoforms in mammalian cells. Biochim. Biophys. Acta 1439, 199–213.

Meneray, M., Fields, T. and Bennett, D. (1997). G_s and $G_{q/11}$ couple vasoactive intestinal peptide and cholinergic stimulation to lacrimal secretion. Invest. Ophthalmol. Vis. Sci. 38, 1261–1270.

Mircheff, A. (1993). Understanding the causes of lacrimal insufficiency: Implications for treatment and prevention of dry eye syndrome. Research to Prevent Blindness Science Writers Seminar, 51.

Mircheff, A. (1994). Water and electrolyte secretion and fluid modification. In: Principles and Practices of Ophthalmology: Basic Sciences. (Albert, D., Jakobiec, F. and Robinson, N., Eds.), pp. 466–472. W.B. Saunders, Philadelphia.

Montagna, W. and Ellis, R. (1959). Cholinergic innervation of the meibomian gland. Anat. Rec. 135, 121–128.

Morris, A. and Malbon, C. (1999). Physiological regulation of G protein-linked signaling. Physiol. Rev. 79, 1373–1430.

Newton, A. (1997). Regulation of protein kinase C. Curr. Opin. Cell Biol. 9, 161–167.

Nicolaides, N. (1986). Recent findings on the chemical composition of the lipids of steer and human meibomian glands. The preocular tear film. In: Health, Disease and Contact Lens Wear. (Holly, F., Ed.), pp. 570–596. Dry Eye Institute, Lubbock, TX.

Nishizuka, Y. (1984). The role of protein kinase C in cell surface signal transduction and tumour promotion. Nature 308, 693–698.

Nishizuka, Y. (1992). Intracellular signaling by hydrolysis of phospholipids and activation of protein kinase C. Science 258, 607–614.

Norn, M. (1992). Diagnosis of dry eye. In: The Dry Eye. (Lemp, M. and Marquardt, R., Eds.). Springer-Verlag, Berlin.

Ota, I., Zoukhri, D., Hodges, R.R., Rios, J.D., Tepavcevic, V., Raddassi, L., Chen, L. and Dartt, D.A. (2003). Alpha1-adrenergic and cholinergic agonists activate MAPK by separate mechanisms to inhibit secretion in lacrimal gland. Am. J. Physiol. Cell Physiol. 84, C168–178.

Perra, M., Lantini, M., Serra, A., Cossu, M., De Martini, G. and Sirigu, P. (1990). Human meibomian glands: A histochemical study for androgen metabolic enzymes. Invest. Ophthalmol. Vis. Sci. 31, 771–775.

Pockley, A. and Montgomery, P. (1990–1991). The effects of interleukins 5 and 6 on immunoglobulin production in rat lacrimal glands. Reg. Immunol. 3, 242–246.

Prydal, J.I., Artal, P., Woon, H. and Campbell, F.W. (1992). Study of human precorneal tear film thickness and structure using laser interferometry. Invest. Ophthalmol. Vis. Sci. 33, 2006–2011.

Rae, J. and Farrugia, G. (1992). Whole-cell potassium current in rabbit corneal epithelium activated by fenamates. J. Membr. Biol. 129, 81–97.

Rafferty, D. and Montgomery, P. (1993). Effects of transforming growth factor beta on immunoglobulin production in cultured rat lacrimal gland tissue fragments. Reg. Immunol. 5, 312–316.

Raina, S., Preston, G., Guggino, W. and Agre, P. (1995). Molecular cloning and characterization of an aquaporin cDNA from salivary, lacrimal, and respiratory tissues. J. Biol. Chem. 270, 1908–1912.

Rich, A., Bartling, C., Farrugia, G. and Rae, J. (1997). Effects of pH on the potassium current in rabbit corneal epithelial cells. Am. J. Physiol. 272, C744–C753.

Rios, J., Zoukhri, D., Rawe, I., Hodges, R., Zieske, J. and Dartt, D. (1999). Immunolocalization of muscarinic and VIP receptor subtypes and their role in stimulating goblet cell secretion. Invest. Ophthalmol. Vis. Sci. 40, 1102–1111.

Rios-Garcia, J., Kessler, T., Ota, I. and Dartt, D. (1997). Regulation of glycoconjugate secretion from conjunctival goblet cells in vitro. Invest. Ophthalmol. Vis. Sci. 38 (suppl.), S154.

Robinson, M. and Cobb, M. (1997). Mitogen-activated protein kinase pathways. Curr. Opin. Cell Biol. 9, 180–186.

Rocha, E., Wickham, L., da Silveria, L., Krenzer, K., Yu, F., Toda, I., Sullivan, B., et al. (2000). Identification of androgen receptor protein and 5a-reductase mRNA in human ocular tissue. Br. J. Ophthalmol. 84, 76–84.

Rothman, J. and Wieland, F. (1996). Protein sorting by transport vesicles. Science 272, 227–234.

Rothman, J. and Sollner, T. (1997). Throttles and dampers: Controlling the engine of membrane fusion. Science 276, 1212–1213.

Ruskell, G. (1985). Innervation of the conjunctiva. Trans. Ophthalmol. Soc., UK 104, 390–395.

Ruskell, G.L. (1985). Innervation of the conjunctiva. Trans. Ophthalmol. Soc., UK 104, 390–395.

Satoh, Y., Oomori, Y., Ishikawa, K. and Ono, K. (1994). Configuration of myoepithelial cells in various exocrine glands of guinea pigs. Anat. Embryol. 189, 227–236.

Seifert, P. and Spitznas, M. (1993). Demonstration of nerve fibers in human accessory lacrimal glands. Graefes Arch. Clin. Exp. Ophthalmol. 232, 107–114.

Seifert, P. and Spitznas, M. (1999). Vasoactive Intestinal polypeptide (VIP) Innervation of the Human Eyelid Glands. Exp. Eye Res. 68, 685–692.

Seifert, P., Spitznas, M., Koch, F. and Cusumano, A. (1994). Light and electron microscopic morphology of accessory lacrimal glands. Adv. Exp. Biol. Med. 350, 19–23.

Seifert, P., Stuppi, S. and Spitznas, M. (1997). Distribution pattern of nervous tissue and peptidergic nerve fibers in accessory lacrimal glands. Curr. Eye Res. 16, 298–302.

Sharif, N., Crider, J., Griffin, B., Davis, T. and Howe, W. (1997). Pharmacological analysis of mast cell mediator and neurotransmitter receptors coupled to adenylate cyclase and phospholipase C on immunocytochemically defined human conjunctival epithelial cells. J. Ocul. Pharmacol. Ther. 13, 321–336.

Sharma, A. (1993). Energetics of corneal epithelial cell-ocular mucus-tear film interactions: Some surface-chemical pathways of corneal defense. Biophys. Chem. 47, 87–99.

Shi, X. and Candia, O. (1995). Active sodium and chloride transport across the isolated rabbit conjunctiva. Curr. Eye Res. 14, 927–935.

Sternberg, P. and Alberola-Ila, J. (1998). Conspiracy Theory: RAS and RAF do not act alone. Cell 95, 447–450.

Sullivan, D.A. (1999a). Immunology of the lacrimal gland and tear film. Dev. Opthalmol. 30, 39–53.

Sullivan, D. (1999). Ocular mucosal immunity. In: Handbook of Mucosal Immunology. (Ogra, P. and Mestecky, J., Lamm, M., et al., Eds.), pp. 1241–1281. Academic Press, Orlando.

Sullivan, D. and Sato, E. (1994). Immunology of the lacrimal gland. In: Principles and Practices of Ophthalmology: Basic Science of the eye. (Albert, D., Jakobiec, F. and Robinson, N., Eds.), pp. 479–486. W.B. Saunders, Philadelphia.

Sullivan, D., Wickham, L., Rocha, E., Kelleher, R., da Silveira, L. and Toda, I. (1998). Influence of gender, sex steroid hormones, and the hypothalamic-pituitary axis on the structure and function of the lacrimal gland. Adv. Exp. Med. Biol. 438, 11–42.

Sunahara, R., Dessauer, C. and Gilman, A. (1996). Complexity and diversity of mammalian adenylyl cyclases. Ann. Rev. Pharmacol. Toxicol. 36, 461–480.

Taussig, R. and Gilman, A. (1995). Mammalian membrane-bound adenylyl cyclases. J. Biol. Chem. 270, 1–4.

Thody, A. and Shuster, S. (1989). Control and function of sebaceous glands. Physiol. Rev. 69, 383–408.

Thompson, H., Beuerman, R., Cook, J., Underwood, L. and Nguyen, D. (1994). Transcription of message for tumor necrosis factor-alpha by lacrimal gland is regulated by corneal wounding. Adv. Exp. Med. Biol. 350, 211–217.

Tiffany, J. (1987). The lipid secretion of the meibomian glands. Adv. Lipid Res. 22, 1–62.

Tiffany, J. (1995). Physiological functions of the meibomian glands. Prog. Retin. Eye Res. 14, 47–71.

Treisman, R. (1996). Regulation of transcription by MAP kinase cascade. Curr. Opin. Cell Biol. 8, 205–215.

Tsunoda, Y. (1993). Receptor-operated Ca^{2+} signaling and crosstalk in stimulus secretion coupling. Biochim. Biophys. Acta 1154, 105–156.

Walcott, B. (1990). Leu-enkephalin-like immunoreactivity and the innervation of the rat exorbital gland. Invest. Ophthalmol. Vis. Sci. 31 (suppl.), 44.

Walcott, B. (1994). Anatomy and innervation of the human lacrimal gland. In: Principles and Practice of Ophthalmology: Basic Sciences. (Albert, D., Jakobiec, F. and Robinson, N., Eds.), pp. 454–458. W.B. Saunders, Philadelphia.

Watsky, M., Jablonski, M. and Edelhauser, H. (1988). Comparison of conjunctival and corneal surface areas in rabbit and human. Curr. Eye Res. 7, 483–486.

Wickham, L., Gao, J., Toda, I., Rocha, E., and Sullivan, D. (2000). Identification of androgen, estrogen and progesterone receptor mRNAs in the eye. Acta Ophthalmol.

Wieczorek, R., Jakobiec, F., Sacks, E. and Knowles, D. (1988). The immunoarchitecture of the normal human lacrimal gland. Relevancy for understanding pathologic conditions. Ophthalmology 95, 100–109.

Yoshino, K., Monroy, D. and Pflugfelder, S. (1996a). Cholinergic stimulation of lactoferrin and epidermal growth factor secretion by the human lacrimal gland. Cornea 15, 617–621.

Yoshino, K., Garg, R., Monroy, D. and Pflugfelder, S. (1996b). Production and secretion of transforming growth factor beta (TGF-beta) by the human lacrimal gland. Curr. Eye Res. 15, 615–624.

Zieske, J., Mason, V., Wasson, M., Meunier, S., Nolte, C., Fukai, N., Olsen, B., et al. (1994). Basement membrane assembly and differentiation of cultured corneal cells: Importance of culture environment and endothelial cell interaction. Exp. Cell Res. 214, 621–633.

Zoukhri, D. and Dartt, D. (1995). Cholinergic activation of phospholipase D in lacrimal gland acini is independent of protein kinase C and calcium. Am. J. Physiol. 268, C713–C720.

Zoukhri, D., Hodges, R., Willert, S. and Dartt, D. (1997a). Immunolocalization of lacrimal gland PKC isoforms. Effect of phorbol esters and cholinergic agonists on their cellular distribution. J. Membr. Biol. 157, 169–175.

Zoukhri, D., Hodges, R., Sergheraert, C., Toker, A. and Dartt, D. (1997b). Lacrimal gland PKC isoforms are differentially involved in agonist-induced protein secretion. Am. J. Physiol. 272, C263–C269.

Zoukhri, D., Hodges, R.R., Sergheraert, C. and Dartt, D.A. (2000). Cholinergic-induced Ca^{2+} elevation in rat lacrimal gland is negatively modulated by PKC delta and PKC epsilon. Invest. Ophthalmol. Vis. Sci. 41, 386–392.

FURTHER READING

Hodges, R.R., Shatos, M.A., Tarko, R.S., Vrouvlianis, J., Gu, J. and Dartt, D.A. (2005). Nitric oxide and cGMP mediate alpha 1D-adrenergic receptor-stimulated protein secretion and p42/p44 MAPK activation in rat lacrimal gland. Invest. Opthalmol. Vis. Sci. 16, 2781–2789.

Mochly-Rosen, D. and Gordon, A. (1998). Anchoring proteins for protein kinase C: A means for isozyme selectivity. FASEB J. 12, 35–42.

THE CORNEA
EPITHELIUM AND STROMA

Niels Ehlers and Jesper Hjortdal

Advances in Organ Biology
Volume 10, pages 83–111.
© 2006 Elsevier B.V. All rights reserved.
ISBN: 0-444-50925-9
DOI: 10.1016/S1569-2590(05)10003-2

INTRODUCTION

The cornea is the anterior, transparent part of the collagenous wall of the eyeball. It is the window of the eye to the outer world. Its properties allow for the formation of an optical image on the light-sensitive retina in the back of the eye. This requires transparency and regularity, but it also demands that the gross dimensions of the eye be kept constant. A regulated hydrostatic pressure within a relatively stiff eyeball accomplishes this.

The aim of this chapter is to describe the human cornea. Regarding dimensions, there are of course large species variations. Functional aspects have often been studied in animal corneas and usually extended to all other species. Unless otherwise stated the text applies to the "standard human cornea".

I. GROSS ANATOMY

In the human eye, the cornea forms a dome-like projection from the spherical shape of the eyeball (Figure 1). The change in curvature is abrupt enough to form a groove just posterior to or outside the corneal periphery. This is also the transition from opaque sclera to clear cornea. The diameter of the adult cornea is about 11 mm, slightly larger horizontally than vertically. The central curvature of the surface is 7.8 mm (44 diopters), slightly larger in the horizontal than in the vertical direction (meridian), corresponding to a toricity of 0.5 diopters axis horizontal. The thickness of the central human cornea is 0.52 mm and increases toward the periphery. In contrast, many animal corneas have the same thickness over a considerable area. The surface area of the human cornea is about 1.3 cm^2, corresponding to 15% of the total surface of the eyeball. In some animals, the proportion is higher (e.g., 25% in rabbit and almost 50% in rat and mouse). The relative size of the cornea varies among animal species living under different conditions, being generally large in night-living animals and small in those active in daylight. A short review on ocular dimensions of some commonly studied animals is given in Table 1.

The cornea is a highly specialized tissue with a multilayered structure. Usually five layers are distinguished from outside: epithelium, Bowman's

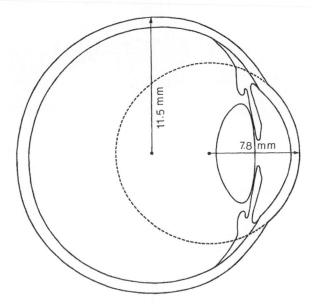

Figure 1. Diagram of the eyeball with protruding cornea. The pericorneal sulcus is created by the dome-like protrusion of the cornea. Reprinted from Histology of the Human Eye, Hogan, Alvarado and Weddell, p. 92, 1971, with permission from Saunders.

Table 1. Average Dimensions of the Cornea of Some Species

Species	Eyeball mm	Diameter Cornea mm	Curvature mm	Cornea Thickness
Man	24	11	7.9	0.52
Cynomolgus	18	9	7	0.4
Rabbit	17	12	8.0	0.4
Rat	6	5.5	3	0.2
Mouse	3.5	3	1.75	0.15
Pig	20	15	10	0.7
Ox	35	20	17.5	0.8

Data collected from various sources and own measurements.

membrane, stroma, Descemet's membrane, and endothelium (Figure 2). The cornea may also be divided into a central part and a peripheral or limbal zone.

This chapter describes the structure of the anterior epithelium and stroma separately. The biochemistry is also considered separately, while the physiology is discussed for epithelium and stroma together, because they exert

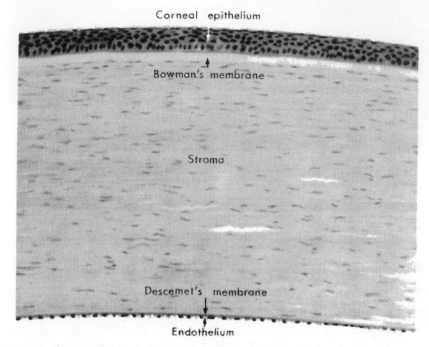

Figure 2. Histological section of the central cornea showing the 5 layers: epithelium, Bowman's membrane, stroma, Descemet's membrane, and endothelium. Reprinted from Ophthalmic Pathology, Hogan and Zimmermann, pg. 278, 1962, with permission from Saunders.

their functions in common. Three physiological aspects will be discussed: (1) The maintenance of the epithelium as a protective barrier to external noxious influences, (3) the biomechanics, which ensures the shape and the stability of the cornea (and the eye), and is the requirement for an image formation on the retina, and finally (2) the transmission of light through the cornea.

Before describing in detail the structure and function of these parts a summary of the embryology may be relevant. For a detailed description of the embryonal and fetal development of the cornea, textbooks in anatomy should be consulted (e.g., Barishak, 2001).

II. EMBRYOLOGY

The cornea can be identified at an early stage of development (6th week) (Figure 3). Waves of cells migrate into the space between the ectoderm and lens vesicle. The cells come in three waves: the first in the 7th week, forming

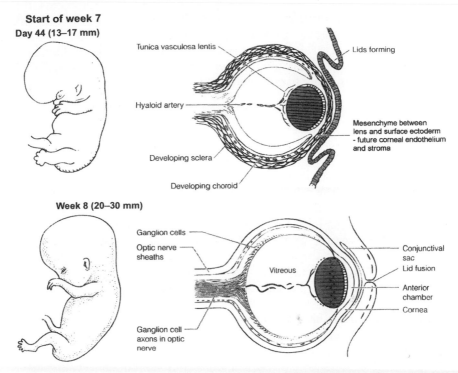

Figure 3. Top: Invasion of neural crest cells to form endothelium and keratocytes at 6th week. Bottom: Fusion of lids in front of the eye, isolating the cornea-conjunctival cavity from 3rd month until shortly before term. (From Forrester, J.D., Dick, A.D., McMenamin, P. and Lee, W.R. (1996). The Eye, Basic Sciences in Practice. Saunders, London.)

the corneal endothelium; the cells of the second wave migrate between the ectoderm, now called the primitive epithelium, and the endothelium; and gives rise to the keratocytes, the cells of the corneal stroma. The third wave of migrating cells enters between endothelium and lens and forms the iris stroma. The endothelial and the stromal cells are believed to be of neural crest origin and to differentiate into mesenchymal cells (Hayashi et al., 1986). Keratan sulphate proteoglycan can be demonstrated in the corneal stroma soon after arrival of these cells, and at the 8th week it is present in the keratocytes and the endothelial cells but not in the epithelial cells (Azuma et al., 1994). The density of keratocytes and endothelial cells is high at birth and decreases steadily during life. Cell division is rare for both cell types.

For a certain period of the intrauterine life (3rd to 6th month), the eyelids fuse in front of the eye isolating the corneal epithelium from the amniotic fluid and possible inductive effects. The specific effect of this arrangement, which is universal among mammals, is still not clear.

At birth the cornea is relatively large compared to the eyeball. The diameter is 10 mm, the curvature similar to the adult, the thickness slightly higher but reduces to adult value within months.

III. STRUCTURE AND COMPOSITION OF EPITHELIUM

The corneal epithelium is a stratified, squamous, nonkeratinized epithelium (Figure 4). Its thickness in man is about 50 μm, corresponding to 10% of the corneal thickness. The cells are arranged in three parts. The innermost or deepest part is formed by the basal cells, which are columnar and closely packed, about 15 μm in diameter. The middle part is made up of 2–3 layers of wing- or goblet-cells, called so because of their shape, becoming increasingly flatter towards the surface. The outermost 2–3 layers of flattened, squamous cells form the superficial part. The total number of cell layers in the human epithelium is therefore 5–7. Mitoses are found almost exclusively in the basal layer.

Smaller mammals have fewer layers, but always a distinguishable basal layer and a superficial flat-cell layer. Large mammals (ox, horse, elephant, whale, etc.) can have 10–20 layers of cells.

The basal cells have a distinct smooth and nonundulating basement membrane, to which the cells are attached by hemidesmosomes. The thickness of the basement membrane is 40–60 nm. It is similar in structure and composition to basement membranes of other squamous epithelia. It contains collagen type IV, but also type VII and XII. Laminin, fibronectin, fibrin, and the proteoglycan perlecan, has also been demonstrated. The primary proteins of the hemidesmosomes are bullous pemphigoid antigen and integrin heterodimer (a laminin receptor). The intracellular part of the hemidesmosomes is linked to keratin filaments. In the basement membrane the hemidesmosomes are linked to anchoring fibrils made of type VII collagen passing through the superficial stroma (Bowman's layer) to a depth of 2 μm, and ending in anchoring plaques composed of laminin. A firm fixation of the cells to the underlying smooth surface is thus secured.

The cells of all the layers of the epithelium interdigitate and are separated by a 10–20 nm intercellular space. The cells of the corneal epithelium have four types of intercellular junctions. The cells are attached to each other by desmosomes and communicate through *gap* junctions, which allow for passage of small molecules. The superficial cells in addition are connected by adherent junctions and tight junctions (zonulae occludens) establishing a diffusion barrier in the surface of the epithelium. Other molecules along the cell membrane also function in the cell to cell adhesion (e.g., cadherins and

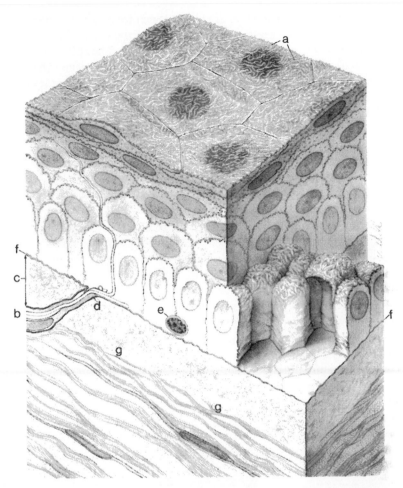

Figure 4. Block diagram of epithelium with penetration of nerves through Bowman's membrane. Reprinted from Histology of the Human Eye, Hogan, Alvarado and Weddell, pg. 113, 1971, with permission from Saunders.

integrins). Thus much is done to keep the cells together as a layer, in good accordance with its function as a protective barrier.

The cytoplasm of the cells appears rather uniform with tonofibrils, which in the more superficial cells run parallel to the surface. Of the three cytoplasmic filament types found within all cells, keratin filaments are the major type within the corneal epithelium. The keratins either type I (acidic) or class II (neutral and basic) belong to a group of about 30 proteins. The keratin filaments are formed by two polypeptide chains, one from each group (Gipson and Joyce, 2000). As the basal cells differentiate into more superficial cells, two keratin pairs are expressed in sequence. K5 and K14 are found in the basal cells, while suprabasal cells express K3 and K12. K12 is believed to be cornea specific (Schermer et al., 1986). The role of the keratin filaments is to form a cytoskeleton anchoring the cells to each other and to the stroma through desmosomes and hemidesmosomes. Of the other types of cytoskeletal filaments, the actin filaments are prevalent as a network along the apical cell membrane, where they extend into the microplicae. At the junction of lateral cell membranes they are associated with junctions of adherens and tight type. The actin filaments are important in providing the cytoskeletal connection of adhesion molecules, such as integrins and cadherins. The third cytoskeletal filament, the microtubules are involved in the spindles of mitotic basal cells, where they provide the framework for chromosomal segregation.

The cytoplasm is poor in organelles, which almost disappear as the cells move to the surface. In the basal cells the Golgi apparatus is relatively well developed while the mitochondria are small. The cell membrane of the outermost cells show projections resembling microvilli, or in flat preparations microcristae. These microexcrescences possibly have a role in maintaining the structure of the tear film. The outer surface of the superficial cells in addition is covered by a glycocalyx. The nuclei are round in the basal cells and become more and more indented and folded with clumped chromatin as the cells move towards the surface.

The corneal epithelium has cellular functions common to other epithelia, so-called housekeeping functions, maintaining the basal cellular functions in renewal and the overall barrier function against the outside. For these functions, the cells have the machinery with nucleus, protein expression, energy production, and transport processes. The cells, in particular the basal cells normally contain large amounts of glycogen.

A. The Tear Film

The tear film covers the surface of the epithelium. This thin layer of fluid maintains the optical quality of the cornea. It is in itself a layered structure. The air–fluid interface is composed of hydrophobic lipids from the Meibomian glands. Polar lipids (a.o. phospholipids) reduce the interfacial tension at the lipid–water border. The fluid layer is composed of an aqueous phase of tears. The transition to the surface cell is again stabilized, by polar molecules, the glycocalyx of the cell membrane, or a thicker mucoid layer. The possibility of species variations is not resolved, in particular considering the large differences in anatomy of eyelids and glands among different species. The exact thickness of the mucoid layer and of the entire precorneal tear film is still a matter of investigation. The usually given value for tear film thickness is 7–10 microns.

The tear film is renewed at every blink and upon opening the eye its structure is immediately rebuilt. Until the next blink there is a drain of aqueous tears by gravity and by capillary attraction exerted by the curvature of the meniscus along the lid margins.

B. Light and Dark Cells

In the basal layer, a marked difference in staining characteristics of the cells may be evident. Many studies have been dedicated to this difference between light and dark cells. The most likely explanation seems to be that these are cells in different stages of development. When studied by scanning electron-microscopy the outer surface also reveals light and dark cells. It is believed that the light cells are those recently exposed to the surface (i.e., the difference here also indicate a process of maturation). However, the implications of light and dark cells in the basal and in the superficial layer are not clear.

C. Langerhans Cells

Langerhans cells are wandering cells of macrophage nature that are found as stellate cells between the epithelial cells, mainly in the periphery but occasionally central and in the stroma.

D. Innervation

The cornea is one of the most densely innervated tissues in the body (Figure 5). A review was recently published by Müller et al. (2003). Most corneal nerves are sensory branches from the trigeminal nerve. It is estimated that there are approximately 7000 nocireceptors/mm^2 in the human cornea. The density of nerve endings is 3–400 times that of epidermis. All mammalian corneas receive sympathetic innervations from the superior cervical ganglion. In the human cornea sympathetic fibers are assumed to be very rare. Parasympathetic innervation has been described in certain animals but it is unclear if the human cornea receives parasympathetic nerves.

When the nerve bundles enter the cornea in the periphery, they soon loose their myelin sheaths to become clinically almost invisible. The nerves are located in the anterior third of the stroma. Keratocytes are often located in close contact with nerve fibers. To reach the epithelium, the nerves turn abruptly 90° and penetrate the Bowman's layer. After penetrating this membrane they turn once again abruptly 90° to continue parallel to the corneal surface. Finally, the nerves turn for a third time now to pass upwards between the epithelial cells. Different types of epithelial nerve fibers and endings can be distinguished morphologically. Epithelial cell membranes come in close contact with nerve terminals. The function of

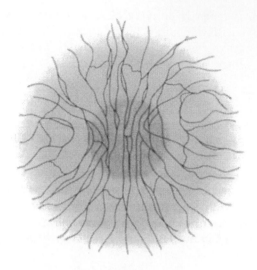

Figure 5. Distribution of nerves within the cornea. Reprinted from Corneal Nerves: Structure, Contents and Function, Vol. 76, Müller et al., Exp. Eye. Res., 521–542, 2003, with the permission from Elsevier.

these epithelial specializations remains unclear. Their appearance suggests a release to or takes up from cells, or extracellullar substances.

The sensory nerves of the cornea express a variety of biologically active substances. Many of the nerves contain substance P (SP) and/or calcitonin gene-related peptide (CGRP) in addition to pituitary adenylate cyclase-activating peptide (PACAP). Also excitatory amino acids, such as glutamate and aspartate, are expressed. The corneal sensory nerves are thought to exert a trophic function on the epithelium. It remains to be determined if the neuro-chemical substance of a corneal nerve correlates with its electrophysiological or trophic functions.

E. Limbal Structure

The corneo–scleral junction or limbus is a transitional zone, where the clear cornea continues into the opaque sclera, corresponding to the corneo–scleral sulcus at the outer eye (Figure 6). Its anterior border can be defined histo-logically as a line connecting the ends of Bowman's and Descemet's mem-branes. The posterior border of the limbus is the line, where the tissue changes from transparent to opaque. At the end of Bowman's membrane the character of the epithelium changes towards more cell layers and less regularity although the cell junctions remain the same. There are more

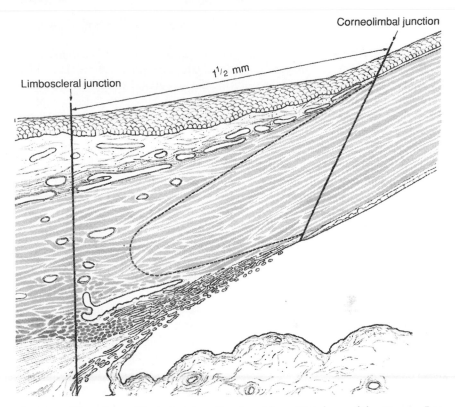

Figure 6. Overview of limbal region. Reprinted from Histology of the Human Eye, Hogan, Alvarado and Weddell, p. 84, 1971, with permission from Saunders.

organelles in the cytoplasm; and the basement membrane assumes a wavy course due to radial folding.

The basal layer of the limbal epithelium contains the stem cells of the epithelium, from which regeneration takes place. Basal cells and stem cells can be distinguished by their expression of Keratins 12 and 3, the latter being a specific marker for differentiation (Edelhauser and Ubels, 2002). A loose subepithelial stromal tissue appears overlying the compact corneal stroma. This tissue is supplied by blood vessels arranged in a superficial and a deep network. The radially oriented crests of tissue are termed the palisades of Vogt. This arrangement increases substantially the basal cell area, a fact that may be of importance for the nutrition and for exchange of metabolites.

At the limbus, the collagen fibrils of the corneal stroma turn and run circumferentially forming an annulus that maintains the curvature of the cornea and the cornea–scleral sulcus.

IV.   STRUCTURE AND COMPOSITION OF STROMA

The stroma makes up about 90% of the corneal thickness. It is a dense tissue consisting of sheets of collagenous material each composed of highly ordered collagen fibrils with interspersed cells, the keratocytes (Figure 7). The matrix between the collagen fibrils contains proteoglycans and other proteins (Table 2). Blood vessels are absent except for the extreme periphery (see limbal structure).

The basic structural unit of the fibrillar collagens is tropocollagen, an asymmetric molecule about 300 nm long and 1.5 nm in diameter. Fibrillar collagens are composed of three amino acid chains coiled in a triple helix. These molecules polymerize to form elongated collagen fibrils with a diameter of 25–30 nm and interspaces about 40–45 nm. Several types of collagen are known. The stroma is composed of collagen fibrils, mainly type I with lesser amounts of collagen V, VI, and XII. Collagen type V is located in the centre of the fibril and type I on the fibril surface. The fibril diameter is regulated by the ratio between type I and type V collagen. The more the type V, the smaller the diameter. The interfibrillar distance is probably regulated by collagen type XII, which form lateral bridges between the fibrils. Collagen type VI forms microfibril networks.

The extrafibrillar matrix contains proteoglycans composed of a core protein molecule, to which glycosaminoglycan sidechains are covalently linked. The proteoglycans bind to the exterior surface of the collagen fibril (Scott and Haigh, 1988). Glycosaminoglycans (old name: mucopolysaccharides) are long carbohydrate molecules composed of repeating disaccharide units. They occur in four main forms: chondroitin sulphate, dermatan sulfate, heparan sulfate, and keratan sulfate. Keratan sulphate proteoglycan and dermatan sulphate proteoglycans are the predominant proteoglycans in the corneal stroma. The proteoglycans of the corneal stroma belong to the family of small leucine-rich repeat proteoglycans. The core proteins of these molecules have a high homogeneity and contain 7–10 leucine-rich motifs with the sequences L-X-X-L-X-L-X-X-N-X-L, where X is any amino acid. The corneal proteoglycans were originally named according to their carbohydrate side chains and are known as chondroitin sulphate/dermatan sulphate (CS/DS) proteoglycans and keratan sulphate (KS) proteoglycans. The genes coding for the core protein of the proteoglycans have now been cloned, and the molecules are now named for their core protein, with lumican, keratocan, mimican, and decorin, having been identified in the cornea.

The cornea contains three closely related keratan sulphate proteoglycans, which are more abundant in the posterior stroma. Lumican is the core protein of keratan sulphate proteoglycan. Lumican is a protein with 342 amino acids and 11 leucine-rich repeats, with a single keratan sulphate side chain. Lumican is abundant in the cornea and is essential to the maintenance of

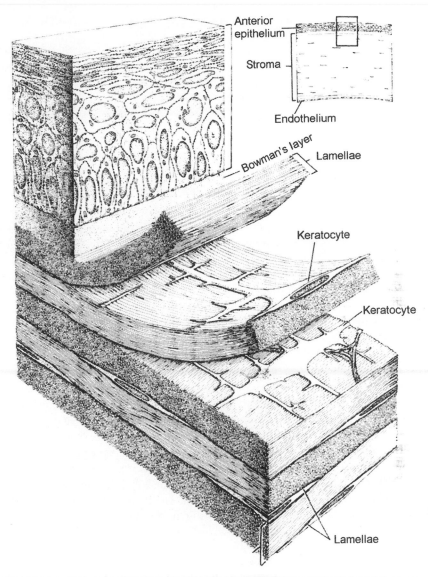

Figure 7. Block-diagram of cornea illustrating stromal lamellae and flat keratocytes.

transparency. Keratocan is the second keratan sulfate proteoglycan in the cornea. It has 10 leucine-rich repeats and carries 3 KS side chains. It exists almost exclusively in the cornea. The third keratan sulfate proteoglycan is mimican. It has five leucine-rich repeats and carries only one keratan sulphate chain. Lumican, keratocan, and mimican are also found in other tissues, but

Table 2. Composition of Stroma

Substance	%
Water	78
Collagen	15
Other proteins	5
Keratan sulphates	0.7
Chondroitin sulphate	0.3
Salts	1

they are present only in the cornea in a highly sulphated form. The sulphation of the sidechains in the cornea increases the water-binding properties.

Decorin is a CS/DS proteoglycan with nine leucine-rich motifs and a single CD/DS side chain. Decorin is the core protein of dermatan sulphate proteoglycan. It is more abundant in the anterior stroma than in the posterior, and is the only CS/DS proteoglycan in the cornea.

The turn-over time for the proteoglycans appears to be several months and for collagen even longer. Lesions in the stromal collagen and in Bowman's layer never regain their original structure.

The collagen fibrils are packed in bundles extending from limbus to limbus, the bundles again being arranged in lamellae. This structure is seen clearly in polarized light. The bundles in the posterior region of the stroma cross in right angles, whereas those in the anterior stroma cross at oblique angles. This simple structure model applies well to lower vertebrates. In mammals and in particular in the human and primate cornea the adjacent lamellae are of different thickness, make various angles with one another, and are limited in width. In the meridional section the bundles in the posterior stroma run in parallel, in the anterior stroma they cross.

The thickness of the lamellae is said to be 1.5–2.5 microns, which would mean 200–250 lamellae in the human cornea. The width of the single lamella is reported in the literature as about 1 mm. In the anterior quarter of the human cornea, the lamellae become much narrower and appear to interweave in an irregular manner. The most anterior portion forms the Bowman's zone in man and primates.

The anterior part of the stroma is acellular and by light microscopy it appears homogeneous. It is known as Bowman's membrane. It is prominent in human and primate corneas, but less so in many animal species. It is a 10-micron zone of randomly orientated collagen fibrils without cells. Its constituents are probably synthesized by both the basal epithelial cells and the superficial keratocytes. It contains several collagen types (I, V, and VII), and proteoglycans. The fibrils in Bowman's membrane are thinner (20 nm) than those of the stroma, and accordingly shows a higher type V to type I collagen ratio. Type VII collagen is found in anchoring fibrils connecting the epithelial hemidesmosomes to plaques located 1–2 microns into the anterior portion of Bowman's membrane.

These anchoring fibrils intertwine with type I fibrillar collagen forming a network that stabilizes the connection between the epithelium and the stroma.

In some lower animals fibers run perpendicular to the surface across the entire thickness of the stroma. They are known as sutural fibers of Ranvier.

The matrix of the stroma is produced and maintained by the stromal cells, the keratocytes. The cells are flattened and arranged between the collagen bundles in the human cornea between and within the lamellae (Figure 8). They are found in all levels of the stroma except Bowman's zone. The cells are rather poorly supplied with organelles, but contain rough endoplasmic reticulum and Golgi apparatus reflecting an active protein synthesizing function. The keratocytes have slender cytoplasmic processes and form gap junctions with neighbouring cells. The activity of the cells is coordinated via these communications. The cells apparently form a syncytium as injected dye can spread from cell to cell. The density of cells have been estimated to be about 5×10^4 cell/mm^3, corresponding to a total of about 2.5 million keratocytes in a cornea (Møller-Pedersen et al., 1994, 1995), and make up almost 10% of the stromal volume. The density decreased with age. The mRNA for lumican, keratocan, and mimican have been identified in keratocytes. Proteoglycans and collagen is produced by the cells, while the final orientation of the molecules takes place extracellularly.

The cells may be activated by damage. Some cells may then differentiate into myofibroblasts

V. ASPECTS OF PHYSIOLOGY

Three main issues will be discussed, (1) Functions of the epithelium as a barrier, (2) image transmission through the cornea, and (3) mechanical stability of the cornea.

A. Function of the Epithelium as a Barrier

The corneal epithelium is the barrier of the cornea to the exterior. Primarily this means to establish the regular surface, upon which the tear film can form the optically perfect refracting surface, secondarily it is to be a barrier to passage of nutrients, solvents, and other substances. The cells in the surface are closely joined. The repeated washing of the surface also supports this barrier because blinking movements wash material off the surface and help to smooth irregularities.

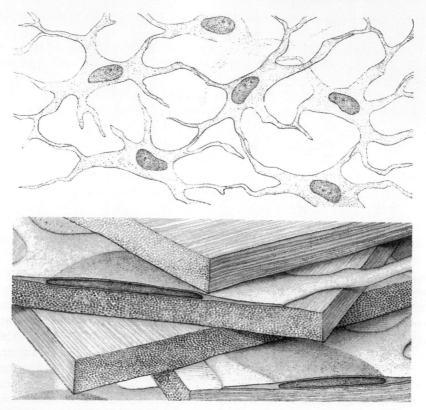

Figure 8.   Top: Keratocyte syncytium in frontal view. Bottom: Lamellar structure of the stroma. Reprinted from Histology of the Human Eye, Hogan, Alvarado and Weddell, p. 61, 1971, with permission from Saunders.

VI.   EPITHELIAL PERMEABILITY AND ELECTROPHYSIOLOGY

The corneal epithelium is a tight epithelium with a relatively low-ionic conductance. If the cornea is mounted between two fluid chambers it will generate a transepithelial potential (stromal side positive). The potentials are maintained by an epithelial transport of sodium inwards and a transport of chloride outwards. The sodium enters the surface cells through channels in the cell surface and is extruded into the intercellular space by an ouabain sensitive Na^+/K^+ ATP ase. The cell interior is maintained at a negative potential of a magnitude of 25–40 mV (Figure 9). The transepithelial resistance about

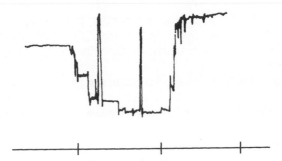

Figure 9. Intraepithelial negative potentials. The potential of the intracellular spaces at the level of the stromal side (Rabbit cornea).

10–20 Ω cm^2 is due to the tight junctions between the superficial cells. When a transepithelial potential is maintained the intercellular spaces are equipotential with the stroma.

The chloride transport in the epithelium results in osmotic transport of water out of the cornea. *In vivo* the importance of the transport for maintaining corneal thickness and transparency is minimal and it is probably primarily involved in epithelial homeostasis. Aquaporins 5 and 3 have been identified in epithelium (Hamann et al., 1998). The role of these water channels in the physiology of the cornea and its epithelium is not yet clear.

The electrophysiological experiments were performed on animal corneas, notably rabbit and cat. Very little specific information exists regarding ion transport in the primate and human cornea, and a universal model for corneal epithelial ion transport has not yet been presented. It seems likely that these transporters are homeostatic and maintain conditions in the epithelium of significance for cell division, migration, and differentiation with the purpose of maintaining the epithelial barrier to the outside world.

VII. NUTRITION AND METABOLIC SUPPLY

The living and reproducing cells of the cornea are metabolically active. The basic metabolites are oxygen and glucose, but vitamins and proteins are important too. The breakdown products to be removed are carbon dioxide, lactic acid, and other waste products.

There are three exchange routes for metabolites and accumulated breakdown products: the anterior epithelial surface, the posterior endothelial surface, and the peripheral limbus. One particular condition distinguishes the corneal metabolism from that of many tissues and organs (i.e., the wide range of temperature, over which the cornea must function). The average temperature of the corneal surface in the open eye is below body temperature. Values of 32–34 °C are given. In the closed eye, the temperature rises to body temperature, while under extreme conditions (e.g., skiing), ice crystals may be formed.

Of the three possible routes of exchange, the peripheral limbal are of the least importance. Early experiments interfering with this route of nutrition by cuts around the periphery, or coagulation of the peripheral vascular blood supply result in no changes in the cornea. By contrast, isolating the anterior surface of the cornea (with a tight and gas-impermeable contact lens) results in oedema of the epithelium and stroma due to interference with the oxygen supply. It is agreed that the cornea gets its oxygen directly from the air by diffusion (partial pressure in the atmosphere 140 mmHg, in venous blood 55 mmHg). The oxygen consumption for the intact isolated cornea is of the order of 0.7 µl/h/mg dry weight, for the isolated stroma the oxygen consumption is lower, around 0.2 µl/h/mg dry weight. Due to lipid solubility of the oxygen molecule, the diffusion across the cornea is sufficient to supply the stromal needs. Glucose enters through the endothelial barrier. Glucose must come from the aqueous, as the concentration in tears combined with the volume in the tear film simply does not allow for enough glucose to be supplied.

Within the stroma, movements take place by diffusion driven by concentration gradients. Molecules move in the stroma in accordance with their molecular weight. Large molecules are restricted in their movement by the limited interfibrillar space. A so-called bulk-flow of water is met with very high resistance. There are no channels or pipelines between the collagen lamellae, as suggested in older literature.

VIII.   METABOLIC PATHWAYS IN THE CORNEA

The enzymes for the glycolytic pathway, the citric acid or tricarboxylic-acid cycle and the hexose monophosphate shunt are present in the cellular layers of the cornea. Glycolysis and the hexose monophosphate shunt are both purely cytosolic enzyme pathways. Glycolysis is an important pathway for glucose utilization and produces lactic acid under anaerobic conditions. In accordance with the relative paucity of mitochondria the greater part of glucose oxidation in the epithelium is by way of the hexose monophosphate shunt and only about one-third enters the tricarboxylic-acid cycle (which requires mitochondria). The sorbitol pathway, which converts glucose to the alcohol sorbitol, has also been identified in the corneal epithelium. In the keratocytes the oxidation is entirely through the tricarboxylic-acid cycle, as 6-phosphogluconate dehydrogenase is not found.

IX. RENEWAL OF EPITHELIUM AND STROMA

The epithelial cells are constantly being renewed by mitotic activity in the basal layers. Usually one cell was assumed to start moving up in the epithelium while the other daughter cell remained basal. Newer observations (Beebe and Masters, 1996) indicate that the two resultant cells of a single division move together toward the apical surface, leaving a need for new cells to divide and mature. Animal experiments with tritium labeled thymidine showed a turn-over time of about 1 week in the rabbit. It is often assumed that the human epithelium needs a similar time span for a cell to move from the basal layer to the very surface, where it is desquamated. Clinical observations of human epithelial pathology, however, suggest a much slower process of turn-over.

The cells in the epithelium are believed to interact in maintaining the homeostasis of the cell layer barrier. Many of these interactions are mediated by cytokines, growth factors, and chemokines. An epithelium to keratocyte interaction also exists. Attention has been paid to different growth factor systems including the EGF system, which has been shown to be up regulated upon injury leading to proliferation.

Regeneration after abrasion involves an initial sliding of surrounding cells to cover the denuded area with a monolayer before the structural regeneration of the multilayered epithelium begins. It has been reported that several months may pass before full ultrastructural integrity is restored.

The limbal epithelium is assumed to be the ultimate reservoir for renewal from its basal stem cells. How the whole physiological process from stem cell to mature superficial epithelial cell occurs, considering also the mitotic activity in the basal cells of the central cornea is not entirely clear. Recent observations on whole corneas in organ culture showed that the limbus-proximal epithelium did not take part in the primary wound closure (Hardarson et al., 2004). This surprising observation may explain the stellate wound shape and pattern of closure lines often observed clinically when epithelium heals after an abrasion.

Compared to epithelium the stroma is a bradytrophic tissue. The cells stay for a long time. Mitoses are not seen. Early experiments with radioactive markers estimated a turn-over for the sulphated polysaccharides of 2–3 months and for collagen even longer turn-over times.

A. Image Transmission Through the Cornea

The interaction between light and matter depends upon nature of the medium, its molecules, and the existence of particles or objects of certain sizes. Dissolved molecules may interact with light energy by absorption at specific wavelengths. The electrons in the chemical bond have a resonance frequency with a kinetic energy corresponding to the quantum energy of the specific wavelength. By removing certain wavelengths the solution becomes colored. By contrast, large particles or objects may be described as having a surface at which light is reflected or refracted. For intermediate-sized particles, the order of the magnitude of the wavelength of light is diffracted. This is explained through the wave nature of light. The particles can be considered as new secondary light sources, from which the propagated light may be in phase leading to amplification, or out of phase leading to a decrease in light intensity.

The main function of the cornea is to allow the formation of an image on the retina. To do this, it must have a regular surface to avoid light scatter, and a curvature that sufficiently refracts light to focus on the retina. In addition, the media and interfaces between them (epithelium, stroma, endothelium, aqueous) must not scatter or reflect light appreciably.

X. THE OPTICAL QUALITY OF THE SURFACE

When light strikes a surface perpendicularly, a fraction of the light is reflected, the specific amount depends on the change in refractive index according to the Fresnel formula:

Reflected fraction $I_r/I_i = (n_1 - n_2)^2/(n_1 + n_2)^2$

I_r being intensity of reflected light, I_i intensity of incident light, n_1 being the refractive index of the first medium, and n_2 the refractive index of the second medium.

For light falling perpendicular to the cornea with its surface lipid layer with a thickness of about 40 nm (Olsen, 1985) and refractive index of 1.5 about 4% will be reflected. The remaining 96% are refracted according to Snell's law:

$$\text{Sin } a_{in}/\text{Sin } a_{out} = n_1/n_2.$$

This formula is used for calculation of image formation in large pupil systems, today using computerized ray tracing programmes (e.g., Zee-max). With a small pupil only the paraxial rays are considered and the formula for image formation in the eye becomes:

$$1/f = 1/f_{cornea} + 1/f_{lens} - d \cdot 1/f_{cornea} \cdot 1/f_{lens},$$

where f is the focal length and d the optical distance between cornea and lens.

XI. TRANSPARENCY OF CORNEAL TISSUE

As mentioned above, efficient propagation of light through the cornea requires that the tissues remain transparent and the interfaces do not scatter light. Traditionally the main interest has been directed toward explaining the transparency of the thick collagenous stroma but actually all layers deserve an explanation regarding transparency. Factors of relevance are summarized in Table 3.

No reflections take place at the epithelial basement membrane, at Bowman's membrane, or at Descemet's membrane because the change in refractive index is negligible. At the endothelium, however, a specular reflection may occur. This is used to obtain an image of the endothelial cell pattern. The reflection takes place at the transition from cell to aqueous humor, because of a change in refractive index (Olsen, 1982a) (Figure 10).

When a narrow slit of light passes through the cornea it reveals the layered structure. The boundary layers, epithelium, and endothelium lights up due to scatter while the stroma is optically almost empty. It is likely that the scatter from the cells occurs from the organelles (i.e., by a diffraction phenomenon caused by differences in refractive index between organelle and cytoplasm). A certain amount of reflection at intracellular membranes cannot be ruled out. The importance of cytoplasmic crystallins in the keratocytes for reducing their visibility was stressed by Jester et al. (1999). Under controlled geometric conditions the width of the optical section can be used to calculate the thickness of the tissue (Ehlers and Hjortdal, 2004).

Table 3. Explaining Transparency of the Corneal Layers

Tear film	Extreme molecular regularity of lipid layer in surface soluble, colorless molecules in lipid film and in water phase
Epithelium	homogenous refractive index, small cytoplasmic organelles
Bowman's membrane	homogenous refractive index, colorless molecules
Stroma	fibril regularity, thin and with regulated distance, homogenous refractive index
Descemet's membrane	homogenous, colorless
Endothelium	homogenous refractive index, colorless molecules

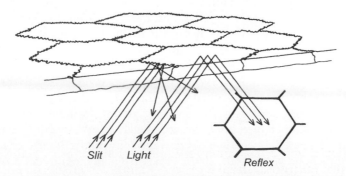

Figure 10. Stromal reflection at transition from endothelial cell to aqueous humour.

XII.   STROMAL TRANSPARENCY

The classical discussion on corneal transparency assumes equal refractive index, or a gradual change in index from fiber to ground substance (Caspersson and Engström, 1945). Most old theories did not account for the different anatomical parts of the cornea but in fact referred only to the stroma. In early physiologic–optical studies, the refractive index was given much attention (Aurell and Holmgren, 1945). It was noted that the overall refractive index of 1.376 could be calculated from the refractive index of collagen and the interfibrillar substance using the law of Gladstone and Dale for an optically composite system:

$$n_{cornea} = n_{collagen}\, d_{collagen} + n_{fluid}\, d_{fluid}$$

with the refractive index for collagen (1.49) and fluid (1.33), and d the tissue volume fraction of fibrils (28%) and fluid (72%). The calculation gives $n = 1.375$ in good agreement with the usually accepted refractive index of corneal stroma. This accordance was considered an argument for a two-compartment theory explaining corneal transparency.

On certain assumptions Maurice (1957) calculated that more than 90% of 500 nm green light would be scattered by the stroma, and stressed that the primary need was to explain the transparency of the normal cornea, and not why a diseased cornea became opaque. The classical *Uniform Refractive Index Theory* would require that the electron-microscopic picture of the stroma with fibrils and interfibril substance was an artefact not present *in vivo*. The collagen fibrils should be swollen and their refractive index reduced to 1.376, the index of the cornea and interfibrillar fluid should be concentrated to reach the same refractive index. With this model it is difficult to explain the opacification of the corneal stroma upon swelling observed also in the *in vivo* cornea (Olsen, 1982b). Maurice (1957) therefore presented his *Interference Theory*. According to the interference theory each collagen fiber scatters light independently of the other fibers (Figure 11). Light waves that meet will destroy, or reinforce each other according to their relative phase. When waves from a number of evenly spaced sources interfere, the combined radiation may be restricted to sharply defined directions at definite angles to the original one. One direction in which recombination takes place irrespective of the fiber spacing is the zero-order direction, the direction of the incoming light. When the spacing is decreased to the wavelength of the light a point is reached when the first-order image is deviated by 90°. With this and closer spacing only the zero-order image remains and radiation in all other directions are completely suppressed by destructive interference and all light energy goes into the constructive interference in the forward direction (Meek et al., 2003). Apart from any loss by reflection and absorption the light will be transmitted unchanged by the grating.

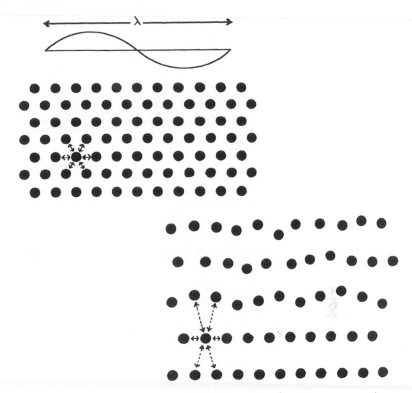

Figure 11. Interference theory for explaining stromal transparency. a. In normal regular stroma destructive interference reduce lateral scatter and promotes forward propagation. b. With an irregular pattern energy is lost by lateral scatter, and the cornea appears opaque. (From Maurice, D.M. (1969). The cornea and sclera. In: The Eye. (Davson, H., Ed.), Vol. 1, Chap. 7, pp. 489–600. Academic Press, London.)

This explanation requires that all collagen fibrils are of equal diameter and are equidistant. Quantitative electron microscopic studies have questioned this regularity but still found evidence for some ordered structure. The small diameter of the corneal collagen fibrils also leads to reduced amount of scatter. Proteoglycans bound to the collagen fibrils maintain the regular fibril pattern. Lumican-deficient mice develop corneal opacification and a disturbed arrangement of the collagen fibrils (Chakravarti et al., 2000). Decorin-deficient mice, in contrast have clear corneas [EK1].

The stromal transparency is now usually explained on the basis of light scatter from the semicrystal lattice arrangement of the collagen fibrils in the lamellae. Each bundle of fibrils will act as a diffraction grating with a spacing less than a wavelength of light and destructive interference will suppress the diffuse scattering of light. A recent review of stromal transparency was published by Meek et al. (2003). It was stressed that all current theories explaining stromal transparency agree upon three points: (1) each stromal fibril is a scatterer, (2) destructive interference must occur, and (3) the cornea is thin.

A. Mechanical Stability of the Cornea

The corneal outer surface supports the shape of the tear film, in which most of the optical refraction of the eye takes place due to the abrupt change in refractive index from air to tissue. Fortunately, the structure of the corneal tissue results in a remarkably stable surface shape, which provides a stable refracting surface (Hjortdal, 1998).

XIII. CORNEAL SHAPE AND TISSUE MECHANICS

The intraocular pressure of the eye sets up the wall tension in the cornea tissue (and in the sclera), and in response to the stress, the stroma strains and elongates (Figure 12). The highly ordered organization of collagen fibrils within the corneal stroma makes the tissue very inextensible. The elongation of the corneal surface arc length is less than 0.5% within physiological and pathological intraocular pressure variations, and intraocular pressure changes, therefore, does not affect the curvature and the corresponding refractive power of the cornea to any significant extent.

In vitro studies of the human cornea have shown that the extensibility of the tissue is very dependent on the level of hydration: a swollen cornea is more extensible than a normally hydrated cornea. This can be understood if the corneal stroma is mechanically treated as a fiber-reinforced material with the collagen fibrils taking up the fibril part, and the ground substance with water forming the "cement". If the hydration is increased, the fibrils at the posterior of the cornea will become wavy, thereby not taking any significant part in strengthening the tissue (Hjortdal, 1998).

While the collagen fibrils normally are responsible for giving corneal tissue strength and stiffness, the ground substance between the fibrils plays a major role for the compressive and the shearing physical properties of the tissue.

XIV. STROMAL SWELLING

The corneal stroma swells when placed in water or aqueous solutions. The expansion takes place perpendicular to the surface, only accompanied by a slight contraction in the plane of the surface. The maximal swelling differs

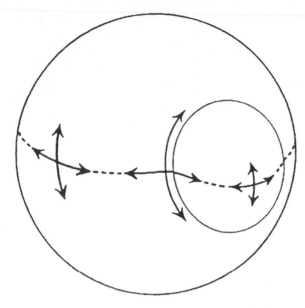

Figure 12. Eyeball with arrows indicating the directions and relative size of mechanical stresses within the ocular tunics (Maurice, D.M. (1969). The cornea and sclera. In: The Eye. (Davson, H., Ed.), Vol. 1, Chap. 7, pp. 489–600. Academic Press, London.)

among animal species, being smallest in man and primates with about 2–2.5 times increase in thickness, compared to about 12 times in the horse.

The stromal swelling, which is accompanied by an opacification, is caused by electrostatic repulsion between negatively charged groups, sulphate and carboxyl in the polysaccharides of the proteoglycan molecules. In accordance with this the swelling is minimal at pH 4, the pK of the carboxyl groups. At neutral pH, the collagen fibrils do not swell. The limitation of the swelling in solutes of different ionic strength is in accordance with this assumption. This expansive force of the stroma can be measured *in vitro* as the swelling pressure. The swelling pressure increases considerably with less corneal thickness, and amounts to approximately 60 mmHg at the normal thickness of 0.52 mm (Olsen and Sperling, 1987) (Figure 13). *In vivo*, a negative intrastromal pressure can even be measured with a cannula (imbibition pressure).

An exception to the swelling is found among certain fishes, the explanation here being the existence of sutural fibers of Ranvier running perpendicular to the surface and thus keeping the anterior and posterior surface together.

In the living eye, the relation between intraocular pressure and corneal thickness is influenced by the permeability of the limiting corneal cell layers and the endothelial fluid pump. Positive as well as negative relations between

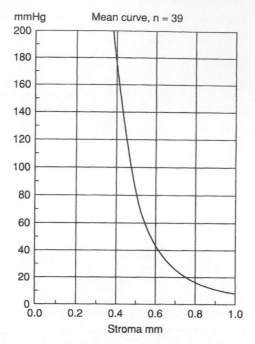

Figure 13. Stromal swelling pressure of human corneal stroma as a function of thickness. At normal thickness 0.5 mm the swelling pressure is 80 mmHg (Olsen, T. and Sperling, S. (1987). The swelling pressure of the human corneal stroma as determined by a new method. Exp. Eye Res. 44, 481–490).

intraocular pressure and corneal thickness have been described. The limiting cell layers are necessary for opposing the expansive force (measured as the swelling pressure) of the corneal stroma, which would otherwise suck water (measured as the imbibition pressure). Thus, even under zero load from the intraocular pressure, the corneal stroma is under a compressive load generated by the endothelial fluid pump. The imbibition pressure decreases from the centre of the cornea to the corneal periphery, possibly reflecting the pumping capacity of the corneal endothelium, the stroma's resistance to water movement, and the local concentration of corneal proteoglycans. The intraocular pressure, the stress distribution between the corneal fibrils, the swelling pressure of the ground substance determined by the local concentration of proteoglycans, the barrier and pumping functions of the limiting cell layers, and the magnitude of and resistance to water flow determine the local corneal hydrostatic pressure. Water, the predominant component of the stroma, will move from regions of high pressure to regions of low pressure, thus leveling out any pressure differences. How this distribution of water determines the shape of the cornea and thereby the refractive properties of the cornea remains to be determined in detail.

The cornea has little resistance to compressing forces along the direction of the collagen fibrils. The tissue will therefore fairly easily bend if an applanating lens is pressed against the surface of the cornea. Some force is, however, needed and this force has been found to increase with corneal thickness. The reading of applanation-based and many other types of tonometers are dependent on corneal thickness (Whitacre and Stein, 1993).

XV. FINAL REMARKS

With the given limitations it has been necessary to omit several fields of importance for corneal homeostasis. However, it has been the aim still to present the structure, composition, and function of the cornea in an open-minded way and to point to the many still unanswered questions and hopefully to stimulate further studies of this simple and still fascinating structure. It was decided to leave out immunological and pharmacological aspects and also to limit the discussion on pathophysiology and pathology, even if this often gives clues to understanding the normal cornea. For information about these subjects the reader is referred to modern textbooks and for a deeper penetration into any problem; the necessity of reading the original publications cannot be overstressed.

REFERENCES

Azuma, N., Hirakate, A., Hida, T. and Kohsaka, S. (1994). Histochemical and immunohisto-chemical studies on keratan sulfate in the anterior segment of the developing human eye. Exp. Eye Res. 58, 277–286.

Barishak, Y.R. (2001). Embryology of the Eye and Its Adnexa, 2nd edn. Karger, Basel.

Beebe, D.C. and Masters, B.R. (1996). Cell lineage and the differentiation of corneal epithelial cells. Invest. Ophthalmol. Vis. Res. 37, 1815–1825.

Edelhauser, H.F. and Ubels, J.L. (2002). The cornea and sclera. In: Adler's Physiology of the Eye. (Kaufman, P.L. and Alm, A., Eds.), 10th edn., pp. 47–114. Mosby, St. Louis.

Forrester, J.D., Dick, A.D., McMenamin, P. and Lee, W.R. (1996). The Eye: Basic Sciences in Practice. Saunders, London.

Gipson, I.K. and Joyce, N.C. (2000). Anatomy and cell biology of the cornea, superficial limbus and conjunctiva. In: Principles and Practice of Ophthalmology. (Albert, D.M., Jakobiec, F.A., Azar, D.T., Gragoudas, E.S., Power, S.M. and Robinsen, N.L., Eds.), Chap. 56, pp. 612–629. Saunders, Philadelphia.

Hamann, S., Zeuthen, T., la Cour, M., Nagelhus, E.A., Ottersen, O.P., Agre, P. and Nielsen, S. (1998). Aquaporins in complex tissues: Distribution of aquaporins 1–5 in human and rat eye. Amer. J. Physiol. 274 (Cell Physiol. 43), C1332–C1345.

Hardarson, T., Hanson, C., Claesson, M. and Stenevi, U. (2004). Time-lapse recordings of human corneal epithelial healing. Acta Ophthalmol. Scand 82(In press).

Hayashi, K., Sueishi, K., Tanaka, K. and Inomata, H. (1986). Immunohistochemical evidence of the origin of human corneal endothelial cells and keratocytes. Graefes Arch. Clin. exp. Ophthalmol. 224, 452–456.

Hjortdal, J.Ø. (1998). On the biomechanical properties of the cornea with particular reference to refractive surgery. Acta Ophthalmol. Scand. 76 (suppl. 225), 1–23.

Hogan, M.J., Alvarado, J.A. and Weddell, J.A. (1971). Histology of the Human eye – An Atlas and Textbook. Saunders, Philadelphia.

Hogan, M.J. and Zimmerman, L.E. (1962). Ophthalmic Pathology – An Atlas and Textbook. Saunders, Philadelphia.

Jester, J.V., Møller-Pedersen, T., Huang, J., Sax, C.M., Kays, W.T., Cacanagh, H.D., Petroll, W.M. and Piatigorsky, J. (1999). The cellular basis of corneal transparency: Evidence for "corneal crystallins." J. Cell Sci. 112, 613–622.

Maurice, D.M. (1957). The structure and transparency of the cornea. J. Physiol. Lond. 136, 263–286.

Maurice, D.M. (1969). The cornea and sclera. In: The Eye. (Davson, H., Ed.), Vol. 1, Chap. 7, pp. 489–600. Academic Press, London.

Meek, K.M., Leonard, D.W., Connon, C.J., Dennis, S. and Khan, S. (2003). Transparency, swelling, and scarring of the corneal stroma. Eye 17, 927–936.

Müller, L.J., Marfurt, C.F., Kruse, F. and Tervo, T.M.T. (2003). Corneal nerves: Structure, contents and function. Exp. Eye Res. 76, 521–542.

Møller-Pedersen, T., Ledet, T. and Ehlers, N. (1994). The keratocyte density of human donor corneas. Curr. Eye Res. 13, 163–169.

Olsen, T. (1982b). Light scattering from the human cornea. Invest. Ophthalmol. Vis. Sci. 23, 81–86.

Olsen, T. (1985). Reflectometry of the precorneal film. Acta Ophthalmol. 63, 432–438.

Olsen, T. and Sperling, S. (1987). The swelling pressure of the human corneal stroma as determined by a new method. Exp. Eye Res. 44, 481–490.

Schermer, A., Galvin, S. and Sub, T.T. (1986). Differentiation-related expression of a major 64K corneal keratin in vivo and in culture suggests limbal location of corneal epithelial stem cells. J. Cell Biol. 103, 49–62.

Scott, J.E. and Haigh, M. (1988). Identification of specific binding sites for keratan sulphate proteoglycans and chondroitindermatan sulphate proteoglycans on collagen fibrils in cornea by the use of cupromeronic blue in 'critical-electrolyte-;concentration' techniques. Biochem. J. 253, 607–610.

Whitacre, M.M. and Stein, R. (1993). Sources of error with use of Goldmann-type tonometers. Surv. Ophthalmol. 38, 1–30.

FURTHER READING

Albert, D.M., Jakobiec, F.A., Azar, D.T., Gragoudas, E.S., Power, S.M. and Robinsen, N.L. (Eds.) (2000). Principles and Practice of Ophthalmology. Saunders, Philadelphia.

Aurell, G. and Holmgren, H. (1945). On the metachromasia of the cornea with special reference to the question of the causes of transparency. (in Swedish) Hygieae 108, 1277–1279.

Caspersson, T. and Engström, H. (1945). The transparency of the corneal tissue. (in Swedish) Hygieae 108, 1279–1282.

Chakravarti, S., Petroll, W.M., Hassell, J.R., et al. (2000). Corneal opacity in lumican-null mice: Defects in collagen fibril structure and packing in the posterior stroma. Invest. Ophthalmol. Vis. Sci. 41, 3365–3373.

Ehlers, N. and Hjortdal, J.Ø. (2004). Corneal thickness: Measurement and implications. Exp. Eye Res. 78, 543–548.

Kaufman, P.L. and Alm, A. (Eds.) (2002). Adler's Physiology of the Eye, 10th edn. Mosby, St. Louis.

Kaufman, H.E., McDonald, M.B., Barron, B.A. and Waltman, S.R. (1988). The Cornea. Churchill Livingstone, New York.

Møller-Pedersen, T. and Ehlers, N. (1995). A three-dimensional study of the human corneal keratocyte density. Curr. Eye Res. 14, 459–464.

Olsen, T. (1982a). Specular microscopic onvestigations on the corneal ebdothelium and its involvement in corneal oedema. Acta Ophthalmol. 70 (suppl. 155), 1–46.

Smolin, G. and Thoft, R.A. (1994). The cornea. Scientific Foundations and Clinical Practice, 3rd Edn. Little, Brown and Company, Boston.

THE CORNEAL ENDOTHELIUM

Jorge Fischbarg

I. INTRODUCTION

The corneal endothelium, also called corneal posterior epithelium, is a comparatively thin ($\sim$4.5 μm height), innermost layer of the cornea. The name of "endothelium" arose from its histological similarity to the arterial

Advances in Organ Biology
Volume 10, pages 113–125.
© 2006 Elsevier B.V. All rights reserved.
ISBN: 0-444-50925-9
DOI: 10.1016/S1569-2590(05)10004-4

endothelium and similarly thin layers, which cover the walls of body cavities. This name is perhaps not what one would choose for it nowadays, as by now we know that functionally it is not a passive endothelial layer but instead a typical fluid transporting epithelium with characteristic leaky tight junctions between the cells. Still, no international convention has rectified this yet. As a result, when writing or talking about the corneal endothelium to audiences outside the eye field, it is always advisable to point out the correct function of the layer.

II. THE ENDOTHELIUM; CORNEAL TRANSPARENCY

In spite of its small thickness, the endothelial cell layer is extraordinarily important, in that its fluid-transporting function is what keeps the cornea transparent. As the outer coat of the eye, the cornea has to be mechanically very sturdy to resist impacts and blows. It is also in the visual pathway, so it has to be transparent. Hence, it is understandable that the corneal stroma has evolved into a relatively thick organ made up of multiple collagen layers, which give it great mechanical resistance. However, for this organ to be transparent, the collagen fibrils that make the layers have to be precisely arranged in space. The collagen is therefore embedded in a ground substance with plenty of molecules that act as spacers; these molecules are mucopolysaccharides. Essential as they are, they also create a complication: correct spacing can only be achieved if these mucopolysaccharides are kept somewhat dehydrated. If left to their own resources, the mucopolysaccharides would imbibe water, and the cornea would become first hazy and ultimately opaque, or barely translucent as comparable pieces of connective tissue such as tendons. In fact, that is precisely what happens in the terminal stages of corneal diseases, states termed corneal decompensation or corneal edema, which result in eventual blindness of corneal origin.

The endothelium is what keeps the ground substance spacers relatively dehydrated. A balance is achieved, at which point fluid leaks slowly from the anterior chamber of the eye into the corneal stroma. As fluid leaks in, the endothelium pumps it out. One might ask "Why does fluid leak in?" "Would it be simpler if evolution had dictated that the endothelial cell layer would be relatively water impermeable?" As repeated below, there are at least two explanations for this: (1) that nutrients have to come from the aqueous into the stroma, and that may be happening at least in part across permeable intercellular junctions; and (2) the layer has to transport fluid, which of lately is being argued to be ferried precisely across these permeable junctions.

III. CELL NUMBER PROGRESSION

In humans, the endothelial cells are precious because they largely do not divide. As people age, they gradually lose corneal endothelial cells. Most people have enough reserve of them to last a lifetime. However, some people have a disease called Fuch's dystrophy, in which, for unknown reasons, corneal endothelial cells die off faster, eventually falling below a critical cell density needed to prevent corneal edema. Likewise, endothelial cells can be lost in other conditions. In general, anything that damages endothelial cells, such as inflammation in the eye (as in uveitis), very high-intraocular pressure (as in acute glaucoma), or trauma from intraocular surgery, can accelerate the loss of endothelial cells. Concerning this last, during cataract removal surgery, small bits of cataract fragments may hit the endothelium, which along with irritating fluid circulating inside the eye can damage the endothelial layer. The end result is the same: when too many corneal endothelial cells have disappeared, corneal edema develops. This edema may reach a point, at which the corneal epithelium forms blisters that rupture intermittently, causing sharp pain (a condition termed "bullous keratopathy"). Eventually, blindness ensues. The only current therapy for this is corneal transplantation, which fortunately under the right conditions tends to work rather well. Still, as transplantation is not a perfect solution, there have been for some time efforts to develop artificial corneas, which have accelerated in recent years.

Why a decline in endothelial cell numbers (called cell count, usually in cells mm^{-2}) leads to loss of transport function has not been clarified as yet. As an interesting observation, declines to comparatively low numbers (perhaps below 500 cells mm^{-2}), sometimes do not have as drastic an effect on function as might be expected. Hence, it is conceivable that a parameter other than cell count might be more informative. In this connection, some calculations of possible use are as follows:

Given a number of cells per unit area (say nca), the area of a cell would be the inverse:

$$Ac = 1/nca.$$

The endothelial cells resemble a six-sided polygon and from tables, the side (cs) of an n-sided polygon is:

$$cs = \frac{2}{n} \cdot \left[n \cdot Ac \cdot \tan\left(\frac{\pi}{n}\right) \right]^{0.5}$$

The cell perimeter (cp) is: $cp = n \cdot cs$, and the total perimeter of cells (tp) is

$$tp = \frac{cp \cdot nca}{2},$$

where the division by 2 corrects for the fact that each cell would be counted twice otherwise. From this, the total perimeter can be verified to be:

$$tp = nca \cdot \left[\frac{n}{nca} \cdot \tan\left(\frac{\pi}{n}\right)^{0.5} \right],$$

and for the endothelial hexagons, $tp = 1.861 \cdot nca^{0.5}$.

The square-root dependence of the perimeter means that its decrease as the number of cells decrease will be somewhat gradual, with a precipitous fall taking place only at very low counts. This would be relevant for both models considered below for the endothelial fluid transport.

IV. SHAPE, FUNCTION, AND TRANSPARENCY OF ENDOTHELIAL CELLS

The corneal endothelium is placed directly in the visual pathway. Therefore, a prime imperative is that this layer be transparent. However, as anticipated above, there are other requirements for it as well. The cornea is one of the very few organs in the body without blood vessels (at least in a healthy eye). Hence, nutrients for the stromal cells (keratocytes) have to come across the endothelium from the aqueous humor in the anterior chamber. Last but not least, as mentioned above, the endothelium has to pump fluid from the stroma to the aqueous. These several requirements have dictated the microscopic anatomy and the physiology of the layer. An exploration of how evolution has solved this quandary provides a fascinating view of epithelial biology.

The corneal endothelium transports fluid at a significant rate similar to that of other epithelia, of the order of $45 \mu m \ hr^{-1}$, that is $4.5 \mu l \ hr^{-1} \ cm^{-2}$. For this transport to take place, the layer has to be metabolically very active, which explains the abundance of endothelial intracellular organelles. As another characteristic of tissues engaged in fluid transport, the endothelium is rich in mitochondria. An excellent description of endothelial histology is given in a classical textbook (Alvarado, Hogan, and Weddell), from which an illustrative cross-sectional scheme of the endothelium has been redrawn (Figure 1). The fact that the layer has to be transparent might perhaps explain its very small thickness (about $4.5 \mu m$) (Figure 2), as a thicker cell with the density of light-scattering structures the endothelium might interfere with light transmission. However, there is one more element that appears to be in place to ensure a suitable optical medium inside these cells. The corneal endothelium shares with other ocular epithelia a remarkable feature: the intracellular presence of large amounts of water-soluble proteins, in this case the enzymes transketolase and aldehyde dehydrogenase class 1. These proteins do not conceivably function in this case as the enzymes that they are but have instead been seemingly expressed because their water solubility results in a cellular content, which appears homogeneous to light, and hence results in relatively

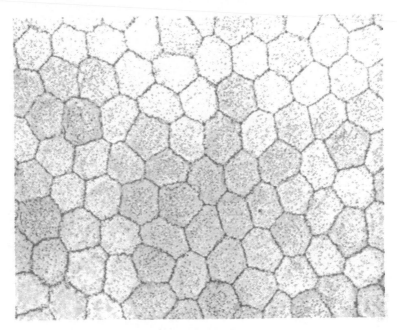

Figure 1. Corneal endothelial layer viewed from the apical side. The characteristic hexagonal pattern of the apical ends of the cells is very well apparent. Figure reproduced from Hogan, Alvarado, and Weddell.

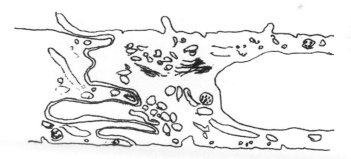

Figure 2. Cross section of a corneal endothelial cell layer. At the bottom, the cells are attached to Descemet's membrane (not shown). A convoluted intercellular space is visible in all its length. The cells have a dense population of organelles. At the apical surface (top), the cells show projections. Scheme redrawn from Hogan, Alvarado, and Weddell.

little light loss. This is similar to the well-known characteristic of lens fibers, which express large amounts of water-soluble crystallins for the same reason, namely, to diminish light scattering by the cytoplasm.

At the apical border, the endothelial cells form a characteristic regular array of hexagons (Figure 1). They are one of the few cells in the body that can be seen quite well in the living condition, as these cells are clearly visible with optical instruments such as the slit lamp and specular microscopes. Aside from this curiosity, their number and shape can be recorded and analyzed, which are useful in clinical settings to evaluate the endothelial condition, to follow the impact of surgeries, and to try to predict future function.

At the basolateral border, the endothelial cells are very different. The lateral membrane increases in size progressively from the apex to the base. As a result, to fit in a given volume, the side of the cell has to assume the form of a heavily pleated skirt. As a result, the cross section of the intercellular spaces at the level of the base of the cell has the form seen in Figure 3.

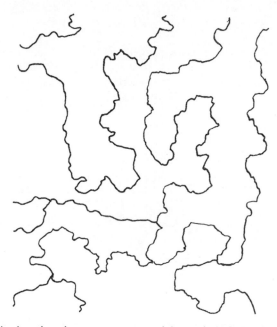

Figure 3. At the basal end, a cross section of the endothelial cells shows a pattern very different to the regular hexagons at the apical end. Each cell lateral surface has acquired the shape of a pleated skirt, and the imprint of the intercellular spaces forms a peculiar puzzle, in which neighboring cells can be about 1μm close to each other. Reprinted from Experimental Eye Research, Vol. 25, Hirsch et al., pp. 281., 1977, with permission from Elsevier.

Measurements taken from electron micrographs indicate that at the cell base the perimeter is 6.6 times longer than that at the apex.

An explanation for this is perhaps that the lateral membrane is the place, where the Na^+ pump molecules are located. As there need to be a large number of them per cell, the membrane area needs to be large enough to accommodate them. The number in the literature for Na^+ pumps per cell is 3×10^6 (in rabbit endothelium). This seems consistent: given a net transendothelial Na^+ flux of 0.53 μEquiv h^{-1} cm^{-2}, one cycle of the pump would require 30 ms, which is in line with the relatively slow rates of turnover of complex transporters such as the present example. Moreover, other representative numbers are also in line: for a cell with an apical side of the hexagon measuring 11.2 μm, the apical perimeter would be 67.2 μm, the basal perimeter 450 μm, the lateral area 3.1×10^3 μm^2, and hence each Na^+ pump would occupy in the average a square of 32 nm to the side. Considering there are other transporters in that same membrane, this space seems adequate for Na^+ pump complexes that are perhaps 5–10 nm wide, but this would of course not apply for a much smaller lateral area. The lateral membrane infoldings appear thus necessary if the cell, given its small volume, is to transport as much as it does.

V. ELECTROLYTE TRANSPORT BY THE ENDOTHELIUM

As shown in Figure 4, the endothelium is polarized in such way that some crucial transporters are restricted to one of the sides of the cell. The Na^+ pump is one good example of it; it is restricted to the lateral membrane. Although it is common to infer that epithelial cells transport "salt," a closer examination shows that these cells transport a good deal of HCO_3^- from stroma to aqueous but no matching cation, at least not through the cell membrane. In fact, the transcellular movement of Na^+ takes place in the opposite direction. In initial estimates, the amount of Na^+ leaking back into the cell via apical epithelial Na^+ channels is as much as 50–70% of that transported by the Na^+ pumps. The rest of the Na^+ inflow would take place via the cotransporters and exchangers indicated in Figure 4, and also via carriers of amino acids and other nutrients that cotransport substrates together with Na^+. These Na^+ pathways are diagrammed in Figure 5.

VI. ELECTRICAL PHENOMENA IN AND AROUND THE ENDOTHELIUM

Of course, electroneutrality cannot be breached. If the movement of HCO_3^- would proceed all by itself, the resulting separation of charges would create an impossibly high electrical potential difference. In practice, the separation

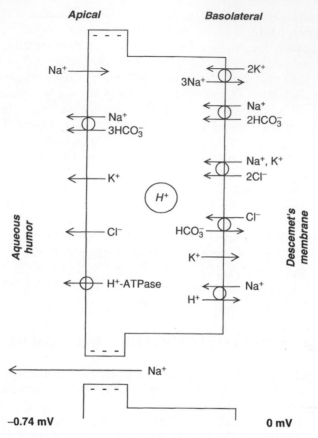

Figure 4. Tangential section across idealized corneal endothelial cells. This illustration lists all the transporters (cotransporters ⇒; exchangers ⇔) and channels known to contribute to either cell homeostasis or translayer transport. Most locations are known; a few are presumed. The electrolytes transported in each case are given. Standard arrows are for transporters, thicker arrows are for the channels. All electrolyte flows shown take place across cell plasma membranes, except for the arrow across the junction, which signifies that Na^+ moves along the paracellular pathway driven by the electrical field created by the cell. The junction is lined predominantly with negative charges that discriminate against anion movements.

of charges occurs up to a point but results in a potential that produces flow of Na^+ ions across the paracellular pathway. The intercellular junctions of the endothelium, although they form a generic tight junction zonula all around the apical cell borders (as in a beer six pack), share with all other fluid transporting epithelia a key characteristic: they are leaky. Hence, the somewhat confusing but otherwise adequate name they receive, namely,

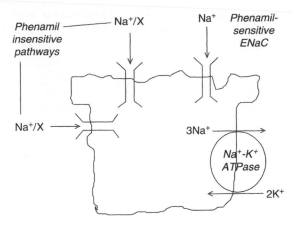

Figure 5. A schematic illustration of Na$^+$ pathways in corneal endothelial cells. At the top, the apical side, with epithelial Na$^+$ channels, for which the acronym is EnaC; phenamil is a specific inhibitor for them. On the lateral cell membranes, the sodium pump, or Na$^+$-K$^+$-ATPase. Other Na$^+$ pathways insensitive to phenamil are shown located on both apical and basolateral sides.

"leaky tight junctions". Estimates obtained from fluid flow driven across them by hydrostatic pressure differences place the junctional width at 35–40 Å, or 3.5–4 nm. From this, one would estimate that compounds with molecular weights as large as 10–15 kD might traverse them. Curiously, there is evidence that a molecule as large as bovine serum albumin (MW 66,430; estimated diameter ~70 Å) traverses the endothelium, albeit very slowly. Whether such migration takes place across the cell body (say, via vesicles), or the junction is unclear; in the last case, junctions may have a mosaic of pores of different sizes, either constantly or in a time-dependent fashion.

The leakiness of the junction translates into a specific electrical resistance of 20–60 ohm cm^{2}[1] for the endothelial layer. By comparison, cell membranes in general including that of the endothelial cell display specific resistances of about 1 Kohm cm^2. Hence, given an electrical potential across the endothelium,[2] the resulting electrical current will circulate practically all through the junctions. Of course, the spontaneous standing current resulting from endothelial activity returns via the endothelial cell membranes, high resistance, and all. The selectivity and numbers of the transporters/

---

[1] The units are correct as stated, ohm cm^2. This can be seen considering the more intuitively clear inverse units, Siemens/cm^2, in which a conductance is normalized by the area considered.

[2] Whether induced by the endothelium itself or by an external voltage source.

channels involved, and the electrochemical gradients that drive them join in a harmonious ensemble.

These cells may not be efficient at transporting salt as such, but they produce a hefty local electrical current. They generate a potential difference of about 700 µV, which when put across a 20 ohm cm^2 resistance translates into a local current of 35 µAmp cm^{-2}. This may not appear much, but as it is all funneled into the junctions, the local current density is 91 milliamps cm^{-2}, a more impressive number.

VII. WATER PATHWAYS

Aside from passage of water across the permeable junctions driven by hydrostatic pressure differences, water can traverse the endothelial cell membranes. Due to the abundance of aquaporin 1 water channel proteins all around the cell membrane, endothelial cells are reasonably permeable to osmotic flows. Their osmotic permeability (abbreviated P_f, for permeability to filtration) is some 90 µm s^{-1}. For comparison, a cell membrane made of a lipid bilayer and devoid of all proteins would exhibit a P_f of some 5 µm s^{-1}, and the most permeable cell membrane reliably characterized so far, the red blood cell, has a P_f of ~250 µm s^{-1}.

What is the functional significance of this cell membrane P_f. Could it account for fluid transport, for instance? We know that this layer can transport fluid at a rate (abbreviated J_v, for volume flow in the Katchalsky nomenclature) of about 45 µm hr^{-1}, in units popular in the corneal research field, or 4.5 µL h^{-1} cm^{-2} in the more usual units found in the general physiological literature. Incidentally, these units are very practical for communication purposes, but do not conform to the Système Internationale. For the purists, the politically correct rate of transport would be 12.5 nm s^{-1}. To return to the question of fluid transport, given the endothelial cell P_f and the J_v noted, one uses the following formulation:

$$Jv = Lp \cdot \Delta\pi = Lp \cdot R \cdot T \cdot \Delta C = \left(\frac{LP \cdot R \cdot T}{Vw}\right) \cdot Vw \cdot C = Pf \cdot Vw \cdot \Delta C,$$

$$\therefore \Delta C = \frac{Jv}{Pf \cdot Vw}.$$

Here, Lp is the hydraulic permeability, $\Delta\pi$ the osmotic gradient, R the gas constant, T the absolute temperature, ΔC the concentration gradient, Vw the water molar volume, and P_f the osmotic permeability mentioned above. It follows that the concentration gradient required is ~8 milliOsmol l^{-1}. Moreover, these gradients would be separately required across both the basolateral and the apical membranes that the flow would have to traverse in series.

VIII. THE MECHANISM UNDERLYING FLUID TRANSPORT: LOCAL OSMOSIS VS. ELECTROOSMOSIS

The mechanism by which fluid would be osmotically transported by way of small local osmotic gradients is known as local osmosis. It is a popular explanation, in fact the predominant one, frequently found in textbooks. However, an in-depth examination of the consequences of such mechanism for the corneal endothelial layer finds problems with it.

The transmembrane gradient of 8 mOsm mentioned immediately above is uncomfortably large. Compared to the osmolarity of the extracellular medium, $\sim$300 mOsm, it represents an anisotonicity of 2.6%. It is doubtful whether it can be maintained in the steady state, as gradients as small as 1% can trigger a volume regulatory cellular response, whereby the cell interior changes its osmolarity to the ambient one. So there is a priori at least an almost threefold discrepancy regarding the mere possibility of such steady-state gradients. A second problem is that they have never been actually detected. Third but not least, aquaporin 1 null mice have their cell membrane P_f cut about in half but transport fluid about as well as their wild type counterparts, which makes the discrepancy above approximately fivefold. Of course, arguments can be made that perhaps P_f values are difficult to measure and may be underestimated, but it seems doubtful whether small adjustments may bridge that wide gap.

Not surprisingly, explanations other than steady-state local osmosis are being sought for fluid transport. One possibility is that the gradients may be transient, and consequently fluid secretion may be cyclical. Another one is that the driving force for the flow would arise from electroosmotic coupling at the leaky tight junctions. Local osmosis and electroosmosis are compared in the schemes of Figure 6.

With reference to that figure, for a steady-state local osmosis mechanism to operate, the cell would have to be 8 mOsm hypotonic. The gradient between the intercellular spaces and the cell inside could be less than that, since the lateral membrane is $\sim$4.5 larger than the apical one, which would cut that gradient down to 1.8 mOsm. Still, this means the intercellular spaces would be hypotonic by $\sim$10 mOsm to the external medium. Whether this can be maintained while solute diffuses via the wide-open basal ends is doubtful. In addition, a hypotonic cell would transfer fluid across the basal membrane to the stromal side, in a direction opposite to that of fluid transport.

By contrast, the elements for electroosmotic coupling appear to fall in place without those inconsistencies above. Net HCO_3^- and Na^+ fluxes have been detected across the layer; there is an important distinction, namely, when the electrical field across the endothelium is nulled in the short-circuit condition, the HCO_3^- flux persists, but the Na^+ flux disappears, which is consistent with its being passive and traversing a leak pathway, in this case

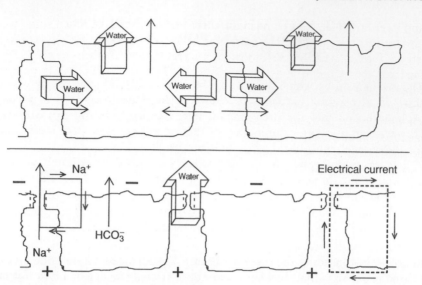

Figure 6. Two competing views of the mechanism of fluid transport across corneal endothelium. In the top diagram, fluid transport across the layer results from local osmotic gradients across basolateral and apical cell membranes. In this view, electrolyte transport across these membranes would result in the two separate osmotic gradients, with consequent separate osmotic flows adding up in series across each of the cell sides. The bottom diagram depicts instead the hypothesis of electroosmotic origin for fluid transport. In this view, cell transport of bicarbonate from the stroma to the aqueous side plus the activity of the Na^+ pump carrying Na^+ from the cell to the lateral space contribute to create an electrical potential difference (aqueous side negative). This gives rise to the local circulating electrical current shown at the right, from aqueous to stroma across the cell, and from stroma to aqueous across the paracellular pathway including the intercellular spaces and the intercellular junctions. This current is carried by Na^+ ions across negatively charged intercellular junctions; electroosmotic coupling between these Na^+ ions and water in the restricted space inside the junctions would create the driving force for the fluid pump.

the intercellular junction. To help carry the local current, much of the Na^+ flux carried by the Na^+ pump recirculates back into the cell via apical Na^+ channels. Electrical fields imposed across the endothelium result in fluid movements, and such coupling depends on the integrity of the intercellular junctions.

IX.   CONCLUSIONS

Historically, the endothelium was thought to be rather unimportant until perhaps the 1940s. The realization that it had a central function came gradually during 1950s and 1960s. The discoveries of its fluid transport and its electrical activity came in the early 1970s. After that, much study went into the membrane proteins that transport electrolytes and into the biochemistry of the cell. Thanks to all the accumulated information, in-depth study of the mechanism underlying fluid transport has become now possible and may reveal that fundamental cell question. Other important areas, such as the regulation of layer development, cell reproduction and their connection with the primary transport function of the endothelium and with the other elements of the cornea, and the eye vicinity, are now coming into focus. The next decades promise to be as full of enterprising and revealing research as the past ones, and with each passing one, the goal of understanding and controlling endothelial function will be tantalizingly closer.

REFERENCES

Hirsch, M., Renard, G., Faure, J.P. and Pouliquen, Y. (1977). Study of the ultrastructure of the rabbit corneal endothelium by the freeze–fracture technique: Apical and lateral junctions. Exp. Eye Res. 25, 27–88.

FURTHER READING

Fischbarg, J. (2003). On the mechanism of fluid transport across corneal endothelium and epithelia in general. J. Exp. Zoolog, Part A Comp. Exp. Biol. 300 (1), 30–40.
Parikh, C.H. and Edelhauser, H.F. (2003). Ocular surgical pharmacology: Corneal endothelial safety and toxicity. Curr. Opin. Ophthalmol. 14 (4), 178–185. Review.
Joyce, N.C. (2003). Proliferative capacity of the corneal endothelium. Prog. Retin. Eye Res. 22 (3), 359–389. Review.
Bonanno, J.A. (2003). Identity and regulation of ion transport mechanisms in the corneal endothelium. Prog. Retin. Eye Res. 22 (1), 69–94. Review.
Laibson, P.R. (2002). Current concepts and techniques in corneal transplantation. Curr. Opin. Ophthalmol. 13 (4), 220–223. Review.
Verkman, A.S. (2002). Aquaporin water channels and endothelial cell function. J. Anat. 200 (6), 617–627. Review.
Liesegang, T.J. (2002). Physiologic changes of the cornea with contact lens wear. CLAO J. 28 (1), 12–27. Review.
Bourne, W.M. (2001). The effect of long-term contact lens wear on the cells of the cornea. CLAO. J. 27 (4), 225–230. Review.
Edelhauser, H.F. (2000). The resiliency of the corneal endothelium to refractive and intraocular surgery. Cornea 19 (3), 263–273. Review.
Streilein, J.W. and Stein-Streilein, J. (2000). Does innate immune privilege exist? J. Leukoc. Biol. 67 (4), 479–487. Review.
Jester, J.V., Moller-Pedersen, T., Huang, J., Sax, C.M., Kays, W.T., Cavanagh, H.D., Petroll, W.M. and Piatigorsky, J. (1999). The cellular basis of corneal transparency: Evidence for "corneal crystallins". J. Cell Sci. 112 (Pt 5), 613–622.
Hogan, M.J., Alvarado, J.A. and Weddell, J.E. (1971). Histology of the Human Eye. W.B. Saunders Co., Philadelphia.

CILIARY BODY AND CILIARY EPITHELIUM

Nicholas A. Delamere

ABSTRACT

The ciliary body is a complex, highly specialized tissue that comprises several cell types. The ciliary muscle is situated at the base of the ciliary body and ligaments originating in the ciliary body attach to the lens. Contraction or relaxation of the muscle alters tension on the lens causing it to alter shape and thus shift focus.

Advances in Organ Biology
Volume 10, pages 127–148.
© 2006 Elsevier B.V. All rights reserved.
ISBN: 0-444-50925-9
DOI: 10.1016/S1569-2590(05)10005-6

The surface of ciliary body is elaborated into a series of ridges named ciliary processes. Each ciliary process contains a complicated network of blood vessels that appear leaky to plasma constituents. The ciliary processes are covered by a specialized epithelium bilayer that comprises two distinct epithelial cell types, pigmented ciliary epithelium (PE) and nonpigmented ciliary epithelium (NPE). The ciliary epithelium bilayer constitutes a diffusion barrier between the blood and the aqueous humor in the interior of the eye. Barrier function depends on tight junctions between adjacent NPE cells. The ciliary body is responsible for the production of aqueous humor, a task that requires the polarized cellular distribution and coordinated function of Na, K-ATPase, Na/K/2Cl cotranspor-ter, Na-H exchanger chloride channels and aquaporins in the NPE and PE. There is evidence suggesting an important role for gap junctions between the NPE and PE layers. The rate of aqueous humor secretion can be modified by ion transport inhibitors, by agents that modify gap junction permeability and by maneuvers that change blood flow in the ciliary processes.

Two striking features of the eyeball are first that it is more or less spherical and second that some parts of it are transparent. If you examine the eye of a cat, for example, it is possible to see how the transparent cornea is an inverted dome positioned above the iris and pupil. The shape and transparency of this "window" onto the visual system are both extremely important. Although there is a lens inside the eye, hidden behind the pupil, it turns out that refraction of light crossing the curved cornea plays a dominant role in the focusing of incoming light onto the retina at the back of the globe. The lens acts only as a fine focus mechanism because its shape can be changed, and this alters the effective focal length of the optical system in order for us to switch our gaze between near and distant objects without losing sharpness of the image. Look-ing again at the cat's eye, one can see that the cornea resembles an optical lens of the type normally made from glass. Now elementary physics tells us that the curvature of the cornea dictates its optical characteristics. How does the cornea come to have its unique shape? Although many different factors are involved, one of the governing issues is the existence of a hydrostatic pressure within the eye. It is this hydrostatic pressure that keeps the eyeball in a roughly spherical shape and keeps the walls of the eyeball taut.

The pressure is approximately 15 mmHg higher inside the human eye than outside. The globe is pressurized with fluid in much the same way as a basket-ball is pressurized with air. In the eye, the pressure is called the intraocular pressure (IOP). The pressure is the direct result of a continual secretion of fluid into the interior cavity of the eyeball. The fluid is watery and transparent—it is aqueous humor. The aqueous humor is produced by the ciliary body.

I.   STRUCTURE OF THE CILIARY BODY

Because the ciliary body is hidden behind the iris, it is impossible to see it no matter how deeply you gaze into someone's eyes. It is a ring of tissue on the inner wall of the eyeball, positioned just behind the rear-facing (posterior) surface of the iris. The base of the ciliary body is home to the ciliary muscle, the contraction of which causes the lens to assume a more rounded shape. This is because the lens is suspended by fine ligaments, called zonules that attach to the ciliary body. When the ciliary muscle contracts, the anchoring point of the zonules moves slightly inward, relaxing tension on the zonules, and the natural elasticity of the lens causes it to take on a more spherical shape. This is how our focus shifts. It is the process of accommodation (Kaufman, 1992). When the ciliary muscle relaxes, there is a slight outward motion that tightens the zonules and flattens the lens.

The surface of the ciliary body is elaborated into a series of ridges named ciliary processes. These ridges on the surface of the ciliary body look rather similar to the fins on a motorcycle engine, and in a way they serve a similar purpose—to increase surface area. On a motorcycle engine, the ridges provide a large surface area to dissipate heat. On the ciliary body, the purpose is to increase the surface area available for fluid secretion.

The ciliary processes have a radial orientation, each ridge pointing toward the pupil (Figure 1). In some species, the ridges extend part way across the posterior surface of the iris. Ciliary body anatomy has been described in great detail by Tamm and Lutjen-Drecoll (1996). The entire surface of the ciliary

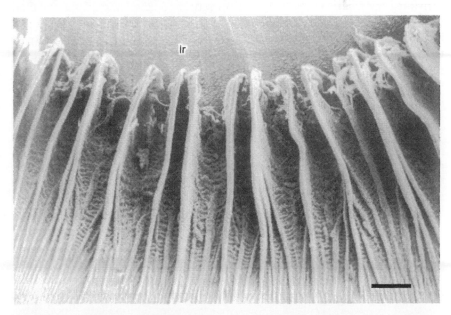

Figure 1. Topography of dog ciliary processes. Shown in this scanning electromicrograph, the ciliary processes resemble fins that are arranged radially at the root of the iris on its posterior surface. Ir, iris. Bar = 0.5 mm. Taken from Morrison et al. (1987). Invest. Opthalmol. Vis. Sci. 28, 1325–1340, J.B. Lippincott Company, used with permission.

body, the ridges as well as the valleys in between, is covered with a specialized epithelium—ciliary epithelium (Figure 2a and b). What is unique, and not very well understood, is the fact that this covering is a bilayer made of two different epithelial cell types. Other epithelial cell barriers in the body are either mono-layers, such as lung airway epithelium, or have multiple layers made up of a single cell type such as corneal epithelium. In the ciliary epithelium bilayer, the two cell layers have different developmental origins. The cell layer farthest from the interior of the eyeball is developmentally related to the retinal pigment epithelium. True to their origin, these cells contain black pigment granules, and not very creatively, this half of the ciliary epithelium bilayer has been termed the pigmented cell layer. The other half of the bilayer, the nonpigmented layer, is developmentally related to the neural retina. The pigmented and nonpigmented ciliary epithelial cells (PE and NPE) have numerous differences other than the presence or absence of pigment. Nonpigmented cells are much larger. Non-pigmented cells also contain many more mitochondria than pigmented cells, and this probably signifies a higher degree of metabolic activity in the non-pigmented cell layer. One remarkable feature of the nonpigmented ciliary epithelial cells is the high degree infolding on that part of their surface, which faces aqueous humor. This invagination of what is the basolateral membrane of the nonpigmented ciliary epithelium, appears to be another anatomical feature providing the ciliary body with an enormous surface area available for fluid secretion: not only are there the foldings of the tissue into the ridges and valleys of the ciliary processes, but there are also tiny invaginations on the surface of each nonpigmented ciliary epithelial cell.

II. BLOOD CAPILLARIES IN THE CILIARY PROCESSES

The interior of each ciliary process is filled with a complicated pipe work of blood vessels (Figure 2). From one standpoint, it could be said that the elaboration of the ciliary body into the ridges of the ciliary processes might be a strategy to pack more capillaries into the tissue. Why would this be advantageous? Well, as we shall see in a later section, the ciliary epithelium requires the constituents of blood plasma in order to produce aqueous humor. Also, it is probably fair to assume that in order to fuel their secretory machinery, the ciliary epithelial cells themselves have a high demand for nutrients, which must be delivered by the blood. Thinking along these lines, one can easily see that if the rate of blood flow to the ciliary body were to be reduced sufficiently, then the rate of aqueous humor secretion will be slowed. Importantly, the body itself is very well equipped to regulate blood flow. In many parts of the body, factors such as norepinephrine and some prostaglandins, constrict blood vessels, while other factors, such as nitric oxide, cause relaxation. At any one moment, blood vessel diameter in a

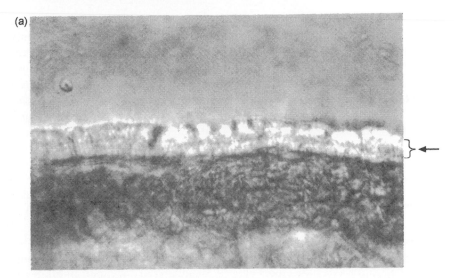

Figure 2a. Light photomicrograph showing the edge of a ciliary process from an albino rabbit. The bilayer of epithelial cells (indicated by an arrow) can be seen covering the blood capillaries within the ciliary process.

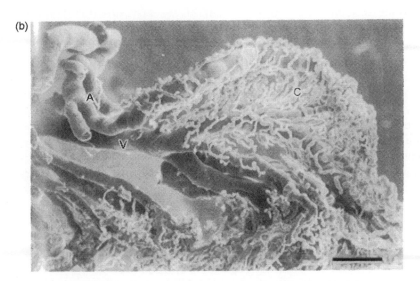

Figure 2b. Scanning electronmicrograph of arterioles (A) and veins (V) entering and exiting the goat ciliary process. These blood vessels serve the very elaborate capillary bed (C). Bar = 0.15 mm. From Morrison et al. (1987). Invest. Opthalmol. Vis. Sci. 28, 1325–1340, J.B. Lippincott Company, used with permission.

tissue is governed by an equilibrium between endogenous relaxing factors and constricting factors. Almost certainly, the body is able to control what goes on in the ciliary process (i.e., the rate of aqueous humor secretion) by controlling the delivery of blood to the ciliary process. Can drugs be applied to the eye to do the same thing? Probably so. Using an ingenious micro-casting technique, investigators have been able to make plastic replicas of the ciliary body vasculature. When blood vessel replicas from the ciliary processes of normal eyes are compared with replicas from eyes treated with catecholamines, such as epinephrine, there is clear evidence of vasoconstriction (Funk and Rohen, 1987).

III. BARRIER FUNCTION OF THE CILIARY EPITHELIUM

For the most part, epithelial cells separate different compartments in the body. In the eye, the ciliary epithelium bilayer forms a barrier between a part of the body that is one of the most densely vascularized and a part of the body that has no blood vessels at all. One of the unique features of the eyeball is the absence of blood in the optical pathway. Neither of the cornea, anterior chamber, or lens, nor vitreous body contain blood. This makes sense. Blood is very messy stuff and not at all transparent. Hold a test tube filled with blood up to the light and simply nothing shines through. The situation is improved if the blood cells are removed, for example, by placing the test tube in a centrifuge and spinning the cells to the bottom to leave just the blood plasma on top. However, the optical transparency through the tube of blood plasma is still poor. Blood plasma contains such a high concentration of large protein molecules that it scatters light. Such an optical disturbance is known as Rayleigh scatter, an effect demonstrated very nicely by shining a flashlight across a smoke-filled room. In these health conscious days when smoke-filled rooms are hard to find, a room with dusty air will do! The more you can see the glow of the light beam as it crosses the air in the room, the greater is the fraction of light being scattered offhyphen;track by the suspended particles instead of continuing in a straight path to shine on the opposite wall. A similar kind of light scatter would interfere with our vision if blood plasma macromolecules were permitted to get into the aqueous humor of the eye. What stops blood proteins from getting into the eye? The barrier seems to be the nonpigmented ciliary epithelium of the ciliary body.

In the tangled network of blood vessels inside each ciliary process, the sheath of vascular endothelium surrounding each capillary is rather leaky. Although blood cells are generally constrained inside the capillaries, plasma constituents leak out. This has been shown very nicely by electron microscopic studies in which a chemical marker, such as horseradish peroxidase (HRP, an enzyme), introduced into the systemic blood supply soon was found to appear in the stroma of the ciliary processes, outside the blood vessels (Raviola, 1977). Since HRP is a protein with a rather high molecular weight (approximately 44 kDa), we are comfortable in assuming it is unable to cross the plasma membrane to get inside a cell. Therefore, one can conclude that in the ciliary process, this large molecule is able to pass easily between the endothelial cells on the capillary walls. Now, what is important is the fact that Raviola observed HRP did not pass into the interior of the eye. It did penetrate between the pigmented ciliary epithelial cells. It also filled up the space between the pigmented cell layer and the nonpigmented cell layer. However, it did not penetrate between the nonpigmented ciliary epithelial cells. This is an important clue—it tells us the site of the diffusion barrier between blood and aqueous humor.

Physiologists coined the term "blood–aqueous barrier" to describe the group of structures that separate the two fluids. The nonpigmented ciliary epithelium layer is one part of this system; the other part of the blood–aqueous barrier is the tight endothelium layer surrounding the blood capillaries in the iris. The tightness of these cell layers effectively isolates the anterior of the eyeball from the blood circulation, leaving easy access only for very small molecules that are highly diffusible. From one standpoint, the system can be said to behave like a molecular sieve. To examine this feature experimentally, a researcher would typically introduce a tracer solute into the bloodstream, generally a molecule "labeled" with a radioactive isotope, then measure its rate of appearance in samples of aqueous humor taken from the eye. Size is a big factor; ions and small solutes, such as sucrose, penetrate into the eye quite quickly, while large molecules and proteins hardly penetrate at all (Davson, 1990). It is important to note that lipid soluble substances do not conform to this pattern. They penetrate much more rapidly than one might predict from their molecular weight. The reason for this is simple; their lipophilic nature enables them to pass through cellular barriers. The phospholipid composition of the cell membrane provides little in the way of a diffusion obstacle to a lipophilic molecule. The characteristics of the blood–aqueous barrier are particularly important when it comes to the development of pharmacological agents. If one's goal is to create an oral anti-inflammatory agent to cure intraocular inflammation, one of the big hurdles is finding a compound that gets into the eye in an effective concentration. It is difficult to establish therapeutic levels of many drugs inside the eye simply because the compounds do not readily cross the blood–aqueous barrier.

The absence of blood in the interior of the eye means that a good deal of generally useful things is missing, like components of the immune system. The result is a situation where the interior of the eye has rather unusual immunological characteristics. However, this can be changed—in response to injury, the blood–aqueous barrier "breaks down," and plasma constituents are allowed to flood into the aqueous humor. Thus, the appearance of protein in aqueous humor is one of the aftermaths of ocular injury and also a telltale sign of ocular inflammation. Such protein entry is detected as "flare," or light scatter (Rayleigh scatter), when a focused beam of light is shone through the aqueous humor compartment. It is easy to see how temporarily sacrificing visual acuity is well worth the advantage of getting clotting factors inside the eye to seal a penetrating corneal wound and the benefits of permitting the entry of immune factors and leukocytes to deal with invading microorganisms. But how does breakdown of the blood–aqueous barrier occur? The molecular mechanism remains slightly mysterious, but the site of breakdown seems to lie within the nonpigmented ciliary epithelium cell layer. It appears that the cell–cell junctions in this layer, the junctions that constitute the diffusion barrier, can be "broken" in response to external stimuli.

IV. TIGHT JUNCTIONS BETWEEN NONPIGMENTED CILIARY EPITHELIUM CELLS

Adjacent cells in the nonpigmented ciliary epithelium layer are joined by protein structures called tight junctions (Figure 3). In a typical tight junction, each cell expresses an assembly of several different protein molecules, including ZO-1 and occludin. Part of the tight junction protrudes from the plasma membrane, and the junctional proteins form a bridge to similar molecules protruding from the neighboring cell. Inside the cell, the tight

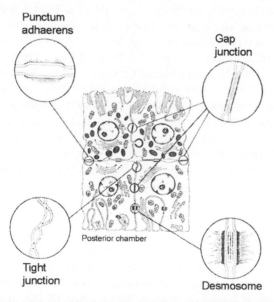

Figure 3. Diagram showing the location of cell–cell junctions in the ciliary epithelium. The pigmented ciliary epithelium is shown at the top with the nonpigmented ciliary epithelium beneath. Note that tight junctions are located in the nonpigmented ciliary epithelium cell layer, the layer which faces the posterior chamber of the eye From Raviola (1977), used with permission.

junction assembly is connected to the actin cytoskeleton. Viewed by scanning electron microscopy, the tight junction is a tangled web of protein fibrils, which forms a band or collar that wraps around one part of the lateral surface of each cell. To one side of this boundary is the cell's apical surface; to the other side is the basolateral surface. Importantly, the junctional complex prevents the diffusion of large molecules in the extracellular space between the cells. The collar of occludin protein threads acts as a seal that large molecules in the extracellular fluid are unable to cross. Thus, it is the tight junctions between cells of the nonpigmented ciliary epithelium, which constitute the barrier to the passage of plasma proteins from the stroma of the ciliary process into the aqueous humor. Importantly, cell biologists have found that the architecture of the tight junction can change very quickly (Schneeberger and Lynch, 1992). In some tissues, there is good evidence that certain stimuli cause the occludin threads to untangle sufficiently for diffusion pathways to open up between the cells. Most probably, this is what takes place in the nonpigmented ciliary epithelium cell layer when the blood–aqueous barrier breaks down.

V.   POLARIZED DISTRIBUTION OF ION TRANSPORTERS

In addition to restricting solute diffusion in the extracellular space between adjacent nonpigmented ciliary epithelium cells, the tight junction complexes constitute a line of demarcation between two very different parts of the nonpigmented ciliary epithelium cell surface. To the one side of the tight junction is the basolateral surface, which as described in an earlier section, is the highly invaginated part of the cell that faces the interior of the eye. To the other side of the tight junction is the apical surface, which faces the apical surface of cells in the pigmented ciliary epithelium layer. In most types of epithelial cells, the ion transport mechanisms (ion channels, pumps, transporters, and exchangers) are not distributed evenly across the cell surface. Instead, some ion transporters are located on the apical surface but not the basolateral surface and vice versa. The molecules are said to be distributed in a polarized fashion. In simple terms, this polarized distribution of ion transport molecules permits the cell to preferentially import certain solutes across one surface and then export them across the other. In this way, epithelial sheets are able to support a net flux of solute across the cell layer. Often, this is accompanied by a flux of water, which due to osmotic forces, is obliged to follow the solute flux. Is this how the ciliary epithelium produces aqueous humor? Probably so, but the situation is likely to be more complicated. Other factors, such as hydrostatic pressure, might also contribute to fluid flow via a process of ultrafiltration across the ciliary epithelium along a paracellular (between the cells) pathway. Some aspects of the process remain somewhat mysterious because in the ciliary body the epithelial sheet comprises two distinct cell layers.

VI.   NONPIGMENTED CILIARY EPITHELIUM

Focusing first on the nonpigmented ciliary epithelium, immunocytochemical studies have shown very clearly that Na, K-ATPase, the active sodium–potassium transport mechanism, is expressed much more densely on the basolateral surface of the cell than the apical surface (Figure 4). Here,

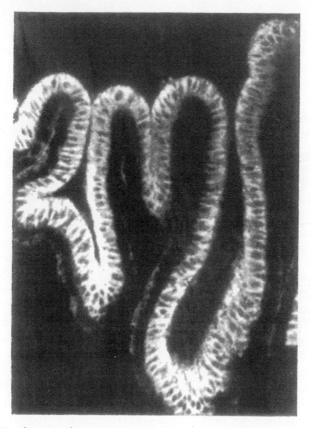

Figure 4.   Distribution of Na, K-ATPase in the ciliary epithelium of the chick. The confocal microscopic image shows the result of an immunolocalization study carried out using an antibody directed against the beta-1 subunit of Na, K-ATPase. As judged by fluorescence intensity, Na, K-ATPase beta-1 polypeptide is abundant in the nonpigmented ciliary epithelium layer, particularly at the basal and lateral surfaces (the NPE layer points upward in this photograph). The Na, K-ATPase beta-1 signal is much more faint in the pigmented cell layer (downward-facing cells in this picture), where it is seen at the basal surface. Taken from an unpublished collaborative study by the authors, Amy E. Moseley (University of Cincinnati, Cincinnati, OH), and Steven Bassnett (Washington University, St. Louis, MO).

the cells express several different isoforms (molecular variants) of Na, K-ATPase, a feature, which suggests the cells are specialized for active sodium–potassium transport (Coca-Prados and Sanchez-Torres, 1998). As in all other cells, the Na, K-ATPase hydrolyzes ATP to produce ADP and uses the energy from each ATP molecule to go through a series of conformational shifts that permit the molecule to translocate three sodium ions outward and two potassium ions inward. Among other things, this active sodium–potassium transport mechanism makes for a relatively low concentration of sodium (<15 mM) in the cytoplasm of the nonpigmented ciliary epithelium. For comparison, the concentration of sodium in the extracellular fluid (blood plasma on one side, aqueous humor on the other side) is approximately 145 mM. As a result, there is a driving force that encourages sodium in the extracellular fluid to cross the plasma membrane and enter the nonpigmented epithelium cell. It seems quite possible that sodium entry could occur across the apical cell surface (the side opposite the Na, K-ATPase), possibly via ion exchange mechanisms, such as Na–H exchange, possibly via cotransporter mechanisms, such as Na/K/2Cl cotransporter, and importantly via gap junctions, which form a conduit to the PE cell (see below). In this way, sodium effectively goes into the NPE cell on one side and out the other. Chloride also crosses the cell, entering apically via gap junctions, and possibly also by an apically located chloride–bicarbonate exchanger or Na/K/2Cl cotransporter that uses the inward sodium gradient as a driving force, and leaving via chloride channels that have been proposed to be polarized, like Na, K-ATPase, on the basolateral membrane. Transepithelial ion movement might be expected to set up a small electrical voltage across the nonpigmented ciliary epithelium layer, and this would tend to energize the movement of additional sodium and chloride ions through the extracellular space between the cells. Because of their size, diffusion of sodium and chloride ions is not likely to be hindered by the tight junctions.

If a model along the lines described above (but perhaps more complex) is true to life, the net transepithelial shift of ions across the nonpigmented ciliary epithelium could feasibly drive the movement of water. Specialized water channels, termed aquaporins, may facilitate the passage of water through the plasma membrane of the cell (Verkman, 2003). Water movement may also be made more efficient by the invaginations on the basolateral surface of the nonpigmented ciliary epithelium. Because these clefts are long and narrow, sodium and chloride ions shifted into the extracellular fluid at the closed end of the invaginations could create localized "pockets" of hyperosmotic solution. Water would tend to move out of the nonpigmented ciliary epithelium cytoplasm to progressively dilute the hyperosmotic solution in the cleft, so promoting the bulk movement of near-isosmotic fluid out of the open end of the invagination. (For a theoretical explanation of this process, see Reuss, 1977). This begins to give insight into the way the ciliary epithelium forms aqueous humor. Recent studies indicate cultured nonpigmented ciliary epithelium monolayers are indeed capable of transporting fluid in an apical to basal direction, and that ouabain, an Na, K-ATPase inhibitor, abolishes the process (Patil et al., 2001). However, the nonpigmented cells represent just half the ciliary epithelium bilayer. There is the role of the pigmented cell layer to be considered.

VII. COOPERATION BETWEEN PIGMENTED AND NONPIGMENTED CELL LAYERS

It seems likely that the two cell types in the ciliary epithelium bilayer work together as something of a functional syncitium. The basis for thinking there could be such a cooperative arrangement was the discovery of gap junctions between adjacent nonpigmented and pigmented ciliary epithelium cells (Raviola and Raviola, 1978). Gap junctions are a type of cell–cell contact that forms a conduit from one cell to the other, often allowing the exchange of fairly large molecules. Gap junctions have a very characteristic appearance when tissues are studied by electron microscopy. The molecular structure of a gap junction is fairly well understood. Membrane-spanning proteins, called connexins, are expressed on the cell surface and connexins from one cell reach out across the extracellular space to link with connexins on the neighboring cell. This protein bridge seems to form a cell–cell pathway that in some respects resembles a giant ion channel (for a detailed review, see Saez et al., 2003). Because charged ions penetrate the connexin channel rather easily, the electrical resistance of the pathway is low and cells joined by gap junctions are electrically coupled. Some researchers have taken advantage of this electrical coupling to study gap junction characteristics by measuring electrical resistance; the resistance between noncoupled cells is large, whereas it is remarkably low when the cells are coupled via gap junctions. Electrophysiological studies on the ciliary epithelium have confirmed gap junction-mediated coupling of the nonpigmented and pigmented ciliary epithelium. Importantly, electrophysiological studies have also led to the demonstration that, as in other tissues, the gap junctions between the nonpigmented and pigmented ciliary epithelium can switch from an "open" to a "closed" conformation (Shi et al., 1996). Adjacent cells might be able to choose whether or not to open their gap junctions and cooperate with their neighbors. Another useful way for researchers to study gap junctions is to use microinjection techniques to deposit fluorescent dye into one cell then record the movement of the dye to adjacent cells as it passes through gap junctions (Oh et al., 1994). The rate of diffusion can be quite rapid.

Gap junctions seem to constitute a quick route for solutes to pass between cells. By means of gap junctions, it is possible to see how the nonpigmented and pigmented ciliary epithelium cells may share cytoplasmic signaling molecules to coordinate cell function. Activation of receptors on the surface of the pigmented ciliary epithelium could potentially cause an increase of cytoplasmic cAMP or cytoplasmic calcium, which diffuses to the nonpigmented cells to modulate the activity of ion transport mechanisms on the opposite side of the ciliary epithelium bilayer. However, bearing in mind that the ciliary epithelium bilayer is essentially a secretory epithelium, it is possible that gap junctions between the nonpigmented and pigmented ciliary epithelium serve a more fundamental purpose. Sodium and chloride ions are probably able to enter the ciliary epithelium bilayer across the basolateral surface of the pigmented epithelium, then diffuse via apically located gap junctions to the nonpigmented epithelium in order to exit through the Na, K-ATPase, ion channels, and cotransporters on the nonpigmented cell's basolateral surface.

VIII. BILAYER MODEL OF ION TRANSPORT

Today, most researchers have adopted the idea that ion (and water) transport across the ciliary epithelium adheres to some sort of syncitial, or cooperative, scheme that involves coupling of the nonpigmented and pigmented cell layers (Jacob and Civan, 1996). This type of model, generally depicts an arrangement of some ion transporters at the basolateral surface of the nonpigmented cells and other ion transporters at the basolateral surface of the pigmented cells (Figure 5). In some respects, the pigmented ciliary epithelium is acting rather like the "apical" side of the nonpigmented cell.

In most schemes, entry of sodium and chloride ions is proposed to occur via Na–H exchange, chloride–bicarbonate exchange, and Na/K/2Cl cotransport at the basolateral surface of the pigmented cell layer. Na, K-ATPase and chloride channels on the basolateral side of the nonpigmented cell shift the sodium and chloride across the opposite side of the bilayer. Potassium channels on the basolateral surface of the nonpigmented cell permit recirculation of potassium ions brought into the cell by the Na, K-ATPase. Recent analysis of ion concentrations in the two cells of the bilayer using an electron microscopic X-ray microanalysis technique suggests that chloride and potassium concentrations within the cytoplasm of the nonpigmented ciliary epithelium might also be sufficient to drive outward ion flow via a Na/K/2Cl exchanger (McLaughlin et al., 2001).

What biological advantage the bilayer arrangement has over a system based on transport across a simple single layer of nonpigmented ciliary epithelium (with apical solute entry) is open to debate. There are puzzling questions still to be answered regarding the role of the two ciliary epithelium cell layers. We do know that the ciliary body shifts some solutes (prostaglandins for example) outward from the eye. Possibly the pigmented cell layer enables solute transport

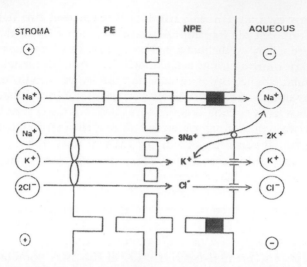

Figure 5. A theoretical model depicting the principal ion transport mechanisms thought to underlie the secretion of aqueous humor by the ciliary epithelium. Non-pigmented cells (NPE) and pigmented cells (PE) communicate via gap junctions. Densely shaded areas between NPE cells represent tight junctions. The model shows sodium, potassium, and chloride entering the PE via a Na/K/2Cl cotransporter and passing via gap junctions to the NPE. There, sodium is exported via Na, K-ATPase (the sodium pump). In the NPE, potassium imported by Na, K-ATPase is recycled out of the cell via potassium channels. Chloride exits the NPE via chloride channels. Taken from Coca-Prados et al., 1995, Am. J. Physiol. 268, C572–C579, The American Physiological Society, used with permission.

to occur in this reverse direction. It is also possible that an important purpose of the pigmented ciliary epithelium is not for solute transport but to absorb stray light—the interior surface of a good optical device is almost always coated with a matte black finish. It needs to be stressed, that theoretical models of ion transport by the ciliary epithelium bilayer are just that—theoretical models. They are "best guesses" based on the data in hand, but there is no certainty that they are correct. In actuality, it turns out to be very difficult to prove how the ciliary epithelium works because it is a tricky tissue on which to do definitive experiments.

IX. DIVIDED CHAMBER STUDIES

If physiologists had their way, the ciliary epithelium bilayer would be a flat sheet that could be placed with ease and precision in an Ussing chamber. This device, named after the Danish physiologist, H.H. Ussing, is a divided chamber, in which an epithelium is arranged between two separate fluid baths. In a typical Ussing chamber experiment, electrodes in each bath are used to measure transepithelial resistance, voltage, and current. Under short circuit conditions, the current is a measure of the rate of net ion transport across the epithelium. Using the same apparatus, actual ion fluxes can be measured by depositing radioisotopes of sodium, chloride, or potassium in the fluid bath on one side of the epithelium and recording the rate of appearance of the isotope in the fluid bath on the opposite side of the cell sheet. To make experiments more informative still, transport inhibitors (e.g., ouabain, amiloride, or bumetanide) or ion channel blockers can be added to one side or the other in order to track down the location of specific transport mechanisms to a particular face of the epithelium.

Unfortunately, the ciliary epithelium is not an ideal flat sheet. Until recently, the most common experimental approach has been to mount the entire iris and ciliary body in a divided chamber, using a plastic disc to occlude the iris and pupil. Because the ciliary body is elaborated into ridges (the ciliary processes) it is hard to know the precise value for the surface area of ciliary epithelium interposed between the two fluid baths. Without a measurement of actual surface area, it is difficult to calculate with certainty the specific electrical resistance of the epithelium. Too, the tissues that remain on the stromal side of the preparation could influence the way the epithelium bilayer works *in vitro*. In spite of these hindrances to interpretation, the iris–ciliary body Ussing chamber preparation has been used with considerable success. The tissue does indeed generate a short circuit current that is sensitive to transport inhibitors like ouabain (Krupin et al., 1984). Under certain conditions, a net flux of ^{14}C-labeled ascorbic acid can be demonstrated in blood-to-aqueous direction (Chu and Candia, 1988). This correlates nicely with the long known fact that the ascorbic acid concentration in aqueous humor is some 20 times higher than the concentration in blood plasma. The ciliary epithelium actively transports ascorbic acid into the aqueous humor and the mechanism still works well in the divided chamber preparation. Other studies with the iris–ciliary body Ussing chamber preparation illustrate the presence of bicarbonate exchange mechanisms in the ciliary epithelium, and this fits with the well-documented slowing in the rate of aqueous humor production elicited by carbonic anhydrase inhibitors (To et al., 2001). Carbonic anhydrase inhibitors are sulfonamide compounds, which reduce the supply of cytoplasmic bicarbonate by inhibiting the reversible, enzyme-catalyzed reaction that generates bicarbonate from carbon dioxide that diffuses into the cell.

In an effort to improve the Ussing chamber preparation, some investigators have successfully developed a way to use enzymic digestion to gently free the ciliary epithelium from the ciliary body (for details on the technique, see Chen and Sears, 1997). The cell bilayer, which appears to remain viable for sometime, can be mounted in a divided chamber and used for studies of transepithelial ion transport. Promising results have been obtained. However, in spite of the advances that have been made, there are still some important questions yet to be answered. One rather fundamental issue is the relatively small magnitude of the short circuit current measured across either the iris–ciliary body or the ciliary epithelium bilayer Ussing chamber preparation. The short circuit current tallies with only a modest net transepithelial ion flow, and this would be a bit too small to provide the osmotic drag needed to account for the rate of fluid secretion that we know takes place *in vivo*. There are many possible explanations for this. For example, it would not be surprising if some cellular functions simply do not survive the dissection of the tissue. Some cellular functions may not work when the blood supply ceases (as it does when the iris–ciliary body is isolated and placed in the divided chamber). It is also fair to say that *in vivo*, the production of aqueous humor may not be linked entirely to active solute transport. Another mechanism, such as ultrafiltration, might contribute.

X. HYDROSTATIC PRESSURE AND ONCOTIC PRESSURE

Blood capillaries are pressurized. It has been estimated that *in vivo*, there is a capillary hydrostatic pressure gradient across the ciliary epithelium bilayer; the pressure is generally estimated at >15 mmHg higher on the ciliary process stromal side. Undoubtedly, this pressure would tend to force fluid between the nonpigmented ciliary epithelium cells. As described in an earlier section, tight junctions between adjacent cells restrict the passage of large molecules, so what emerges on the other side of the epithelium should be a solution lacking macromolecules, but with an ionic composition rather similar to that of plasma. Put simply, the epithelial barrier could act rather like a filter through which fluid is pushed by the force of a pressure gradient. This process of fluid formation is termed ultrafiltration. In actuality, it has been something of a challenge to sort out the extent to which ultrafiltration might contribute to the overall process of aqueous humor production. As yet, there is no clear answer. A third and still different factor to be considered is the oncotic pressure gradient across the ciliary epithelium. There is a tendency for water to move across a semipermeable barrier (in this case, the ciliary epithelium) from a fluid compartment, which has a low concentration of dissolved macromolecules (such as the aqueous humor) to a fluid compartment, which has a high concentration of dissolved macromolecules (blood plasma). It is important to note that the oncotic pressure gradient (estimated at ~14 mmHg) and hydrostatic pressure gradient are in opposite directions. The oncotic pressure gradient will tend to cause absorption of water from aqueous humor back into the ciliary process stroma, more or less balancing out the inwardly directed flow of fluid caused by the hydrostatic pressure gradient.

The argument for the contribution of active ion secretion as a driving force for aqueous humor production originates from compositional studies of aqueous humor. Chemical analysis of newly formed aqueous humor revealed that the concentration of some solutes, such as sodium and bicarbonate, is slightly but significantly different from that expected in an ultrafiltrate of blood plasma (Davson, 1990). In addition, physiologists confirmed the need for active sodium transport by demonstrating that it is possible to slow the rate of aqueous humor production in anesthetized animals by introducing ouabain (a specific inhibitor of Na, K-ATPase) either into the blood or into the interior cavity of the eyeball (see Davson, 1990).

XI. VOLUME REGULATION AND WATER MOVEMENT

In the human eye, aqueous humor is produced at a rate of 1–2 µl/min. At first glance, this rate seems rather slow. However, it has been calculated that with this rate of flow, the throughput of fluid for each nonpigmented ciliary epithelium cell is equivalent to changing the cell's own volume every three minutes. It is easy to see how small mismatches of water entry and exit could cause swelling or shrinkage. To survive, the ciliary epithelial cells must be good at volume regulation.

The mechanisms that cells use to regulate their volume have been studied in a number of tissues. First, we should understand that eukaryotic cells are at the mercy of osmotic forces since they have little in the way of solid architecture to prevent shape changes. Cells shrink if subjected to a hyperosmotic external solution and swell if subjected to hypoosmotic external solution. The rate at which this occurs is a function of the permeability of the cell membrane to water molecules. Some cells are more water permeable than others, and recently researchers have come up with an explanation of why this might be so. Some cells express water channels. Termed aquaporins, water channels are plasma membrane-spanning proteins that seem to provide a conduit for water molecules to enter or exit the cell. Other membrane proteins, such as glucose transporters, also are capable of conducting the passage of water across the plasma membrane but with a much slower throughput (for review, see Fischbarg and Vera, 1995). As might be expected from its role in shifting water, the nonpigmented ciliary epithelium cell layer expresses aquaporin molecules (see Verkman, 2003). The cells are probably highly permeable to water. Consistent with this notion, microscopic studies with intact, living, ciliary processes have shown the NPE responds with rapid volume changes when external osmolarity is changed (Farahbakhsh and Fain, 1987). Now cells tend to dislike volume changes—in many tissues swelling or shrinkage triggers a volume recovery response, during which the cells attempt to return to their original size. There is still debate as to how a cell senses its own volume; perhaps it is able to detect a change in the concentration of cytoplasmic macromolecules or possibly it is able to sense changes in the cytoskeleton caused by swelling or shrinkage. In any case, cell shrinkage often causes a compensatory episode of swelling driven by the activation of solute entry mechanisms, such as the import of sodium chloride via the coupled action of a Na–H exchanger and chloride–bicarbonate exchanger; this response is called a regulatory volume increase (RVI). Cell swelling causes a different response, generally a compensatory episode of shrinkage, driven by the loss of solute via ion channels that are allowed to open; this response is termed a regulatory volume decrease (RVD).

Several researchers have directed their attention to the study of volume regulatory responses to seek answers to questions regarding the function of ion channel and ion exchange mechanisms in nonpigmented and pigmented ciliary epithelial cells. When responses in the two different cell types were compared, something rather interesting became apparent. Pigmented ciliary epithelium cells appear to have a good RVI response but weak RVD response—they seem to be adapted for solute and water uptake (Edelman et al., 1994). Just the opposite, nonpigmented ciliary epithelium cells seem better at RVD than RVI—they appear to be adapted for solute and water loss. These characteristics matchup nicely with the bilayer cooperativity models of ion transport across the ciliary epithelium, which in simple terms, are based on solute and water entry through the pigmented cell layer, exit via the nonpigmented cell layer.

XII. CONTROLLING THE RATE OF AQUEOUS HUMOR FORMATION

The human ciliary body continually produces 1–2 µl of fluid every minute, day in, day out, as long as we live. Is the rate of aqueous humor production always the same? The answer is no. This explains why textbooks generally quote the rate as a range instead of a single value (e.g., 1.5 µl/min). It turns out that human aqueous humor formation rate follows a circadian rhythm, higher in the daytime hours and lower during the nighttime. Certain species, like the rabbit, have the opposite pattern—the rate of production is highest at nighttime. The existence of a circadian rhythm tells us that there are endogenous pathways that control the rate of aqueous humor production. How does this occur? Well, the ciliary body receives both sympathetic and parasympathetic innervation, and there is evidence that aqueous humor production rate is subject to neural control. It has been shown, for example, that the circadian pattern of aqueous humor flow is changed considerably in experimental animals that have been subjected to ganglionectomy (removal of the cervical ganglion) (Gregory et al., 1985). Some mechanistic questions regarding circadian control of aqueous humor flow are still unanswered. While we know that the rate of aqueous humor production can be 40% lower at night in humans, it has proved tricky to define what aspect of ciliary body function is actually responsible slowing fluid production during the dark hours. Aqueous humor formation could be slowed as a consequence of reducing blood flow in the ciliary process capillaries. But equally, the rate of formation could be altered as a result of changes in the function in the ion transport machinery (ion channels, cotransporters, or the Na, K-ATPase) in the ciliary epithelium bilayer or even by opening or closure of gap junctions between the pigmented and nonpigmented cells. Some responses are difficult to interpret because it turns out that peptides, such as endothelin-1, and neurotransmitters, such as norepinephrine, probably elicit changes in both ciliary epithelium function as well as in blood flow to the ciliary processes (Pang and Yorio, 1997). Moreover, the ciliary body itself appears capable of synthesizing then releasing regulatory peptides and neuropeptides that may act in an autocrine manner to alter aqueous humor production (Coca-Prados et al., 1999).

Working on the rationale that most neurotransmitters, neuropeptides, and hormones work via cell surface receptors, researchers have sought to identify the types of receptors expressed in the ciliary body. There is good evidence that adrenergic alpha and beta receptors, dopamine receptors, prostaglandin receptors as well as peptide receptors are localized on the ciliary epithelium. Since most cell surface receptors bring about changes inside the cell by causing the generation of cytoplasmic second messengers, some research groups have sought to determine how the appearance of second messengers, such as cAMP, tally with changes in aqueous humor formation. It turns out that cAMP levels in the ciliary body and aqueous humor change with a circadian rhythm (Yoshitomi et al., 1991). Moreover, the rate of aqueous humor production can be lowered dramatically by cholera toxin and forskolin, two compounds, which cause elevation of cAMP in the ciliary body (see Sears, 1984). On the surface, this suggests that the machinery of aqueous humor formation is controlled by receptors that elevate cAMP in the ciliary epithelium. This could be so, but the system is tantalizingly complex. While some agents that elevate ciliary body cAMP slow aqueous humor production, one of the most widely used clinical strategies to reduce aqueous humor flow in humans is to use beta-adrenergic antagonists (Zimmerman, 1997) and these are drugs that tend to decrease cAMP in the ciliary body. It turns out that the rate of aqueous humor production can be reduced both by beta-adrenergic agonists and alpha-adrenergic agonists, and this goes against the normal situation, where activation of alpha and beta receptors triggers opposing physiological responses. There are other pharmacological agents that slow aqueous inflow (for review, see Zimmerman, 1996). A well-known class of compounds used clinically to reduce aqueous humor production is the carbonic anhydrase inhibitor family of drugs. These compounds are thought to reduce the availability of cytoplasmic bicarbonate ions in the ciliary epithelium bilayer with the result that bicarbonate-dependent ion transport mechanisms are inhibited (To et al., 2001). Still in the experimental stage, researchers are considering whether aqueous humor formation can be suppressed using drugs that inhibit specific ion transport mechanisms such as the Na–H exchanger (Avila et al., 2002). Some clinically proven drugs, like pilocarpine and latanoprost, act on the ciliary body but target the ciliary muscle, the contraction of which can cause mechanical changes that increase the ability of aqueous humor to exit the eye. These compounds lower IOP even though they do not slow the rate of aqueous humor formation.

Why is there a need for pharmacological tools to slow aqueous humor production? Such agents can be used in the therapy of glaucoma. Glaucoma is a blinding disease, which causes vision loss as the result of premature death of ganglion cells in the retina. Persons with glaucoma often have high IOP. There is plenty of evidence to suggest that in many (but perhaps not all) instances, retinal ganglion cell death could be the result of tissue damage caused at the optic nerve head by abnormally high pressure within the eye. Lowering IOP by slowing the production of aqueous humor is often an effective means of retarding progression of the disease. The quest for new and improved glaucoma therapies creates a pressing need for more research to explain how the various different cells in the ciliary body do their job and to determine how we might target the development of new drugs that trick the ciliary body into changing the rate of aqueous humor secretion.

ACKNOWLEDGMENTS

USPS Research Grant #EY06915, the Kentucky Lions Eye Foundation and an unrestricted grant from Research to Prevent Blindness, Inc.

REFERENCES

Avila, M.Y., Seidler, R.W., Stone, R.A. and Civan, M.M. (2002). Inhibitors of NHE-1 Na^+/H^+ exchange reduce mouse intraocular pressure. Invest. Ophthalmol. Vis. Sci. 43, 1897–1902.

Chen, S. and Sears, M. (1997). A low conductance chloride channel in the basolateral membranes of the non-pigmented ciliary epithelium of the rabbit eye. Curr. Eye Res. 16, 710–718.

Chu, T.C. and Candia, O.A. (1988). Active transport of ascorbate across the isolated rabbit ciliary epithelium. Invest. Ophthalmol. Vis. Sci. 29, 594–599.

Coca-Prados, M., Anguita, J., Chalfant, M.S. and Civan, M.M. (1995). PKC-sensitive Cl⁻ channels associated with ciliary epithelial homologue of pI_{Cln}.. Am. J. Physiol. 268, C572–C579.

Coca-Prados, M., Escribano, J. and Ortego, J. (1999). Differential gene expression in the human ciliary epithelium. Prog. Ret. Eye Res. 18, 403–429.

Coca-Prados, M. and Sanchez-Torres, J. (1998). Molecular approaches to the study of the Na^+, K^+-ATPase and chloride channels in the ocular ciliary epithelium. In: The Eye's Aqueous Humor: From Secretion to Glaucoma. (Civan, M.M., Ed.), pp. 25–53. Academic Press, San Diego.

Davson, H. (1990). Aqueous humor and the intraocular pressure. In: Physiology of the Eye. (Davson, H., Ed.), pp. 9–81. Academic Press, New York.

Edelman, J.L., Sachs, G. and Adorante, J.D. (1994). Ion transport asymmetry and functional coupling in bovine pigmented and nonpigmented ciliary epithelium cells. Am. J. Physiol. 266, C1210–C1221.

Farahbakhsh, N.A. and Fain, G.L. (1987). Volume regulation of non-pigmented cells from ciliary epithelium. Invest. Ophthalmol. Vis. Sci. 28, 934–944.

Fischbarg, J. and Vera, J.C. (1995). Multifunctional transporter models: Lessons from the Transport of Water, Sugars, and Ring Compounds by GLUTs. Am. J. Physiol. 268, C1077–C1089.

Funk, R. and Rohen, J.W. (1987). SEM studies on the functional morphology of the rabbit ciliary process vasculature. Exp. Eye Res. 45, 579–595.

Gregory, D.S., Aviado, D.G. and Sears, M.L. (1985). Cervical ganglionectomy alters the circadian rhythm of intraocular pressure in New Zealand white rabbits. Curr. Eye Res. 4, 1273–1279.

Jacob, T.J.C. and Civan, M.M. (1996). Role of ion channels in aqueous humor formation. Am. J. Physiol. 271, C703–C720.

Kaufman, P.L. (1992). Accommodation and Presbyopia: Neuromuscular and biophysical aspects. In: Adler's Physiology of the Eye. (Hart, Jr., W.M., Ed.), pp. 391–411. Mosby, St. Louis.

Krupin, T., Reinach, P.S., Candia, O.A. and Podos, S.M. (1984). Transepithelial electrical measurements on the isolated rabbit iris-ciliary body. Exp. Eye Res. 38, 115–123.

McLaughlin, C.W., Zellhuber-McMillan, S., Peart, D., Purves, R.D., Macknight, A.D. and Civan, M.M. (2001). Regional differences in ciliary epithelial cell transport properties. J. Memb. Biol. 182, 213–222.

Morrison, J.C., DeFrank, M.P. and Van Buskirk, E.M. (1987). Comparative microvascular anatomy of mammalian ciliary processes. Invest. Ophthalmol. Vis. Sci. 28, 1325–1340.

Oh, J., Krupin, T., Tang. L.Q., Sveen, J. and Lahlum, R.A. (1994). Dye coupling of rabbit ciliary epithelial cells *in vitro*. Invest. Ophthalmol. Vis. Sci. 17, 2509–2514.

Pang, I.H. and Yorio, T. (1997). Ocular actions of endothelins. Proc. Soc. Exp. Biol. Med. 215, 21–34.

Patil, R.V., Han, Z., Yiming, M., Yang, J., Iserovich, P., Wax, M.B. and Fischbarg, J. (2001). Fluid transport by human nonpigmented ciliary epithelial layers in culture: A homeostatic role for aquaporin 1. Am. J. Physiol. (Cell). 281, C1139–C1145.

Raviola, G. (1977). The structural basis of the blood-ocular barriers (Review). Exp. Eye Res. 25 (suppl.), 27–63.

Raviola, G. and Raviola, E. (1978). Intercellular junctions in the ciliary epithelium. Invest. Ophthalmol. Vis. Sci. 17, 958–981.

Saez, J.C., Berthoud, V.M., Branes, M.C., Martinez, A.D. and Beyer, E.C. (2003). Plasma membrane channels formed by connexins: Their regulation and functions. Physiol. Rev. 83, 1359–1400.

Schneeberger, E.E. and Lynch, R.D. (1992). Structure, function and regulation of cellular tight junctions. Am. J. Physiol. 252, L647–L661.

Sears, M.L. (1984). Autonomic Nervous System: Adrenergic Agonists. In: Handbook of Experimental Pharmacology, Pharmacology of the Eye. (Sears, M.L., Ed.), Vol. 69, pp. 193–248. Springer-Verlag, Berlin.

Shi, X.P., Zamudoi, A.C., Candia, O.A. and Wolosin, J.M. (1996). Adreno-cholinergic modulation of junctional communications between the pigmented and nonpigmented layers of the ciliary body epithelium. Invest. Ophthalmol. Vis. Sci. 37, 1037–1046.

Tamm, E.R. and Lutjen-Drecoll, E. (1996). Ciliary body. Microsc. Res. Tech. 33, 390–439.

To, C.H., Do, C.W., Zamudio, A.C. and Candia, O.A. (2001). Model of ionic transport for bovine ciliary epithelium: Effects of acetazolamide and HCO_3. Am. J. Physiol. (Cell). 280, C1521–C1530.

Verkman, A.S. (2003). Role of aquaporin water channels in eye function. Exp. Eye Res. 76, 137–143.

Yoshitomi, T., Horio, B. and Gregory, D.S. (1991). Changes in aqueous norepinephrine and cyclic adenosine monophosphate during the circadian cycle in rabbits. Invest. Ophthalmol. Vis. Sci. 32, 1609–1613.

Zimmerman, T.J. (1997). Textbook of ocular pharmacology. (T.J. Zimmerman, Ed.) Lippincott-Raven, Philadelphia.

FURTHER READING

Reuss, L. (1997). Epithelial Transport. In: Handbook of Physiology. (Hoffman, J.F., Ed.), pp. 309–388. Oxford University Press, Oxford, U.K.

THE LENS

Guido A. Zampighi

Advances in Organ Biology
Volume 10, pages 149–179.
© **2006 Elsevier B.V. All rights reserved.**
ISBN: 0-444-50925-9
DOI: 10.1016/S1569-2590(05)10006-8

I. INTRODUCTION

The principal function of the lens and the cornea is to focus light on the retina, thus allowing the central nervous system to receive complex representations of the outside world. In all vertebrate lenses, this function is achieved by highly elongated cells called "fibers" grouped together in a mass, which is transparent, deformable, and capable of maintaining homeostasis for the life of the animal. To meet these stringent requirements, nature developed ingenious evolutionary adaptations to increase the cytoplasms' refraction index and to transport ions and nutrients to fibers within the lens interior.

To attain transparency, the lens must accomplish apparently contradictory functions. First, it must increase the refractive index in the cytoplasm by expressing high concentrations of soluble proteins, called crystallins (>90% of the total protein mass). Crystallins are surprisingly diverse proteins that are actually no more transparent than other soluble proteins per se. In fact, many crystallins are enzymes or stress proteins that are used in tissues for nonrefractive purposes (Piatigorsky, 1998). Transparency results from the high concentration and long-range order that they attain in the cytoplasm of lens fibers. Some crystallins are present in all vertebrate lenses (alpha, beta, gamma), while a growing number of others are found only in selected species (taxon-specific). Several excellent reviews have been published about crystallins that function as chaperones (Jaenicke and Slingsby, 2002; Narberhaus, 2002; Horwitz, 2003) and heat shock proteins (Arrigo and Muller, 2002).

Transparency also requires the elimination of light-scattering elements, such as capillaries, and the reduction of the number of cytoplasmic organelles responsible for normal cell functions. The elimination of capillaries is particularly taxing because nutrients, such as glucose, remain within a narrow range of concentration in the interstitial fluid. If organs were to rely exclusively on diffusion, a layer with few cells would consume all available nutrients and force organs to be small (~50 μm in radius) (Fischbarg et al., 1999). Yet the lens radius reaches thousands of micrometers and instead of a central cavity, it contains a mass of the highly elongated fibers that somehow are able to maintain homeostasis for the life of the animal. A current hypothesis proposes that the lens creates an "internal circulatory system" that links the transport of ions and water to the movement of nutrients and waste products (the "circulating fluxes" hypothesis, Mathias et al., 1997). These fluxes balance the conflicting requirements of attaining a large spherical shape, transparency, and maintaining homeostasis of fibers placed far from their blood supply. Another hypothesis (Fischbarg et al., 1999) proposes that the lens epithelium generates a fluid transport from the aqueous humor into and through the lens, which contributes to nutrient transport and waste removal.

II. THE CELLULAR ARCHITECTURE
OF THE LENS EPITHELIUM

A. Polarity: The Developmental Viewpoint

A brief review of the development of the lens facilitates an understanding of the structural and functional properties needed to attain homeostasis and transparency. The lens develops from a thickening of the ectoderm covering the optic vesicle (the plate or placode), which contains a depression called the lens pit. This pit widens to form the lens vesicle, which is composed of a single-cell layer surrounding a liquid-filled central cavity (Coulombre and Coulombre, 1963). The cell layer comprising the vesicle is polarized with the apical surface facing the vesicle's lumen, the basal surface attached to the basal lamina and the lateral surfaces facing each other. This peculiar orientation of the apical, basal, and lateral surfaces is maintained in the mature lens. To secure transparency, however, the lumen of the original vesicle must be replaced with a medium of high index of refraction.

The elimination of the vesicle's lumen starts with the elongation of the cells in the posterior pole (the "primary fibers") and continues throughout life with the differentiation of the "secondary fibers" at the equator (Coulombre and Coulombre, 1963; Kuwabara, 1975). The cells at the anterior pole of the original vesicle remain undifferentiated and are called the "anterior epithelium". The continuous differentiation of secondary fibers at the equator (the "bow region") creates a geometrical problem because the surface of a sphere increases with the square—but its volume with the cube—of its radius. Formation of the lens "nucleus" neatly solves this problem but places fibers in the interior far from the aqueous and vitreous humors. An important question is: how does the lens maintain the homeostasis of the fibers located in the interior. To answer this question, it is necessary to characterize the pathways followed by nutrients and waste products throughout the lens.

B. The Paracellular and Transcellular Pathways

The lens relies on "paracellular" and "transcellular" pathways to maintain homeostasis of the nuclear fibers. The paracellular pathway embodies the narrow (10- to 20-nm wide) intercellular spaces between neighboring epithelial cells and fibers and is regulated principally by tight junctions. In the lens, the transepithelial resistance (i.e., the measurement of the transport through the pathway) is $\sim$60 ohms/cm^2. This low resistance indicates that epithelial cells and fibers either lack tight junctions altogether or are connected by "leaky" tight junctions. In either case, the low resistance of the paracellular pathway along the anterior–posterior axis indicates that ions and nutrients in the humors can move readily within the lens interior. On the other hand, to enter the transcellular pathway, nutrients and ions must first cross the barrier represented by the phospholipid bilayer of the plasma membrane through a variety of proteins that function as channels and transporters. Important research venues involve the identification of the proteins mediating the transport of ions, small molecules, and water in the different regions of the lens.

C. The Cellular Architecture of the Lens

The cellular architecture of the lens is easy to understand when considered from the vantage point of epithelial polarity (i.e., the location of the apical, basal, and lateral surfaces of the cells and fibers comprising the lens). From the polarity vantage point, the lens is divided into the "cortex" and the "nucleus". The cortex is composed of the anterior epithelial cells and superficial fibers with their basal surface attached to the capsule. A critical characteristic is that cortical cells and fibers are bathed by the aqueous and vitreous humors. The nucleus, on the other hand, is composed of fibers in the interior that have lost their connection to the humors. While it is possible to describe the morphology of cells and fibers in greater detail, this simple classification underscores the crucial problem of how to maintain the homeostasis of the nuclear fibers. The next sections contrast the structure of the apical, lateral, and basal surfaces of the cortex and nucleus.

D. The Cortex

While all cortical cells share the common property of being bathed by the humors, they differ in the degree of elongation along the apical–basal axis. At the anterior pole, the epithelial cells (orange, Figure 1) have a short apical–basal axis (dotted lines, Figure 2A and B), while at the equator, the anterior epithelial cells elongate into the aforementioned "fibers" (yellow region, Figure 1; and dotted lines Figures 5 and 6). At the posterior pole, the fibers achieve the longest apical–basal axis, extending all the way from the posterior capsule to the anterior pole (blue fibers, Figure 1; opposed triangles, Figures 2, 4, and 5).

One problem created by this peculiar organization of the cortex is that of studying the paracellular pathway. Labeling the intercellular space with electron dense deposits, such as lanthanum, delineated the pathway and allowed the identification of the tight junctions. These specialized junctions appear as points of intimate plasma membrane contact between adjacent cells (arrow, Figure 2C). In the lens, tight junctions are found among some cells in the anterior epithelium (Figures 2C and 3), but are absent in the cortical and nuclear fibers (Zampighi et al., 2000). Therefore, the paracellular pathway is

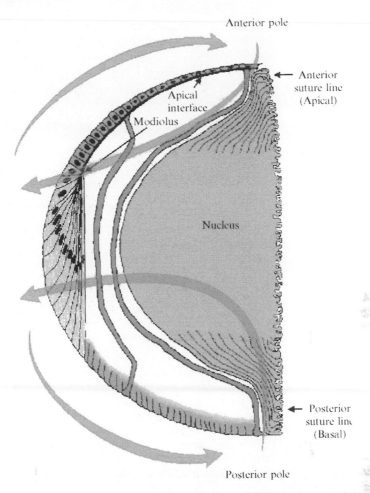

Figure 1. The lens is depicted as a simple folded epithelium. The apical–basal axis of the cells elongates along the anterior–posterior direction. At the anterior pole, the cells exhibit a cuboidal shape and are called the "anterior epithelium" (orange). At the equator (yellow), the cells elongate to form "fibers". At the posterior pole, the fibers are fully developed and extend to the anterior pole (blue). The lens "cortex" is composed of the cells and fibers with their basal surface attached to the capsule and their apical surface forming the "apical interface" and the "modiolus". The apical interface is formed by the interaction of epithelial cells (orange) with the posterior cortical fibers (blue). The modiolus is formed by the apical ends of the equatorial fibers (yellow) only. The lens "nucleus" (gray) is composed of fibers with their apical and basal surfaces forming the anterior and posterior suture lines. The arrows indicate the direction followed by the fluxes carrying nutrients into the nucleus. Reprinted from Experimental Eye Research, Vol. 75, Micro-domains of AQP0 in lens, Zampighi et al., pp. 505–579., 2002, with permission from Elsevier.

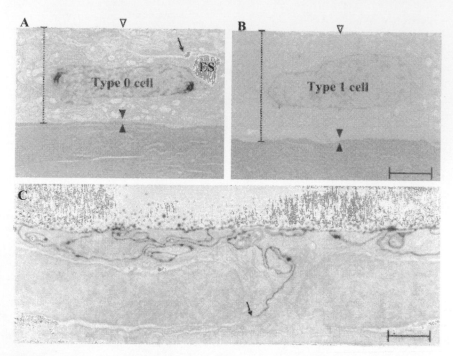

Figure 2. Thin sectioning electron microscopy of anterior epithelial cells (A) and
(B). The anterior epithelial cells have the shortest apical–basal axis (dotted lines). The
basal surface is attached to the capsule (open arrowheads), and the apical surface
forms the apical interface through association with the posterior fibers (opposed filled
arrowheads). The extracellular spaces of Type 0 Cell (arrow) are wide (ES), while
those separating Type 1 Cells are narrower. (C) Lanthanum precipitates (dark
lines) reveal plasma membrane folds in the basal and lateral surfaces and tight
junctions in Type 1 Cell (arrow). Scale bars: (A) and (B), 2.6 μm; (C), 1.3 μm (from
Zampighi et al., 2000).

represented principally by the long and tortuous intercellular spaces separating cells and fibers. Tight junctions appear restricted to a few cells in the
anterior epithelium only.

To gain access to the transcellular pathway, ions and nutrients in the
humors must first cross the plasma membrane of the cells and fibers in the
lens cortex using a variety of specialized channels and transporter proteins.
Once in the cytoplasm, these ions and nutrients move from the cytoplasm of
one cell to that of a neighboring cell through a network of specialized gap
junction channels constructed from proteins of the connexin family. The
ability of gap junction channels to promote direct communication with the
cytoplasm of cortical cells and fibers is supported by electrical and dye

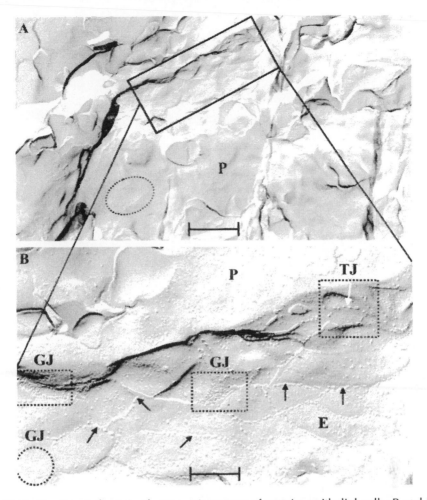

Figure 3. Freeze-fracture electron microscopy of anterior epithelial cells. P and E indicate the protoplasmic and external fracture faces of the plasma membrane. (A) Low magnification view of the junctional complex composed of tight and gap junctions. The dotted ellipse encircles a gap junction between an anterior epithelial cell and a posterior cortical fiber. (B) Higher magnification view of the region enclosed in the rectangle. Tight junctions (TJ) are composed of strands of particles in the P face (white arrow inside TJ dotted square) and complementary furrows in the E face (black arrows). Gap junctions (areas inside dotted rectangles) appear as plaques of particles in the P face (area inside the GJ dotted rectangles) and complementary pits in the E face (area inside the dotted circle). Scale bars: (A), 0.6 μm; (B), 0.2 μm (from Zampighi et al., 2000).

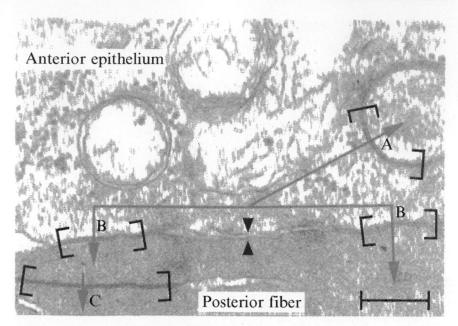

Figure 4. The network of gap junctions at the anterior pole. The gap junctions appear as dark bands (brackets) formed by the close apposition of neighboring plasma membranes. Gap junctions allow communication among the anterior epithelial cells (A) and between them and posterior cortical fibers (B). In addition, gap junctions allow communication among the posterior cortical fibers (C). The red arrows indicate a possible direction of the fluxes. Scale bar: 0.26 μm (from Zampighi et al., 2000).

coupling experiments (Rae et al., 1990; Miller et al., 1992; Mathias et al., 1997; Baldo et al., 2001).

Immunofluorescence and electron microscopy observations have provided information about the location and density of gap junctions as well as the connexin isoforms constructing the channels. In the cortex, gap junctions are present in three regions: (a) at the anterior pole (Figures 1, 3, and 4); (b) at the equator (Figures 5–7.); and (c) at the extensive lateral surfaces (Figure 14) (Zampighi et al., 2000, 2002). Gap junctions at the anterior pole allow the posterior cortical cells to communicate with the anterior epithelial cells. Gap junctions at the lateral surfaces allow the fibers in the same layer to communicate with each other and with fibers in the layers immediately above and below (Figure 14). The highest density of gap junction channels couple fibers at the lens equator (Figure 7). This distribution of gap junction channels is consistent with the electrical coupling measured using electrophysiological methods (Mathias and Rae 1985).

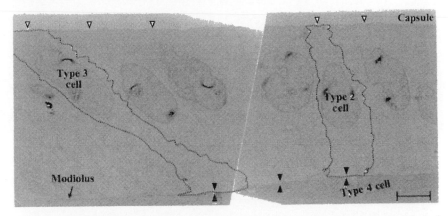

Figure 5. Thin sectioning of fibers at the lens equator (yellow, Figure 1). The dotted lines indicate the location of the plasma membrane. The orientation of the apical–basal axis changes from being almost perpendicular to the capsule (Type 2 Cell) to a steeper angle (Type 3 Cell) with increasing elongation. The basal surface is in contact with the lens capsule (open arrowheads). The apical surface associates with the posterior fibers (double fill arrowheads, Type 4 Cell). Scale bar: 4.0 μm (from Zampighi et al., 2000).

In conclusion, the cortex is well suited to move ions and nutrients from the aqueous and vitreous humors into the lens interior. The paracellular pathway is open and appears regulated by "leaky" tight junctions in only a limited region of the anterior epithelium. The extensive network of gap junction channels creates a direct cell-to-cell pathway into the lens interior.

E. The Nucleus

The nucleus is composed of a group of omega-shaped fibers located in the lens interior. In contrast to fibers forming the cortex, nuclear fibers arrange the apical and basal surfaces in novel structures called "suture lines" that appear as dense demarcations of precise shape to the naked eye. Apical-to-apical associations in the anterior pole and basal-to-basal associations in the posterior pole form the suture lines (Figure 8). The sutures establish complex linear patterns that change with age and are characteristics of the animal specie under study (Kuszak, 1995). For the purpose of this presentation, however, suture lines are considered important components of the transcellular and paracellular pathways of the lens nucleus.

Morphologically, the suture lines are characterized by elaborate folding of the plasma membrane, the presence of cellular organelles, including

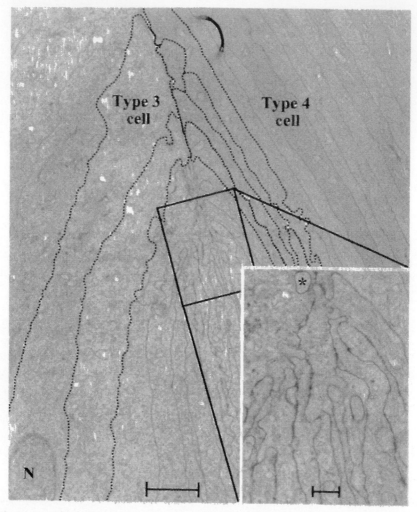

Figure 6. The lens equator. The dotted lines indicate the plasma membrane of two fibers (Type 3 Cell) and several cortical fibers (Type 4 Cell). The equatorial fibers comprising the modiolus exhibit highly folded plasma membranes (area inside rectangle). The inset shows the area inside the rectangle at higher magnification to demonstrate organelles, such as annular gap junctions (asterisk). N, nucleus. Scale bars: 2.6 μm; inset 1.2 μm (from Zampighi et al., 2000).

vesicles and mitochondria, and a wide (~20 nm) extracellular space that allows lanthanum to move readily deep into the nucleus (Zampighi et al., 1992). Curiously, lanthanum precipitates remain in the extracellular space at the suture lines and they do not permeate into the intercellular spaces

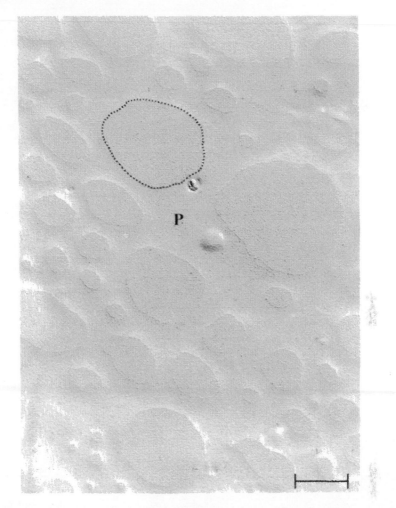

Figure 7. Gap junctions at the equator are larger in diameter and present at a higher density than gap junctions in coupling anterior epithelial cells (Figure 3). The dotted line indicates the boundary of a plaque. Scale bar: 0.66 μm (from Zampighi et al., 2000).

separating the lateral surfaces of nuclear fibers. This compartmentalization of the space probably results from the presence of "tongue-and-groove" junctions.

The elaborate plasma membrane folding and the presence of cellular organelles in suture lines are puzzling. Cells rely on folds to increase the area of the plasma membrane in an effort to accommodate a higher density of transporter proteins, such as the Na,K-ATPase, for example. The

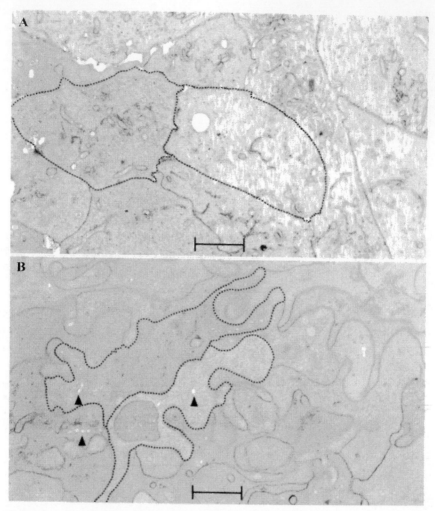

Figure 8. Morphological characteristics of the posterior suture line. (A) The suture line close to the capsule is composed of fibers with a rectangular shape, few plasma membrane folds and cellular organelles. (B) The suture line deeper inside the nucleus has elaborated plasma membrane folds (dotted lines) and the cytoplasm also contains some mitochondria, flattened cisternae, and vesicles with clear lumen (arrowheads). Scale bars: (A) and (B), 1.2 μm (from Zampighi et al., 2000).

presence of cellular organelles, such as mitochondria, appears to support the hypothesis of active metabolism at the suture lines. Yet one question not resolved by morphological observations is whether suture lines contain active Na,K-ATPase transporters and functional organelles deep into the nucleus. This question can only be resolved through further experimentation using immunocytochemistry and biochemical approaches.

Morphological studies have limited success in the identification of the gap junctions in the nucleus due to the progressive proteolysis that connexins undergo with aging (Lin et al., 1998). Information about the critical functional role of the gap junction network in the nucleus has been unveiled by mouse genetic experiments (Baldo et al., 2001; White et al., 2001; Chang et al., 2002). These studies show that the ablation of connexins expressed in fibers induces cataract formation or alters lens development (see in later sections). Yet, the mechanisms responsible for the pathological and developmental changes in the eye remain undetermined.

In summary, the lens must be transparent and possess a large spherical shape to focus images on the retina. These requirements lead to the creation of a tissue that cannot be sustained metabolically solely by diffusion from the aqueous and vitreous humors. Therefore, pathways must be established to move ions and nutrients into the interior. The next section deals with the channels and transporter proteins comprising the machinery responsible for transport into the lens.

III. THE MOLECULAR MACHINERY

A. Channels

Channels are integral proteins that enclose small water-filled cavities for the movement of ions and water across the plasma membrane. Channels move large numbers (10^8/sec) of ions along their electrochemical gradients with great fidelity. For example, K^+ channels select K^+ over Na^+ with a precision of 1 in 10^4 and water channels select water over ions and even protons. The aquaporin and the gap junction connexin channels are the best understood protein-forming channels in the lens. The proteins mediating the Na^+ and Cl^- conductance are undetermined (Mathias et al., 1997).

B. Aquaporins

Since water is the principal constituent of bodily fluid and tissues (55 M), the characterization of its transport in cells and tissues is crucial to the understanding of the normal and pathological conditions of the lens. For many years, it was assumed that water moved within and out of cells through the lipid bilayer of the plasma membrane. This view changed by the identification of a 28-kDa molecular weight integral protein (originally called CHIP-28) found in tissues exhibiting high-water permeability such as kidney and red cells. When expressed in *Xenopus* oocytes, CHIP-28 induced swelling in hypotonic buffers. Surprisingly, the large capacity for water transport induced by CHIP-28 is not concomitant with a large capacity for the transport of any other charge ion, including protons. This special selectivity for water suggested the name "aquaporin," and the original CHIP-28 become aquaporin-1 (AQP1). The family contains more than 150 proteins that have a strict selectivity for water (the aquaporins), or are permeable to water and neutral solutes, such as glycerol (the aquaglyceroporins) (see Agre et al., 2002, for recent review).

Understanding of the mechanisms of water transport through aquaporins started with the identification of two tandem repeats in the 28-kDa monomer, each formed by three transmembrane alpha helices with two highly conserved loops containing the signature motif, asparagine, proline, alanine (NPA). A large body of information led to the hypothesis that each AQP1 subunit contains an aqueous pore shaped as an "hourglass," a structure formed by the NPA boxes reentering the channel from opposite sides of the membrane (Jung et al., 1994). This hypothesis was confirmed and refined with the determination of the atomic structure of AQP1 (Sui et al., 2001) and of the aquaglyceroprorin from *E. coli* named the glycerol facilitator (Fu et al., 2000). The atomic structure of these channels demonstrates that each 28-kDa monomer is composed of six transmembrane alpha helices and two half-spanning helices containing the NPA box. In cells, four channels are assembled in a single particle (Figure 10C).

A pore at the center of the monomer (Figure 10A) is composed of vestibules at the external and cytoplasmic surfaces connected by a narrow section of 2.8 Å in diameter close to the middle of the lipid bilayer (blue region, Figure 10A). Charged residues at the vestibules exclude ions from the pore. At the narrowest part, the pore arranges the carboxyl oxygen of the polypeptide chain to resemble the contacts that a water molecule makes with other water molecules while in solution. In a sense, the pore's walls look like water and trick the molecules to interact with the walls of the pore and move across the membrane in accordance with the direction of the osmotic gradient: a solution that is as simple as it is beautiful.

Lens Aquaporins. The lens expresses AQP1 in the anterior epithelium and aquaporin-0 (AQP0) in cortical and nuclear fibers (Stamer et al., 1994). Aquaporin-1 is expressed in many epithelia and thus it is not unexpected that the undifferentiated anterior epithelium also expresses this water channel. On the other hand, cortical and nuclear fibers express AQP0 (the former MIP), a channel that represents $\sim$50% of the integral protein of the plasma membrane. The importance of the AQP0 water channel in lens physiology is underscored by the demonstration that mutations in the gene are linked to genetic cataract in mice and humans. Four mouse AQP0 mutations including point mutation at amino acid 51 (A51P), replacement of the last 61 amino acids (203–263) at the carboxyl terminus by a transposon sequence (CatFr), deletion of the AQP0 amino acids 121–175 (Hfi) and deletion of amino acids 46–49 (CatTohm) resulted in autosomal dominant cataracts (Shields and Bassnett, 1996; Sidjanin et al., 2001; Okamura et al., 2003). In humans, genetic linkage studies in two families identified mutations, such as Glu-138-Gly that induce lamellar opacities throughout the lens and an ensuing loss in vision. The atomic structure of AQP1 indicates that these point mutations reside in important structural areas of the channel. Therefore, the water permeability function of AQP0 appears to be critical to lens transparency.

Expression in heterogeneous systems has contributed to our understanding of the water permeability function of lens AQP0 and AQP1 (Mulders et al., 1995; Zampighi et al., 1995). Expression in *Xenopus laevis* oocytes is particularly convenient because the water permeability of the cell (P_f) and the number of AQP0 and AQP1 channels are measured in the *same oocyte*. These studies demonstrated that while AQP1 and AQP0 increase the water permeability of oocytes (P_f), the permeability of the single AQP1 channel (p_f) was $\sim$40, larger than the permeability of the AQP0 channel (Zampighi et al., 1995; Chandy et al., 1997). Such a difference in single-channel water permeability raises the question of why AQP0 and AQP1 are expressed in different regions of the lens.

A partial answer to the question was obtained by studying the distribution, density, and state of aggregation of the AQP0 channels in the native fiber plasma membrane (Zampighi et al., 2002). The studies used freeze-fracture-immuno-labeling (FRIL), a method based on the observation that frozen biological membranes fracture down the middle of the phospholipid bilayer-producing complementary protoplasmic (P) and external (E) fracture faces (Figures 9 and 10). Proteins spanning the lipid bilayer, such as AQP0, appear as intramembrane particles on the P fracture face, a fact that allows labeling with immuno-gold complexes (Fujimoto, 1995; Rash and Yasumura, 1999; Zampighi et al., 2002). In this manner, the distribution, density, and arrangement of the AQP0 channels can be deduced from the patterns established by the immuno-gold complexes on the fracture faces.

Studies using FRIL unveiled an elaborated microdomain of AQP0 distribution in the plasma membrane of equatorial fibers. In one microdomain, the AQP0 channels were seen intermingled with the endogenous complement of integral proteins in the plasma membrane (red and dense dots, Figure 11). This microdomain is consistent with AQP0's function in water permeability because the channels are exposed to wide ($\sim$20 nm) extracellular spaces. The density of AQP0 tetramers ($\sim$720 μm^{-2}), as well as the single channel permeability (5×10^{-16} cm^{-3}/sec) estimated in oocytes predict that the AQP0 comprising this microdomain increases the water permeability of the fiber over the water permeability of the phospholipid bilayer (Zampighi et al., 2002).

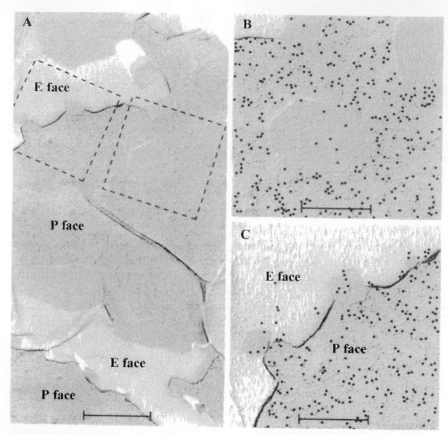

Figure 9. Distribution of AQP0 tetramers (dark particles) and gap junctions (reddish areas) in the protoplasmic (P) and external (E) fracture faces of the plasma membrane of equatorial fibers (yellow, Figure 1). (A) Low magnification view showing 18 gap junction plaques surrounded by AQP0 tetramers (dark particles). The gap junction plaques varied from small (8–10 channels) to large (~5000 channels). (B) and (C) Higher magnification views of the regions inside the squares in panel A. In this region, the AQP0 tetramers (dark particles) are located outside the gap junction plaques. Scale bars: (A), 1.0 μm; (B) and (C), 400 nm. Reprinted from Experimental Eye Research, Vol. 75, Zampighi et al., Micro-domains of AQP0 in lens, pp. 505–579., 2002, with permission from Elsevier.

The AQP0 channels also infiltrate the gap junction plaques (Figure 9) indicating that AQP0 and gap junction channels colocalize in the same plaque (Figures 12 and 13). Such colocalization of both types of channels raises the possibility that AQP0 has a secondary function in cell-to-cell communication. Finally, the AQP0 channels also form specialized "fiber junctions" composed of square arrays abutted against a protein-free plasma

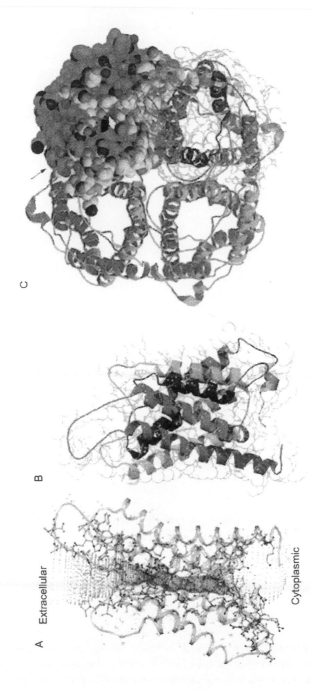

Figure 10. Atomic structure of the AQP1 water channel. (A) The pore that passes water (blue) is composed of vestibules at the extracellular and cytoplasmic surfaces and a narrow pore with a constriction close to the middle of the bilayer. (B) Six transmembrane helices form a right-handed bundle across the lipid bilayer. The long helix 1 in front (red) has the amino terminus and helix 6 (magenta) the carboxyl-terminus. (C) In cells, four monomers (24 transmembrane alpha helices) form a particle (the tetramer). Each monomer is colored differently, and one is represented with a space filled model (from Sui et al., 2002).

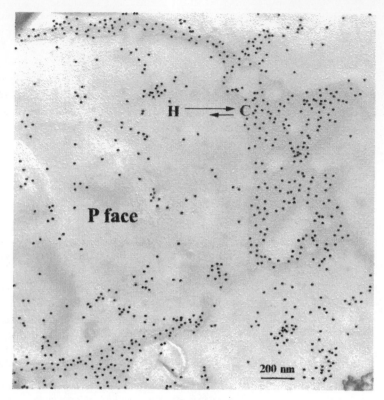

Figure 11. Immunolocalization of Cx50 (dark particles) at the lens equator. The red particles in the vicinity of the immuno-gold particles in the P face represent potential intramembrane particles corresponding to Cx50 hemichannels. H and C indicate the equilibrium of hemichannels and channels in the plasma membrane of equatorial fibers. Scale bar: 200 nm (from, Zampighi, 2003).

membrane, which are arranged in long ribbons (yellow regions, Figure 14; Zampighi et al., 2002). Such a particular arrangement of fiber junctions is thought to function in fiber adhesion.

In conclusion, the water channel AQP0 is the most abundant integral protein in lens plasma membrane of cortical and nuclear fibers. Genetic linkage indicates that mutations in AQP0 result in cataracts and loss of vision. While there is no doubt that a significant amount (~50%) of AQP0 function as water channels, some AQP0 appear to play a role in fiber adhesion and perhaps also in communication through association with the gap junction channels. How these novel functions of AQP0 impact lens homeostasis and cataract formation, however, remains a topic for further investigation.

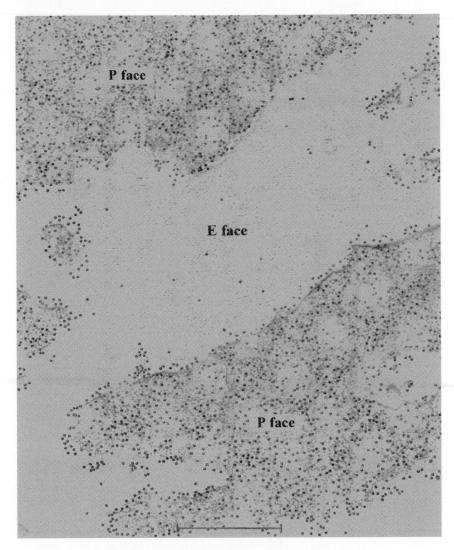

Figure 12. Distribution of AQP0 (dark particles) in equatorial fibers. The red particles represent the integral proteins of the fiber plasma membrane. The AQP0 channels in this region of the plasma membrane are thought to function in water permeability. Scale bar: 500 nm. Reprinted from Experimental Eye Research, Vol. 75, Zampighi et al., Micro-domains of AQP0 in lens, pp. 505–579., 2002, with permission from Elsevier.

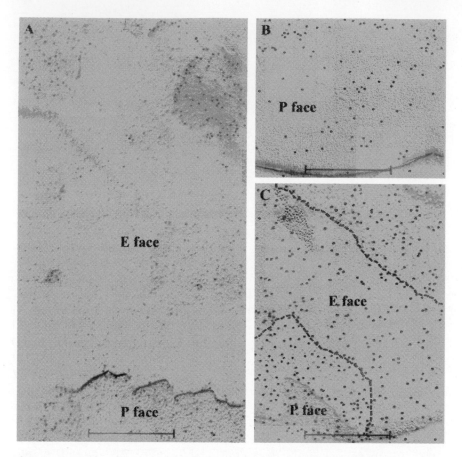

Figure 13. Colocalization of AQP0 (dark particles) and gap junctions (red areas). (A) Five gap junction plaques (red regions) are labeled with immuno-gold particles representing the AQP0 water channel. (B) Higher magnification views of a P face of a large gap junction. (C) Higher magnification of the E face of the gap junction plaque composed of complementary pits. Scale bars: (A), (B), and (C), 400 nm. Reprinted from Experimental Eye Research, Vol. 75, Zampighi et al., Micro-domains of AQP0 in lens, pp. 505–579., 2002, with permission from Elsevier.

C. Connexins

The connexin family encompasses 21 proteins that function as cell-to-cell channels (dodecamers) and as hemichannels (hexamers). Connexin channels span the plasma membranes of adjacent cells, bridge the intervening extra-cellular space, and aggregate to form the aforementioned "gap junctions". Each channel encircles a large water-filled pore of ~2 nm diameter at the

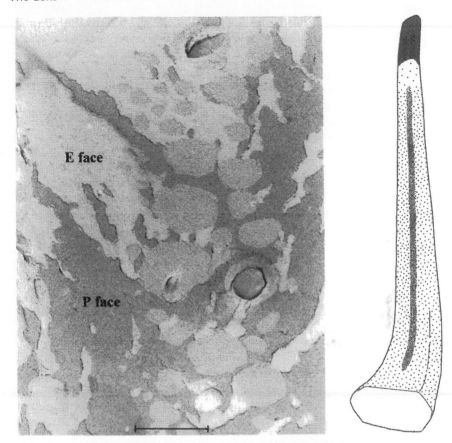

Figure 14. Overall distribution of connexins and AQP0 channels in the plasma membrane of equatorial fibers. Low magnification view (~28 μm²) shows the arrangement of AQP0 channels in the P face of the nonjunctional plasma membrane (dark particles) and the distribution of "fiber junctions" (yellow areas) composed of gap junction channels intermingled with AQP0. Model of an equatorial fiber with the approximate distribution of AQP0 and gap junctions along their apical–basal axis. The apical end (the modiolus) contains the gap junction plaques (red areas) and the diffuse AQP0 water channels (green dots). The green band indicates the distribution of mixed junctions composed of connexin and aquaporin channels. Scale bar: 1.5 μm. Reprinted from Experimental Eye Research, Vol. 75, Zampighi et al., Micro-domains of AQP0 in lens, pp. 505–579., 2002, with permission from Elsevier.

geometrical center (Unwin and Zampighi, 1980). A pore of such dimensions should allow the passage of ions and, at least in principle, of most second messenger molecules between the cytoplasm of adjacent cells. Consequently, the functions attributed to the gap junction channels are diverse. In excitable tissues, they synchronize smooth and cardiac muscle contraction and the fight-or-flight reflexes in arthropods and fish. In nonexcitable tissues, such as the lens, gap junctions are thought to set the resting voltage and to provide a pathway for nutrients and second messenger molecules to move within the lens interior (Mathias et al., 1997). They are also thought to control fiber maturation and lens size (White et al., 2001).

In contrast, connexin hemichannels span a single plasma membrane, which give them a possible role in cell permeability. In fact, when expressed in oocytes, Cx50 hemichannels function as nonselective with high conductance and their activity is dependent on the extracellular calcium concentration and intracellular pH (Zampighi et al., 1999). While the contribution of hemichannels to the permeability of single cells would certainly be physiologically significant, no experimental data directly supports such a role in the lens or any other tissue in the body.

Lens Connexins. The lens is an extremely well-coupled tissue. Electrophysiological experiments indicate that the cortex of the lenses of control mice exhibits a specific conductance per unit area of plasma membrane (G_i) of ~ 1 S/cm^2. Although lower than the cortex, the specific conductance in the nucleus is also sizable (0.4 S/cm^2) (Baldo et al., 2001). Three connexin isoforms, which are encoded by different genes, construct the channels that mediate this coupling. The anterior epithelium expresses connexin43 (Cx43), while the cortical and nuclear fibers express connexin46 (Cx46) and connexin50 (Cx50). The functional properties of these connexins are actively investigated using expression systems and biophysical methods.

Mouse genetics has contributed to the characterization of the gap junction network in the lens cortex and nucleus (Baldo et al., 2001). The knockout approach used in these experiments involves the systematic ablation of selective connexins and the study of the lenses by a variety of methods, including electrophysiology, morphology, and biochemistry. The ablation can be complete ($^{-/-}$, homozygote) or partial ($^{+/-}$, heterozygote). Since fibers express Cx46 and Cx50 in equimolar amounts (Koenig and Zampighi, 1995), one would have expected that the ablation of one isoform should have decreased the gap junction conductance, G_i, by one-half in both the cortex and nucleus. Measurement of G_i in lenses with complete ablation of Cx46 ($^{-/-}$) revealed, however, that the coupling in equatorial fibers remained at $\sim 30\%$, while the coupling in nuclear fibers fell to 0%. Lenses with partial ablation of Cx46 ($^{+/-}$ heterozygote) remained at $\sim 70\%$ of G_i in equatorial fibers and $\sim 50\%$ in the nuclear fibers. The partial ablation of Cx50 also remained at $\sim 70\%$ of the coupling in the equatorial fibers but rose to 100% in the nuclear fibers. In conclusion, Cx46 appears responsible for the entire coupling in nuclear fibers and for most of the coupling in cortical fibers. Unexpectedly, Cx50 plays a role in lens development rather than coupling because its partial, or total ablation induces microphthalmia. Lenses exhibiting complete ablation of Cx50 (homozygotes) were difficult to study because of their extreme fragility.

The changes resulting from the genetic ablation of Cx46 and Cx50 can be understood by assuming that the gap junctions contain mixtures of homomeric (constructed of a single isoform) and heteromeric (constructed of two or more isoforms) gap junction channels. Biochemical evidence supports the existence of heteromeric channels composed of Cx46 and Cx50 in bovine lenses, yet this hypothesis fails to explain why Cx46 is solely responsible for coupling in the lens nucleus. Therefore, while the mouse genetics approach underscores the importance of the gap junction network in lens homeostasis and development, the mechanism, by which these channels establish their action, remains undetermined.

A possible answer for explaining the need for multiple isoforms involves determining the distribution of Cx46 and Cx50 in the surface of fibers. Immunofluorescence experiments using antibodies against Cx46 and Cx50 indicate that these proteins are distributed along the entire length of the fiber (Lo et al., 1996). Unfortunately, immunolocalization has limited spatial resolution and cannot establish whether Cx46 and Cx50 are distributed in gap junctions, arranged as hemichannels or associated with other proteins, such as AQP0, in the plasma membrane.

To determine the arrangement as well as the oligomeric states (channels or hemichannels) of these connexins, several investigators have used freeze-fracture and FRIL methods. These studies have also indicated to an extensive gap junction network in the lens. Gap junctions are present in the anterior epithelium (Figure 3), and the apical interface connecting the anterior epithelium with the posterior cortical fibers (Figure 6). More importantly, however, these studies demonstrated that the lens equator exhibits the highest density of gap junctions (Figures 7 and 13). The extensive lateral surface of cortical and nuclear fibers contained ribbons of "fiber junctions" consisting of mixtures of AQP0 and gap junction channels (Figure 14). On the other hand, the distribution of connexin hemichannels is still undetermined. Preliminary observations suggest that the density of Cx50 hemichannels is substantially lower than the density of gap junction channels. Channels and hemichannels appear to be at an equilibrium that can be altered by phosphorylation (Zampighi, 2003). In summary, a single connexin isoform (Cx46) can explain coupling in lens cortical and nuclear fibers. The contribution of Cx50 appears limited to the equatorial fibers and the GATING induced by cytoplasmic acidification. An unexpected observation resulting from the partial and total ablation of Cx50 is its critical involvement in lens development. The mechanisms responsible for these changes remain undetermined.

Regulation of the connexin pathway. The ability of fibers to regulate Gi under physiological stimulation is a key requirement for maintaining homeostasis of nuclear fibers. One obvious advantage of regulation is the avoidance that damage to a single fiber can propagate to extensive regions of the lens. Also, regulation of the gap junctional conductance might induce subtle changes in the magnitude and direction of the fluxes of nutrients and waste from the lens surface to the interior.

Different mechanisms can regulate the conductance of the gap junction pathway. One mechanism involves triggering channel gating by changing the transjunctional voltage, the intracellular calcium, or the hydrogen concentration. In the lens, this mechanism is thought to prevent that the damage resulting from the death of a single fiber spreads to its neighbors. Another mechanism involves changing the total number of gap junction channels by modifying protein synthesis, or by altering the rate of insertion, or retrieval from the plasma membrane. The paucity of organelles, however, complicates our understanding of how these fibers can regulate the intracellular calcium concentration, control protein synthesis, or orchestrate the traffic of hemi-channels into the plasma membrane. There is mounting experimental evidence suggesting that the number of gap junction channels can be regulated by their disassembly into individual hemichannels. According to this evidence, hemichannels gather in microdomains of the plasma membrane called "lipid rafts" instead of being retrieved into the cytoplasm. These rafts are small ($\sim$70 nm diameter) patches—rich in cholesterol and sphingomyelin—which assemble temporarily in the membrane (Simons and Toomre, 2000; Razani et al., 2002). Since raft formation is reversible, hemichannels can be recycled back into gap junction channels when influenced by cellular signals, such as phosphorylation. The regulation of gap junctional conductance by the cycling of hemichannels between gap junctions and lipid rafts is a hypothesis that awaits further experimentation.

D. Potassium Conductance

Studies in the intact lens indicate that the K^+ conductance appears associated exclusively with cells in the lens surface, predominantly near the anterior epithelium (Rae, 1994). K^+ channels contribute to the maintenance of the fiber resting potential, as we know from experiments where blockers of these channels produced small depolarization when applied to the whole lens. The channels belong to at least three major types: inward rectifiers, large-conductance calcium-activated channels, and delayed rectifiers (Rae et al., 1990; Rae and Shepard, 2000). Inward rectifier channels allow large inward currents at voltages negative to the K^+ equilibrium (E_k) and outward currents over a limited range of voltage positive to E_k but are blocked at more positive voltages. The inward and delayed rectifiers are opened at −60, −40, and −30 mV, respectively, and are thought to contribute to the lens resting potential. The large-conductance calcium-activated channels (BK) are closed at normal intracellular calcium concentration, but open if the concentration reaches the micromolar range. The BK channels might be responsible for the increase in calcium observed in cataracts.

E. Sodium and Chloride Conductance

Epithelia express a special type of Na^+ channel (the "epithelial sodium channel," ENaC) that is characterized by its sensitivity to the drug amiloride. This channel is thought to play a critical role in epithelial Na^+ homeostasis. Mutations that delete or disrupt a C-terminal motif increase renal Na^+ absorption, causing a genetic form of hypertension. In spite of their importance in epithelial physiology, however, ENaC channels have not been identified in the lens. The particular protein responsible for this conductance is not known and the inflow of Na^+ is thought to occur through a leak conductance present in each fiber (Mathias et al., 1985). Since the number of fibers is very large, the magnitude of this conductance is also large and represents one of the critical components of the circulating fluxes model.

Several lines of experimental evidence, including the presence of Cl^- channels in vesicles isolated from plasma membranes of internal fibers, as well as ion substitution studies from isolated lenses, suggest that Cl^- channels might be present in inner cortical and nuclear fibers. This finding is important because, in diabetes, the osmolyte sorbitol accumulates in the cytoplasm. The osmotic insult can be compensated by Cl^- efflux from the cytoplasm. In the cortical fibers, the membrane voltage favors efflux but in the deeper fibers the membrane voltage favors a Cl^- influx and further favors fiber swelling. In this model, osmotic cataract might result from the inability of deeper fibers to regulate their volume leading to their breakdown and precipitation of the crystallins (Donaldson et al., 2001). Yet, it is not clear whether the Cl^- influx into the deeper fibers is the sole cause of the fiber cataract or whether other mechanism, such as Ca^{++} entrance might also be involved in the process.

F. Transporters

Transporters are present in all types of cells and occur in many forms. Each transporter protein is designed to move a different class of compounds—ions, sugars, and amino acids—into the cytoplasm. Some move a single solute across the membrane (uniports). Others function as cotransporters, in which the movement of one solute depends on the transfer of a second one, either in the same direction (symport), or in the opposite direction (antiport). While it is expected that the lens contain a variety of transporters, only a few types have been identified by molecular biology and biochemical methods. The following paragraphs discuss the Na,K-ATPase antiporter that uses ATP to translocate Na^+ and K^+ against their electrochemical gradients, and the glucose facilitated uniporters of the GLUT family.

G. The Na,K-ATPase Antiporter

From early on, it was realized that lenses exhibit currents in their surface resulting from the activity of the Na,K-ATPase transporter. Experiments performed by placing isolated lenses in an Ussing type chamber revealed currents that enter at the anterior pole exit at the posterior pole. The direction of these currents indicated a large density of Na,K-ATPase molecules in the anterior epithelium. When the currents were measured in freestanding lenses using a vibrating probe, however, the direction of the currents changed; they entered at the anterior and posterior poles and exited at the equator (Robinson and Patterson, 1983; Candia and Zamudio, 2002). The direction of the currents measured with the vibrating probe indicates the presence of active Na,K-ATPase pumps at the equator. This observation has revolutionized our understanding of the structure and function of the lens.

The catalytic properties of the Na,K-ATPase (3 Na^+ out, $2K^+$ in per ATP) are provided by a 100-kDa polypeptide, called the alpha subunit. The pumps also have an accessory $\sim$40-kDa polypeptide, the beta subunit, the function of which is not well understood. There are four alpha subunit isoforms: alpha1, alpha2, alpha3, and the testis-specific alpha4 (for review, see Kaplan, 2002), but where they are found depends on the tissue under study. In the frog lens epithelium, RNA protection assays show the presence of alpha1 and alpha2, but not alpha3 isoforms (Gao et al., 2000). In the ovine lens, immunofluorescence experiments indicate that all three isoforms are present in the anterior epithelium and the equatorial fibers. In spite of the presence of the Na,K-ATPase proteins, the activity of the enzyme (measured in plasma membrane fractions isolated from different regions of the rabbit lens) appears restricted to the lens cortex (Delamere and Dean, 1993). In the deeper cortical and nuclear fibers, the alpha subunit is present, but the activity of the pump is undetectable.

In summary, a series of experimental observations indicate that active Na, K-ATPase pumps located in the lens cortex maintain the Na^+ and K^+ homeostasis of the entire lens. These observations underscore the importance of polarization and the establishment of extensive pathways allowing communication between the surface and the fibers occupying the lens interior.

Glucose transporter. Sugars, such as D-glucose, D-fructose, and D-galactose, are the principal sources of carbon and energy in cell metabolism. Due to their large size, sugars require specialized transporters to cross the phospholipid bilayer. In contrast to Na^+,K^+-ATPase transporter that requires energy to translocate cations, sugars enter the cell passively, driven only by the concentration gradient. Of the five isoforms of the GLUT family of proteins, GLUT1, GLUT3, and GLUT4 are relatively high-affinity glucose transporters (20, 10, and 2 mM, respectively). GLUT2 has a much lower affinity ($\sim$40 mM) and GLUT5 does not transport glucose at all. The amino acid sequence of the transporters predicts a secondary structure composed of 12 membrane-spanning alpha helices with the amino and carboxyl termini at the cytoplasmic surface. The transport mechanism is not well understood, but kinetic studies suggest the transporters possess outward- and inward-facing sugar-binding sites. Reorientation of the transporter-sugar complex requires a transition state, in which neither binding site is exposed. Sugar translocation is thought to occur as the protein oscillates between the two conformations.

In the lens, facilitated transport of glucose takes place at both the anterior epithelium where GLUT1 is expressed and at the fibers where GLUT3 is expressed (Merriman-Smith et al., 1999). There is no explanation of why the anterior epithelium should express a different isoform than the fibers. GLUT3 has a lower K_m than GLUT1 ($\sim$10 mM vs. $\sim$20 mM). This is important because GLUT3 receptors appear to be upregulated in diabetes, a condition known to induce osmotic cataract. Whether the upregulation of glucose transporter is directly involved in cataract formation, however, remains to be determined.

Na–nHCO$_3$$^-$ cotransporter. Wolosin et al. (1990), measuring the pH regulation in the toad lens epithelium, concluded that lens epithelial cells possess a DIDS-sensitive electrogenic Na–HCO$_3$$^-$ cotransporter. The presence of such a cotransporter has now been confirmed immunohistochemically by Bok et al. (2001) using an antibody to the pancreatic form of the Na–HCO$_3$$^-$ cotransporter (pNBC1). This transporter is expressed in basolateral and apical membranes of the anterior and equatorial lens epithelium.

IV. THE CIRCULATING FLUXES MODEL

The circulating fluxes model is an attempt to integrate a large body of information obtained from electrophysiological, morphological, and molecular biology methods (Mathias et al., 1997). The model will suffer the inevitable fate of being proven wrong in some or many details by advances in cell and molecular biology. The hope is that nevertheless, it will stimulate investigators who search for a synthesis of the diverse and sometimes contradictory experimental observations obtained by studying the lens.

The circulating fluxes model is based on the observation made almost two decades ago that freestanding lenses of several species exhibit currents directed inward at the poles and outward at the equator (Robinson and Patterson, 1983). The fundamental proposal of the model is that these currents represent fluxes of ions, small molecules, and water that circulate throughout the lens and maintain the homeostasis of the nuclear fibers that lack direct access to the aqueous and vitreous humors.

The fluxes originate with the influx of Na^+, the principal extracellular cation ($\sim$150 mM), through the paracellular pathway composed of the intercellular spaces separating the cells and fibers at the anterior and posterior poles. At the poles, the suture lines create an intercellular path into the lens interior (Figure 7). When Na^+ enters, Cl^- follows to preserve electroneutrality and water to relieve osmotic pressure.

Inside the lens, Na^+, Cl^-, and water cross the plasma membrane and enter the cytoplasm of the fibers. The molecular basis of the Na^+ influx is undetermined and the circulating fluxes model proposes that it is mediated by a leak conductance. The molecular basis of the Cl^- conductance is also poorly understood, although some evidence suggests that Cl^- channels might be located in the cortical fibers. The influx of water can be explained by the permeability of the phospholipid bilayer ($\sim$1 $\times$ 10^{-4} cm/sec) and the abundance of AQP0 water channels in the plasma membrane (Figures 10 and 11). K^+, the principal cytoplasmic cation, is thought to exit through K^+ channels located in the cells of the anterior epithelium. In summary, the inflow of Na^+, Cl^-, and water and the outflow of K^+ are principally responsible for the fluxes that move nutrients within the lens interior.

The balance of Na^+ and K^+ from fibers is returned to the original conditions by the activity of the Na,K-ATPase, which is located preferentially in the anterior epithelial cells and the equator. To reach the cells and fibers containing active Na,K-ATPase, Na^+ relies on an extensive gap junction pathway composed of Cx43, Cx46, and Cx50. The Cl^- efflux is proposed to occur in the superficial cortical fibers due to their more negative membrane voltage. The role of other important cations, such as Ca^{++} remains undetermined.

In conclusion, the principal components of the circulating fluxes model are the asymmetric location of the Na,K-ATPase at the lens equator, the extensive network of gap junction channels and the difference in resting membrane potential between fibers in the lens interior and surface. The crux of the model is that the movement of ions establishes a net flow of water that allows nonelectrolytes, such as glucose, to circulate throughout the lens. This primitive yet effective circulatory system permits the lens to grow many times larger than would be allowed if nutrients had to reach the interior strictly by diffusion from the aqueous and vitreous humors. In essence, the circulating fluxes model is a modification of the more general phenomena of transport through epithelia.

A second mechanism of fluid transport through the lens has been described by Fischbarg et al. (1999). These authors observed that both (a) cultured layers of lens epithelial cells, and (b) the in vitro rabbit lens epithelium moved fluid from aqueous side into and through the lens. In the in vitro rabbit lens the fluid transport reported was of the order of 10 μL/h·lens, which would be sufficient to exchange the extracellular fluid volume of the lens in 1.5–2 h and in the process bringing in nutrients and removing metabolites.

ACKNOWLEDGMENT

This work was supported by a grant from NIH-EY04110.

REFERENCES

Agre, P., King, L.S., Yasui, M., Guggino, WmB., Ottersen, O.P., Fujiyoshi, Y., Engel, A. and Nielsen, S. (2002). Aquaporin water channels—From atomic structure to clinical medicine. J. Physiol. 5421, 3–16.

Arrigo, A.P. and Muller, W.E.G. (2002). Small Stress Proteins. Springer, Berlin.

Baldo, G., Gong, X., Martinez-Wittinghan, F., Kumar, N., Gilula, N. and Mathias, R. (2001). Gap junctional coupling in lenses from alpha8 connexin knockout mice. J. Gen. Physiol. 118, 447–456.

Bok, D., Schibler, M.J., Pushkin, A., Sassani, P., Abuladze, N., Naser, Z. and Kutz, I. (2001). Immunolocalization of electrogenic sodium-bicarbonate cotransporters pNBC1 and kNBC1 in the rat eye. Am. J. Physiol. Renal Physiol. 281, F920–F935.

Candia, O.A. and Zamudio, A. (2002). Regional distribution of the Na^+ and K^+ currents around the crystalline lens of rabbits. Am. J. Physiol. 282, C252–C262.

Chandy, G., Zampighi, G.A., Kreman, M. and Hall, J.E. (1997). Comparison of the water transporting properties of MIP and AQP1. J. Membr. Biol. 159, 631–664.

Chang, B., Wang, X., Hawes, N.L., Ojakian, R., Davisson, M.T., Lo, W.K. and Gong, X. (2002). A Gja8 (Cx50) point mutation causes an alteration of alpha3 connexin (Cx46) in semi-dominant cataracts of Lo10 mice. Hum. Mol. Gen. 11, 507–513.

Coulombre, J.L. and Coulombre, A.J. (1963). Lens development: Fiber elongation and lens orientation. Science 142, 1489–1490.

Delamere, N.A. and Dean, W.L. (1993). Distribution of lens sodium-potassium-adenosine triphosphatase. Invest. Ophthalmol. Visual Sci. 34, 2159–2163.

Donaldson, P., Kistler, J. and Mathias, R. (2001). Molecular solutions to mammalian lens transparency. New Physiol. Sci. 16, 118–123.

Fischbarg, J., Diecke, F.P.J., Kuang, K., Yu, B., Kang, F., Iserovich, P., Li, Y., Rosskothen, H. and Koniarek, J.P. (1999). Transport of fluid by lens epithelium. Am. J. Physiol. 276, C548–C557.

Fujimoto, K. (1995). Freeze-fracture replica electron microscopy combined with SDS digestion for cytochemical labeling of integral membrane proteins. Application to the immunogold labeling of intercellular junctional complexes. J. Cell Sci. 108, 3443–3449.

Fu, D., Libson, A., Miercke, L.J., Weitzman, C., Nollert, P., Krucinski, J. and Stroud, R.M. (2000). Structure of a glycerol-conducting channel and the basis for its selectivity. Science 290, 481–486.

Gao, J., Sun, X., Yatsula, V., Wymore, R. and Mathias, R. (2000). Isoform-specific function and distribution of Na/K pumps in the frog lens epithelium. J. Membrane Biol. 178, 89–101.

Horwitz, J. (2003). Alpha-crystallin. Exp. Eye Res. 76, 145–153.

Jaenicke, R. and Slingsby, C. (2002). Lens crystallin and their microbial homologs: Structure, stability and function. Crit. Rev. Biochem. Molec. Biol. 36, 435–499.

Jung, J.S., Preston, G.M., Smith, B.L., Guggino, W.B. and Agre, P. (1994). Molecular structure of the water channel through aquaporin CHIP. The hourglass model. J. Biol. Chem. 269, 14648–14654.

Kaplan, J.H. (2002). Biochemistry of Na,K-ATPase. Annu. Rev. Biochem. 71, 511–535.

Koenig, N. and Zampighi, G.A. (1995). Cell-to-cell channels from bovine lens are assembled from connexin 46 and connexin 50. J. Cell Sci. 108, 3091–3098.

Kuszak, J.R. (1995). The development of lens sutures. In: Progress in Retinal and Eye Research pp. 567–591. Elsevier Science Ltd.

Kuwabara, T. (1975). The maturation of the lens cell: A morphological study. Exp. Eye Res. 20, 427–443.

Lin, J.S., Eckert, R., Kistler, J. and Donaldson, P. (1998). Spatial differences in gap junction gating in the lens are a consequence of connexin cleavage. Eur. J. Cell. Biol. 76, 246–250.

Lo, W.K., Shaw, A.P., Takemoto, L.J., Grossniklaus, H.E. and Tigges, M. (1996). Gap junction structures and distribution patterns of immunoreactive connexins 46 and 50 in lens regrowths of Rhesus monkeys. Exp. Eye Res. 62, 171–180.

Mathias, R.T. and Rae, J.L. (1985). Transport properties of the lens. Am. J. Physiol. (Cell Physiol.) 18, C181–C190.

Mathias, R.T., Rae, J.L. and Baldo, G.J. (1997). Physiological properties of the normal lens. Physiol. Rev. 77, 21–50.

Merriman-Smith, R., Donaldson, P. and Kistler, J. (1999). Differential expression of facilitative glucose transporters GLU-1 and Glu-3 in the lens. Invest. Ophthalmol. Vis. Sci. 40, 3224–3230.

Miller, A.G., Zampighi, G.A. and Hall, J.E. (1992). Single membrane and cell-to-cell permeability of dissociated embryonic chick lens cells. J. Membr. Biol. 128, 91–102.

Mulders, S.M., Preston, G.M., Dean, P.M.T., Guggino, W., van Os, H. and Agre, P. (1995). Water channel properties of major intrinsic protein of the lens. J. Biol. Chem. 270, 9010–9016.

Narberhaus, F. (2002). Alpha-crystallin-type heat shock proteins: Socializing minichaperons in the contect of a multichaperone network. Microbiol. Mol. Biol. Rev. 66, 64–93.

Okamura, T., Miyoshi, I., Takahashi, K., Mototani, Y., Kon, Y. and Kasai, N. (2003). Bilateral congenital cataract result from a gain-of-function mutation in the gene of the aquaporin-0 in mice. Genomics 81, 361–368.

Piatigorsky, J. (1998). Multifunctional lens crystallins and corneal enzymes. More than meet the eye. Ann NY Acad. Sci. 842, 7–15.

Rae, J.L., Dewey, J., Rae, J.S. and Cooper, K. (1990). A maxi calcium-activated potassium channel from chick lens epithelium. Curr. Eye Res. 9, 847–861.

Rae, J.L. (1994). Outward rectifying potassium currents in lens epithelial cell membranes. Curr. Eye Res. 13, 679–686.

Rae, J. and Shepard, A. (2000). Kv3.3 Potassium channels in lens epithelium and corneal endothelium. Exp. Eye Res. 70, 339–348.

Rash, J.E. and Yasumura, T. (1999). Direct immunogold labeling of connexins and aquaporin-4 in freeze-fracture replicas of liver, brain and spinal cord: Factors limiting quantitative analysis. Cell Tiss. Res. 296, 307–321.

Razani, B., Woodman, S.C. and Lisanti, M.P. (2002). Caveolae: From cell biology to animal physiology. Pharmacol. Rev. 54, 431–467.

Robinson, K.R. and Patterson, J.W. (1983). Localization of steady currents in the lens. Curr. Eye Res. 2, 843–847.

Shields, A. and Bassnett, S. (1996). Mutations in the founder of the MIP gene family underlie cataract development in the mouse. Nat. Genet. 12, 212–215.

Sidjanin, D.J., Parker-Wilson, D.M., Neuhauser-Klaus, A., Prestsch, W., Favor, J., Dean, P.M., Ohtaka-Maruyama, C., Lu, Y., Bragin, A., Skach, W.R., Chepelinsky, A.B., Grimes, P.A. and Stambolian, D.E. (2001). A 76-bp deletion in the Mip gene causes autosomal dominant cataract in Hfi mice. Genomics 74, 313–319.

Simons, K. and Toomre, D. (2000). Lipid rafts and signal transduction. Nat. Rev. Mol. Cell Biol. 1, 31–41.

Stamer, W.D., Snyder, R.W., Smith, B.L., Agre, P. and Regan, J.W. (1994). Localization of aquaporin CHIP in the human eye: Implications in the pathogenesis of glaucoma and other disorders of ocular fluid balanced. Invest. Ophthalmol. Vis. Sci. 35, 3865–3872.

Sui, H., Han, B.-G., Lee, J.K., Wallan, P. and Jap, B.K. (2001). Structural basis of water-specific transport through AQP1 water channel. Nature 414, 872–878.

Unwin, P.N.T. and Zampighi, G. (1980). Structure of junctions between communicating cells. Nature 283, 545–549.

White, T.W., Sellito, C., Paul, D.L. and Goodenough, D.A. (2001). Prenatal lens development in connexin43 and connexin50 double knockout mice. Inv. Ophthamol. Vis. Sci. 42, 2916–2923.

Wolosin, J.M., Alvarez, L.J. and Candia, O.A. (1990). HCO_3^- dependent transport in the toad lens epithelium is mediated by an electronegative Na^+-dependent symport. Am. J. Physiol. Cell Physiol. 258, C855–C861.

Zampighi, G., Simon, S.A. and Hall, J.E. (1992). The specialized junctions of the mammalian lens. Int. Rev. Cytol. 136, 185–225.

Zampighi, G.A., Kreman, M., Boorer, K.J., Loo, D.D.F., Bezanilla, F., Chandy, G., Hall, J.E. and Wright, E.M. (1995). A method for determining the unitary functional capacity of cloned membrane protein expressed in *Xenopus* oocytes. J. Membr. Biol. 148, 65–78.

Zampighi, G.A., Loo, D.D.F., Kreman, M., Eskandari, S. and Wright, E. (1999). Functional and morphological correlates of connexin50 expressed in *Xenopus laevis* oocytes. J. Gen. Physiol. 113, 1–17.

Zampighi, G.A., Eskandari, E. and Kreman, M. (2000). Epithelial organization of the mammalian lens. Exp. Eye Res. 71, 415–435.

Zampighi, G.A., Eskandari, E., Hall, J.E., Zampighi, L. and Kreman, M. (2002). Microdomains of AQP0 in the lens. Exp. Eye Res. 75, 505–519.

Zampighi, G.A. (2003). Distribution of connexin 50 channels and hemi-channels in lens fiber cells: A structural approach. Cell Communication and Adhesion. 10, 265–271.

THE VITREOUS

Henrik Lund-Andersen, J. Sebag, Birgit Sander and
Morten La Cour

I. INTRODUCTION

The vitreous body occupies a volume of about 4.5 ml and is the largest single structure in the eye, making up approximately 80% of its total volume. Anteriorly, the vitreous is delineated by, and adjoins the ciliary body, the zonules and the lens; posteriorly, it adjoins the retina (Figure 1).

Advances in Organ Biology
Volume 10, pages 181–194.
© 2006 Elsevier Inc. All rights reserved.
ISBN: 0-444-50925-9
DOI: 10.1016/S1569-2590(05)10007-X

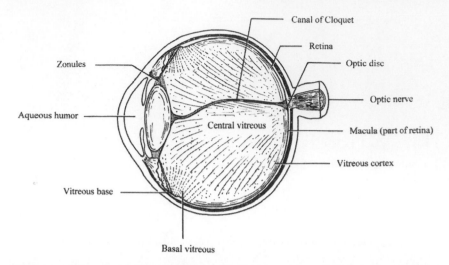

Figure 1. Diagram of the anatomy of the human vitreous. The orientation of the collagen fibrils and the attachment zone in the vitreous base (basal vitreous) is shown, from Bishop (2000).

The vitreous body supports the retina, and is probably necessary for the maintenance of the clarity of the lens (Harocopos et al., 2004). Via anomalous posterior vitreous detachment (PVD) (Sebag, 2004) vitreoretinal pathology can have devastating consequences for visual function, and removal of pathological vitreous by vitrectomy is a common surgical procedure. In the following paragraphs, we will discuss the anatomy, and molecular organization of the vitreous, as well as its development and aging changes.

II. GROSS ANATOMY

The mature vitreous body is a transparent gel, which consists of glycosaminoglycans and collagen fibrils. In the adult, emmetropic eye, the mean axial length of the vitreous is 16.5 mm (Luyckx-Bacus and Weekers, 1966). The vitreous body occupies the vitreous cavity, and has an almost spherical appearance, except for the anterior part, which is concave corresponding to the presence of the crystalline lens. Anatomically, the vitreous can be subdivided into a number of anatomical regions including the basal, central, and cortical vitreous (Figure 1).

The main bulk of the vitreous is composed of the central vitreous, which contains the canal of Cloquet. This structure is a remnant of the fetal hyaloid vascular system, and it joins the optic disc and the posterior surface of the

Figure 2. Dark field slit microscopy of the posterior and central vitreous of a 59-year-old male. The longitudinally oriented central fibrils around the canal of Cloquet, as well as the condensed cortical vitreous are clearly visible, from Sebag (2002).

lens. In individuals below 35 years of age, the canal of Cloquet, and through it the anterior vitreous, is firmly attached to the posterior lens capsule. This attachment is denoted the ligament of Wieger. In the central vitreous, the collagen fibrils mainly course in anterior–posterior directions, parallel to the canal of Cloquet (Figure 2).

The vitreous base is an annular zone that straddles the ora serrata (i.e., the junction between the retina and pars plana of the ciliary body). The vitreous base is a zone of very firm attachments between the vitreous, the retina, and pars plana (Figure 3). It extends from 1.5–2 mm anterior to the ora serrata to 0.5–2.5 mm posterior to it (Bishop, 2000; Wang et al., 2003).

The vitreous cortex is a 100- to 300-μm thick layer that surrounds the central vitreous. The density of collagen fibrils is higher than in the central vitreous, and they course parallel to the surface of the vitreous. Posterior to the vitreous base, the cortical vitreous is less firmly attached to the retinal surface. However, focal points of more firm attachment exist around major retinal vessels, at the fovea, and, most notably, around the optic disc. Clinically, the term anterior and posterior hyaloid is used to denote the anterior and posterior vitreous cortex. Sometimes the term hyaloid membrane is used for the vitreous cortex, but this is a misnomer, since the vitreous cortex is not a true membrane (Heegaard, 1997; Bishop, 2000). The term "hyaloid" should be avoided except to refer to the central artery during vitreous embryogenesis. The vitreoretinal interface is an

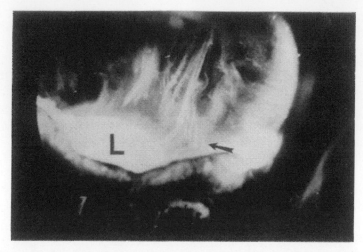

Figure 3. Dark field slit microscopy of the vitreous base region in a 58-year-old female. L denotes the crystalline lens. Anteriorly fibers are seen to fan out prior to insertion in the vitreous base (arrow), from Sebag (2002).

ultrastructurally defined region that includes the outer part of the vitreous cortex and the inner part of the internal limiting lamina of the retina (Heegaard, 1997). The molecules that are located between these structures include laminin and fibronectin, as well as other typical components of an extracellular matrix (Sebag, 1991).

A. Ultrastructure and Biochemistry

The vitreous is a gel that contains more than 99% water. The gel is stabilized by collagen fibrils, glycosaminoglycans, and proteoglycans. The collagen composition in the vitreous resembles that in cartilage. The main collagen type is collagen II, which constitutes approximately 75% of the total vitreous collagen. Smaller amounts of collagen type V, IX, and XI are also found (Sebag, 1998a; Bishop, 2000; Ihanamäki et al., 2004). Collagens II, V, and XI are fibrillar collagens, whereas collagen IX is a so-called fibril-associated collagen with interrupted triple helices, or FACIT collagen (Ihanamäki et al., 2004). Collagen II exists in two forms due to posttranscriptional splicing. Collagen IIA predominates in the vitreous, whereas collagen IIB is more abundant in cartilage. Collagen IIB is a smaller molecule because it lacks exon 2 due to alternative posttranslational splicing (Bishop et al., 1994; Donoso et al., 2003).

All collagens incorporate the characteristic polypeptide α-chains, in which every third amino acid is the small glycine residue. The α-chains are synthesized as preprocollagen, which, after some posttranslational modifications, are assembled into the triple-helical procollagen molecule. The triple-helix of procollagen is tightly twisted, with an almost rod-like structure. The tightness of procollagen is made possible by the primary structure of collagen, where every third amino acid is the small glycine residue, which fits into the narrow core of the triple helix. Procollagen is the secreted entity. Once in the extracellular space, specific proteases remove N- and C-terminal globular proteins, and procollagen subsequently undergo self assembly into collagen fibrils. Procollagen II is a homotrimer, consisting of three identical α-chains that form the rod-like triple helix, which is the building block of the collagen II fibrils. For the other vitreous fibrillar collagen, type V–XI, the procollagen is a heterotrimer that consists of α-chains both collagen V and collagen XI; the stoichiometry of the two is unknown (Bishop et al., 1999; Ihanamäki et al., 2004).

The major collagen fibrils in the vitreous are assembled from more than one type of procollagen, and are thus heterotypical (Bishop, 2000; Ihanamäki et al., 2004). The FACIT collagen IX molecules are regularly aligned along the surface of the vitreal fibrils. Vitreal collagen IX is a proteoglycan, which probably prevents aggregation, or fusion, of these fibrils (Bishop et al., 2004).

Defects in the genes encoding the vitreal collagens have been shown to result in vitreoretinal degenerations. Stickler syndrome is a disease characterized by an optically "empty" vitreous and a high frequency of rhegmatogenous retinal detachments. Extraocular manifestations in Stickler syndrome include flattened facies, arthropathy, hearing loss, and cleft palate. The systemic phenotype might vary considerably, but two main vitreous phenotypes, membranous and beaded, have been described (Snead and Yates, 1999; Richards et al., 2000; McLeod et al., 2002; Donoso et al., 2003). Mutations in the type II collagen gene COL2A1 have been shown to result in the membranous phenotype of Stickler syndrome. Interestingly, mutations in exon 2 of COL2A, which is expressed in the eye, but not in (adult) cartilage, result in the membranous vitreal phenotype, but only in minimal or absent extraocular manifestations (Richards et al., 2000; Donoso et al., 2003). Mutations in the type XI collagen gene COL11A1 have been linked to the beaded phenotype of Stickler syndrome (Donoso et al., 2003).

The other main extracellular matrix component of the vitreous is glycosaminoglycans or GAG's. The major GAG of mammalian vitreous is hyaluronan (formerly denoted as hyaluronic acid). Hyaluronan is a non-sulphated GAG polyanion, which is capable of binding large amounts of water. The volume of the unhydrated hyaluronan is $0.66 \text{ cm}^3/\text{g}$, whereas the volume of hydrated hyaluronan is $2000–3000 \text{ cm}^3/\text{g}$ (Sebag, 1998a). The molecular weight of hyaluronan in human vitreous is probably in the order of 3 million. The molecule coils and forms loose left handed helices in

aqueous solution. In contrast to true gels, hyaluronan has viscoelastic properties which indicates that it does not form stable intermolecular associations
(Almond et al., 1998; Bishop, 2000). Chondroitin sulphate is a sulphated
GAG chain that forms part of two vitreous proteoglycans: type IX collagen
and versican. These molecules probably link the hyaluronan to the fibrillar
skeleton of the vitreous (Reardon et al., 1998; Bishop, 2000).

The structure of the vitreous is critically dependent on its collagen constituents, and proteolytic enzymes are under investigation as adjunct to
vitrectomy, so-called Pharmacologic Vitreolysis (Sebag, 1998b; Tanaka
and Qui, 2000; Sebag, 2002a). Depolymerization of vitreal hyaluronan on
the other hand results in gel wet weight reduction, but not in gel destruction
(Bishop et al., 1999).

The vitreous contains few cells, the so-called vitreocytes or hyalocytes
(Figure 4). These cells have some monocyte markers, and are more numer-
ous in the anterior vitreous (Lazarus and Hageman, 1994; Haddad and
Andre, 1998).

B. Biophysical Aspects

The intact vitreous gel structure acts as a barrier against bulk movement of
solutes. High concentrations of antioxidants, such as ascorbic acid, can
therefore accumulate in the vitreous, and this might help to protect the
lens against oxidative damage (Takano et al., 1997). It is well known that

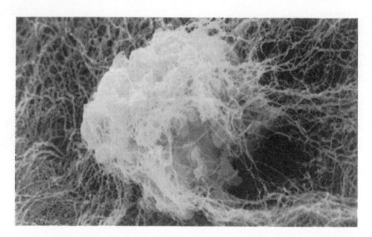

Figure 4. Scanning electron micrograph of the inner surface of the anterior retina
and ciliary body in a rabbit. A free cell, a presumptive vitreocyte, with a warty
surface is seen in the cortical vitreous close to the retina. The network of collagen
fibrils in the cortical vitreous is also clearly seen, from Haddad and André (1998),
with permission from Elsevier.

lack of the vitreous gel, congenital or acquired, is associated with an increased risk of cataract formation (Takano et al., 1997; Snead and Yates, 1999; Richards et al., 2000; Harocopos et al., 2004). The vitreous gel is also hypothesized to act as a sink for potassium ions siphoned out of the Müller cell end-feet during light-induced changes in retinal metabolism (Newman et al., 1984).

The ability of the vitreous to prevent bulk movement of solute depends on the degree of liquefaction of the gel. Since the vitreous by design is invisible, it is difficult to assess the degree of liquefaction by ophthalmic imaging techniques (Sebag, 2002). However, diffusion within the vitreous of a tracer substance, fluorescein, can be illustrated *in vivo* in humans by vitreous fluorophotometry (Lund-Andersen et al., 1985). After intravenous injection of fluorescein a small amount passes through the ocular barriers into the anterior chamber and the vitreous body. In the vitreous, the fluorescein concentration can be measured noninvasively by vitreous fluorophotometry, and the distribution of this tracer within the vitreous can be followed over time. Figure 5 shows measurements with vitreous fluorophotometry from two individuals. The individual in panel A has an intact posterior vitreous, whereas the individual in panel B has a posterior vitreous detachment with mobile liquefied posterior vitreous that allows equalization of the fluorescein concentration by convective fluid movements (Moldow et al., 1998).

C. Development

The development of the vitreous is classically described in three stages (Duke-Elder and Cook, 1963). The primary vitreous constitutes the vascular mesenchyme around the hyaloid vessels that invade the vitreous cavity through the inferior fissure in the optic cup during the 5th–7th week of development (Azuma et al., 1998). The secondary vitreous develops around the primary vitreous, and compresses it to the axial vitreous cavity. Meanwhile, the hyaloid vessels continue to develop, and eventually form a rich vascular network around the lens. However, from the 5th gestational month, the hyaloid vascular system undergoes apoptotic regression, and at birth its only remainder is the canal of Cloquet (see Figure 1) (Duke-Elder, 1963).

Incomplete regression of the primary vitreous causes a variety of ophthalmologic clinical entities, among them congenital cataracts, falciform retinal detachments, and persistent hyperplastic primary vitreous (Goldberg, 1997; Mullner-Eidenbock et al., 2004). The secondary vitreous extends into the area of the future zonules and posterior chamber. From the 6th gestational month, the zonular fibrils start to develop, and at the same time the anterior, triangular part of the secondary vitreous regresses. Because the zonules, and the posterior chamber, replaces a part of the secondary vitreous, these

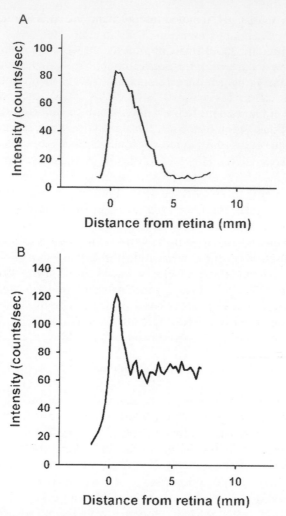

Figure 5. Intravitreal fluorescein profiles as measured *in vivo* with vitreous fluorophotometry 30 minutes after an intravenous injection of fluorescein. Panel (A) shows the fluorescein concentration profile in an individual with an intact vitreous, and the inward diffusion of fluorescein through the gel is reflected by the decreasing concentration towards the center of the vitreous. Panel (B) shows the fluorescein concentration profile in an individual with a complete posterior vitreous detachment. Apart from the fluorescence peak derived from the retina, there is a uniform distribution of fluorescein in the posterior vitreous, probably caused by convective movements in the liquefied vitreous, data from Moldow et al. (1998).

structures are classically denoted the tertiary vitreous (Duke-Elder and Cook, 1963).

In all stages of development, hyaluronan is the major GAG in both primary and secondary vitreous (Azuma et al., 1998). In very early development, most vitreal collagen is type III; however, at the 8th gestational week, vitreal type III collagen is replaced by type II, which subsequently predominates (Azuma et al., 1998). The vitreal extracellular matrix is synthesized by cells in the retina, presumably Müller cells, by the cells in the hyaloid vascular system, and presumably also by cells in the ciliary body (Azuma et al., 1998; Bishop et al., 2002).

III. AGING OF THE VITREOUS

In young individuals, the vitreous is optically homogenous when examined *in vivo* with the slit lamp or, by dark-field microscopy, in enucleated eyes (Sebag, 1989, 2002a). Aging results in structural changes in the vitreous (Sebag, 1987), as well as characteristic alterations in the strength of vitreoretinal adhesion (Sebag, 1991) (Figure 6).

Figure 6. Human vitreous body of a 9-month-old child. The sclera, choroids, and retina are dissected away from the posterior segment, although the retina is still attached to the anterior vitreous. This young vitreous gel is firm, and retains its shape despite the specimen being situated on a surgical towel in room air, from Sebag (2002).

After the first few decades of life, fine parallel fibers appear coursing in the anterior–posterior directions, and mainly attached in the vitreous base and optic nerve head regions. Other fibers occur in the vitreous cortical region and take a circumferential course (Figures 2 and 3). The appearance of visible fibers in aging vitreous is due to aggregation of the collagen fibrils (Sebag, 1989). Collagen IX is present on the surface of vitreal collagen fibers, and its chondroitin sulphate side chain is believed to prevent aggregation of the fibrils, and perhaps thereby maintain the integrity of the gel itself (Bishop, 2000; Bishop et al., 2004). It has recently been shown that aging is associated with a progressive loss of collagen IX from the surface of the vitreous fibrils, and at the same time with an increasing availability of collagen II epitopes on the surface of these fibrils (Bishop et al., 2004). This supports the hypothesis that the "sticky" collagen II is "painted" with "slippery" collagen IX that prevents the collagen II from aggregating. Aging results in the "paint flaking off," and the fibrils consequently sticking together and forming aggregated, thicker, and visible fibers (Bishop, 2000).

Concurrent with the formation of visible fibrils in the vitreous, optically empty spaces, or lacunae, start to evolve in the central vitreous (Figure 7). Within these spaces, the vitreous gel is liquefied (Los et al., 2003). Eventually, the smaller lacunae of liquefied vitreous coalesce into larger cavities or cisterns. In the central, and posterior vitreous a large cistern, named by Jan Worst as the premacular bursa, may form (Worst, 1977; Sebag, 1998a). The mechanisms of vitreous liquefaction remain poorly understood. It is not due to selective loss of hyaluronan within the liquefied lacunae, since the concentration of hyaluronan is the same in liquid as in gel vitreous (Larsson and Osterlin, 1985; Sebag, 1998a; Bishop, 2000). Breakdown of collagen fibrils and loss of chondroitin sulphate proteoglycans have been found both within and adjacent to areas of vitreal liquefaction. This suggests an enzymatic rather than mechanical cause of the age-related vitreous liquefaction (Los et al., 2003).

Aging also results in alterations in the strength of vitreoretinal adhesion that characteristically are unevenly distributed across the vitreoretinal interface. In the posterior part of the eye, there is a progressive weakening of the vitreoretinal adhesion. In the anterior part of the eye, increasing age is associated with remarkable changes in the vitreal base. There is a broadening of the vitreal base, which is caused by a backward shift of its posterior border. At the same time this border becomes more irregular. At 20 years of age, the posterior border of the vitreous base extends only 0.5 mm posterior to the ora serrata; however, beyond 60 years of age this border extends more than 1.5 mm further posteriorly (Bishop, 2000; Wang et al., 2003). Within the vitreous base, the vitreoretinal adhesion increases in strength with age. However immediately posterior to it, the vitreoretinal adhesion diminishes with age (Wang et al., 2003). Thus in older eyes, the posterior border of the vitreous

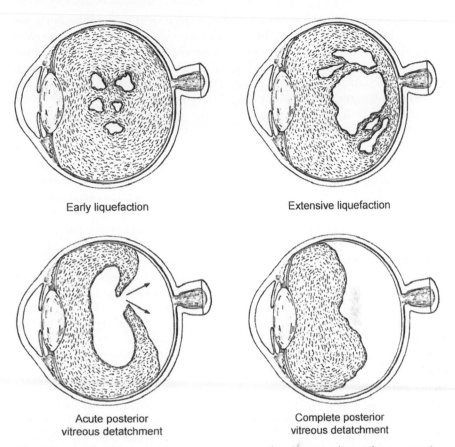

Early liquefaction   Extensive liquefaction

Acute posterior   Complete posterior
vitreous detatchment   vitreous detatchment

Figure 7. Age changes in the human vitreous. The diagram shows the progressive liquefaction of the vitreous gel, which eventually results in a complete posterior vitreous detachment. In a complete posterior vitreous detachment, the vitreous is separated from the posterior retina including the optic disc. This usually evolves during the sixth and seventh decade, from Bishop (2000).

base represents a marked discontinuity in the strength of vitreoretinal adhesion, and it is therefore a predilection site for the development of retinal breaks (Wang et al., 2003).

The age-related weakening of the posterior vitreoretinal adhesion eventually results in the detachment of posterior vitreous cortex from the retina. In the posterior retina, the vitreoretinal adhesion is strongest at the optic disc. Hence, clinically, the posterior vitreous detachment is not considered complete before the vitreous is detached from the disc. This is very often visible biomicroscopically because the "imprint" of the optic nerve head can be seen as an annular opacity, the so-called Weiss' ring, in the detached posterior cortical vitreous (Sebag, 1989, 1991; Akiba et al., 2001). The average age of onset of complete posterior vitreous detachment has been found to be in the seventh decade both in clinical and postmortem studies (Foos and Wheeler, 1982; Yonemoto et al., 1994). However, after the introduction of optical coherence tomography, which enables high resolution studies of the vitreo-retinal interface, it has become apparent that posterior vitreous detachment starts already in the fourth decade with a vitreoretinal separation in a sector of the parafoveal macula. Normally, this subtle separation then proceeds slowly until the posterior vitreous is adherent only at the foveal center and at the optic disc. Finally, in the sixth and seventh decade, the vitreoretinal separation becomes complete as the vitreous is separated first from the foveal center and then from the optic disc (Uchino et al., 2001). Clinically this is referred to as PVD. An innocuous PVD requires concurrent vitreous liquefaction and weakening of vitreoretinal adhesion.

When the degree of vitreous liquefaction exceeds the extent of weakening at the vitreoretinal interface, anomalous PVD results (Sebag, 2004). Failure of the vitreous to separate from the foveal center may cause visual loss due to the formation of a macular hole, or due to distortion of the macula caused by vitreomacular traction (Haouchine et al., 2001). Eventually, the posterior vitreous detachment extends to the posterior border of vitreous base. Because of the sudden increase in vitreoretinal adhesion encountered in this region, peripheral retinal tears may accompany a posterior vitreous detachment, particularly in cases where it evolves quickly, and is symptomatic (Tanner et al., 2000; van Overdam et al., 2001). Symptomatic, peripheral retinal breaks represent an ophthalmologic emergency, because they may cause a retinal detachment with ensuing visual loss (Tanner et al., 2000).

REFERENCES

Akiba, J., Ishiko, S. and Yoshida, A. (2001). Variations of Weiss's ring. Retina 21, 243–246.
Almond, A., Brass, A. and Sheehan, J.K. (1998). Deducing polymeric structure from aqueous molecular dynamics simulations of oligosaccharides: Predictions from simulations of hyaluronan tetrasaccharides compared with hydrodynamic and X-ray fibre diffraction data. J. Mol. Biol. 284, 1425–1437.
Azuma, N., Tajima, S., Konomi, H., Hida, T., Akiya, S. and Uemura, Y. (1998). Glycosamino-glycan and collagen distribution in the developing human vitreous. Graefes Arch. Clin. Exp. Ophthalmol. 236, 679–687.
Bishop, P.N. (2000). Structural macromolecules and supramolecular organisation of the vitre-ous gel. Prog. Retin. Eye Res. 19, 323–344.
Bishop, P.N., Holmes, D.F., Kadler, K.E., McLeod, D. and Bos, K.J. (2004). Age-related changes on the surface of vitreous collagen fibrils. Invest. Ophthalmol. Vis. Sci. 45, 1041–1046.

Bishop, P.N., McLeod, D. and Reardon, A. (1999). Effects of hyaluronan lyase, hyaluronidase, and chondroitin ABC lyase on mammalian vitreous gel. Invest. Ophthalmol. Vis. Sci. 40, 2173–2178.

Bishop, P.N., Reardon, A.J., McLeod, D. and Ayad, S. (1994). Identification of alternatively spliced variants of type II procollagen in vitreous. Biochem. Biophys. Res. Commun. 203, 289–295.

Bishop, P.N., Takanosu, M., Le Goff, M. and Mayne, R. (2002). The role of the posterior ciliary body in the biosynthesis of vitreous humour. Eye 16, 454–460.

Donoso, L.A., Edwards, A.O., Frost, A.T., Ritter, R., III, Ahmad, N., Vrabec, T., Rogers, J., Meyer, D. and Parma, S. (2003). Clinical variability of Stickler syndrome: Role of exon 2 of the collagen COL2A1 gene. Surv. Ophthalmol. 48, 191–203.

Duke-Elder, D. and Cook, C. (1963). System of Ophthalmology. Vol III, part I. Embryology, pp. 1–313. Henry Kimpton, London.

Foos, R.Y. and Wheeler, N.C. (1982). Vitreoretinal juncture. Synchysis senilis and posterior vitreous detachment. Ophthalmology 89, 1502–1512.

Goldberg, M.F. (1997). Persistent fetal vasculature (PFV): An integrated interpretation of signs and symptoms associated with persistent hyperplastic primary vitreous (PHPV). LIV Edward Jackson Memorial Lecture. Am. J. Ophthalmol. 124, 587–626.

Haddad, A. and Andre, J.C. (1998). Hyalocyte-like cells are more numerous in the posterior chamber than they are in the vitreous of the rabbit eye. Exp. Eye Res. 66, 709–718.

Haouchine, B., Massin, P. and Gaudric, A. (2001). Foveal pseudocyst as the first step in macular hole formation: A prospective study by optical coherence tomography. Ophthalmology 108, 15–22.

Harocopos, G.J., Shui, Y.B., McKinnon, M., Holekamp, N.M., Gordon, M.O. and Beebe, D.C. (2004). Importance of vitreous liquefaction in age-related cataract. Invest. Ophthalmol. Vis. Sci. 45, 77–85.

Heegaard, S. (1997). Morphology of the vitreoretinal border region. Acta Ophthalmol. Scand. (Suppl.) 1–31.

Ihanamäki, T., Pelliniemi, L.J. and Vuorio, E. (2004). Collagens and collagen-related matrix components in the human and mouse eye. Prog. Retin. Eye Res. 24, 403–434.

Larsson, L. and Osterlin, S. (1985). Posterior vitreous detachment. A combined clinical and physiochemical study. Graefes Arch. Clin. Exp. Ophthalmol. 223, 92–95.

Lazarus, H.S. and Hageman, G.S. (1994). In situ characterization of the human hyalocyte. Arch. Ophthalmol. 112, 1356–1362.

Los, L.I., van der Worp, R.J., van Luyn, M.J. and Hooymans, J.M. (2003). Age-related liquefaction of the human vitreous body: LM and TEM evaluation of the role of proteoglycans and collagen. Invest. Ophthalmol. Vis. Sci. 44, 2828–2833.

Lund-Andersen, H., Krogsaa, B., la Cour, M. and Larsen, J. (1985). Quantitative vitreous fluorophotometry applying a mathematical model of the eye. Invest. Ophthalmol. Vis. Sci. 26, 698–710.

Luyckx-Bacus, J. and Weekers, J.F. (1966). Etude biometrique de l'oeil humain par ultrasonographie. Bull. Soc. Belge. Ophtalmol. 143, 552–567.

McLeod, D., Black, G.C. and Bishop, P.N. (2002). Vitreous phenotype: Genotype correlation in Stickler syndrome. Graefes Arch. Clin. Exp. Ophthalmol. 240, 63–65.

Moldow, B., Sander, B., Larsen, M., Engler, C., Li, B., Rosenberg, T. and Lund-Andersen, H. (1998). The effect of acetazolamide on passive and active transport of fluorescein across the blood-retina barrier in retinitis pigmentosa complicated by macular oedema. Graefes Arch. Clin. Exp. Ophthalmol. 236, 881–889.

Mullner-Eidenbock, A., Amon, M., Moser, E. and Klebermass, N. (2004). Persistent fetal vasculature and minimal fetal vascular remnants: A frequent cause of unilateral congenital cataracts. Ophthalmology 111, 906–913.

Newman, E.A., Frambach, D.A. and Odette, L.L. (1984). Control of extracellular potassium levels by retinal glial cell K^+ siphoning. Science 225, 1174–1175.

Reardon, A., Heinegard, D., McLeod, D., Sheehan, J.K. and Bishop, P.N. (1998). The large chondroitin sulphate proteoglycan versican in mammalian vitreous. Matrix Biol. 17, 325–333.

Richards, A.J., Martin, S., Yates, J.R., Scott, J.D., Baguley, D.M., Pope, F.M. and Snead, M.P. (2000). COL2A1 exon 2 mutations: Relevance to the Stickler and Wagner syndromes. Br. J. Ophthalmol. 84, 364–371.

Sebag, J. (1987). Age-related changes in human vitreous structure. Graefes Arch. Clin. Exp. Ophthalmol. 225, 89–93.

Sebag, J. (1989). The Vitreous—Structure, Function, and Pathobiology. Springer-Verlag, New York.

Sebag, J. (1991). Age-related differences in the human vitreo-retinal interface. Arch. Ophthalmol. 109, 966–971.

Sebag, J. (1998). Macromolecular structure of the corpus vitreus. Prog. Polym. Sci. 23, 415–446.

Sebag, J. (1998). Pharmacologic vitreolysis (Guest Editorial). Retina 18, 1–3.

Sebag, J. (2002). Is Pharmacologic Vitreolysis brewing (Guest Editorial). Retina 22, 1–3.

Sebag, J. (2002). Imaging vitreous. Eye 16, 429–439.

Sebag, J. (2004). Anomalous posterior vitreous detachment: A unifying concept in vitreo-retinal disease. Graefes Arch. Clin. Exp. Ophthalmol. 242, 690–698.

Snead, M.P. and Yates, J.R. (1999). Clinical and molecular genetics of Stickler syndrome. J. Med. Genet. 36, 353–359.

Takano, S., Ishiwata, S., Nakazawa, M., Mizugaki, M. and Tamai, M. (1997). Determination of ascorbic acid in human vitreous humor by high-performance liquid chromatography with UV detection. Curr. Eye Res. 16, 589–594.

Tanaka, M. and Qui, H. (2000). Pharmacological vitrectomy. Semin. Ophthalmol. 15, 51–61.

Tanner, V., Harle, D., Tan, J., Foote, B., Williamson, T.H. and Chignell, A.H. (2000). Acute posterior vitreous detachment: The predictive value of vitreous pigment and symptomatology. Br. J. Ophthalmol. 84, 1264–1268.

Uchino, E., Uemura, A. and Ohba, N. (2001). Initial stages of posterior vitreous detachment in healthy eyes of older persons evaluated by optical coherence tomography. Arch. Ophthalmol. 119, 1475–1479.

van Overdam, K.A., Bettink-Remeijer, M.W., Mulder, P.G. and van Meurs, J.C. (2001). Symptoms predictive for the later development of retinal breaks. Arch. Ophthalmol. 119, 1483–1486.

Wang, J., McLeod, D., Henson, D.B. and Bishop, P.N. (2003). Age-dependent changes in the basal retinovitreous adhesion. Invest. Ophthalmol. Vis. Sci. 44, 1793–1800.

Worst, J.G.F. (1977). Cisternal systems of the fully developed vitreous body in the young adult. Trans. Ophthalmol. Soc. UK 97, 550–554.

Yonemoto, J., Ideta, H., Sasaki, K., Tanaka, S., Hirose, A. and Oka, C. (1994). The age of onset of posterior vitreous detachment. Graefes Arch. Clin. Exp. Ophthalmol. 232, 67–70.

FURTHER READING

Kishi, S., Demaria, C. and Shimizu, K. (1986). Vitreous cortex remnants at the fovea after spontaneous vitreous detachment. Int. Ophthalmol. 9, 253–260.

Sebag, J. and Balazs, E.A. (1989). Morphology and ultrastructure of human vitreous fibers. Invest. Ophthalmol. Vis. Sci. 30, 1867–1871.

THE RETINA

Morten la Cour and Berndt Ehinger

Advances in Organ Biology
Volume 10, pages 195–252.
© 2006 Elsevier Inc. All rights reserved.
ISBN: 0-444-50925-9
DOI: 10.1016/S1569-2590(05)10008-1

I. OVERVIEW OF THE FUNCTIONAL ARCHITECTURE OF THE RETINA

The retina is the light-sensitive part of the eye. It is a thin film of tissue covering most of the inner wall of the eye (Figure 1), and is formed from nerve cells and glial cells, most of which are of the so-called Müller cell type. The outermost part of the retina is a single layer of pigmented cuboidal epithelial cells, the so-called retinal pigment epithelium (RPE). The functions of the RPE and the retina are tightly coupled. Also developmentally, the pigment epithelium is a part of the retina, as it is derived from the same optic vesicle as the remaining retina. However, the pigment epithelium is in many respects quite different from the rest of the retina, and it is of such importance that it is dealt with in a separate chapter in this book. The rest of the retina, the so-called the neuroretina, is the subject of this presentation.

All vertebrate eyes have a retina, but its structure can vary immensely, and there can be significant differences, even between animal species as closely related as different primates. The emphasis is here on the human retina. It must be understood that knowledge gained in mice, or cats, or even monkeys cannot be assumed to show anything definitive about the human retina. Nevertheless, much of what is known about the human retina has first and most extensively been studied in experimental animals, and then shown to be valid in the human retina also.

In postmortem eyes, the retina is gray with a yellow spot in the center, the so-called macula lutea. In such eyes, the retina is often detached from the pigment epithelium, lying like a grayish cobweb around the vitreous, and hence the term retina (rete = net). Only after the invention of the ophthalmoscope by von

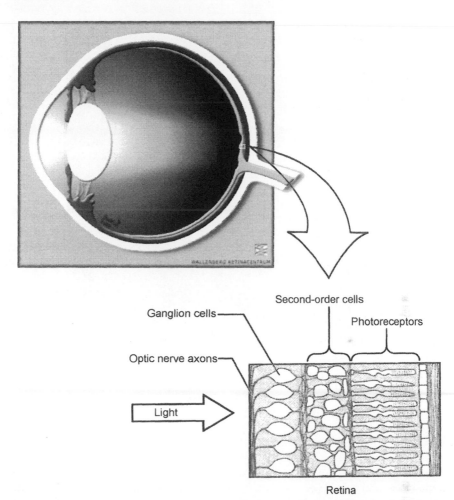

Figure 1. The retina, consisting of photoreceptors, second-order cells, and ganglion cells line the inside of the back of the eye. The axons of the third-order ganglion cells run along the surface of the retina forming the optic nerve where they exit the eye. Light enters the eye via the transparent cornea and is focused on the retina by the lens. [After Dowling (1998).]

Helmholtz in 1851, it became possible to view the posterior part of the living human retina (Figure 2), with its orange-red color, the optic nerve head, and the major blood vessels clearly visible. In the living eye, the macula lutea does not appear yellow, but rather more brownish and darker than the surrounding retina. Although not apparent *in vivo*, the macula does contain special yellow carotenoid pigments (lutein and zeaxanthin),

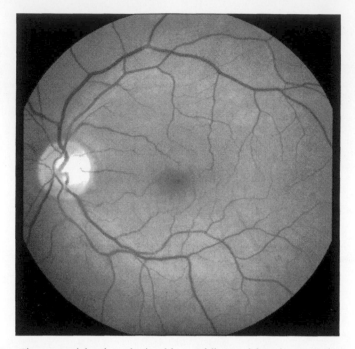

Figure 2. The normal fundus of a healthy middle aged female. Note the optic disc from which the central retinal arteries and veins emerge. Note also the slightly darker macula in the center.

predominantly in the cone axons of the Henle fiber layer, but also further inward in the retina. These pigments are thought to act as a short wavelength light filter, and may also have antioxidant and other chemical protective actions. The central 1 mm of the human retina is thinned, forming the fovea centralis (Figure 3) with a central pit only a few hundred microns in diameter, the foveola. Among mammals, only primates have a fovea, but deep, pitted foveas can be found in certain other animal species, such as predatory birds, which may have visual acuities upwards of five times higher than humans. Some birds even have two foveas.

All vertebrate retinas are constructed according to the same general scheme (Figures 1, 3, and 4). In invertebrates, the blueprint for the retina is quite different, and will not be dealt with here. A nice review of eyes and vision in the animal kingdom has recently appeared (Land and Nilsson, 2002). The major light-sensitive cells in the vertebrate retina, the photoreceptors, do not communicate directly with the brain. Instead, significant signal processing occurs within the retina. Apart from the photoreceptors, five major classes of neurons are found in the retina: horizontal cells, bipolar

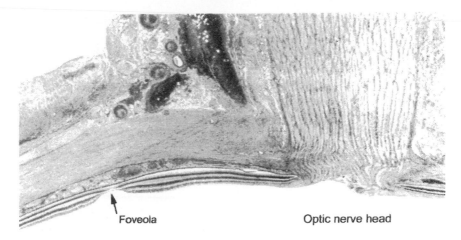

Foveola   Optic nerve head

Figure 3. Section from the posterior part of the human eye, showing the macula region (with its foveola) and the optic nerve head. Adapted from Eugene Wolff's *Anatomy of the Eye and Orbit* (ed. R. J. Last), 6th ed. 1968, p. 141.

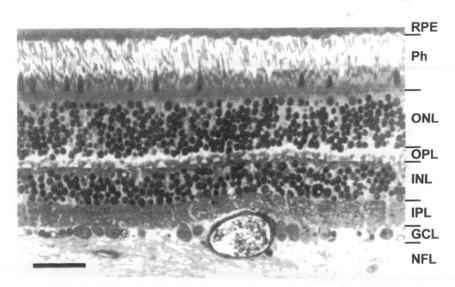

RPE
Ph

ONL
OPL
INL

IPL
GCL
NFL

Figure 4. Phase contrast micrograph of the human retina about 1.25 mm from the center of the fovea, demonstrating the different layers. RPE, retinal pigment epithelium; Ph, photoreceptors (rods and cones); ONL, outer nuclear layer (the cell bodies of the rods and cones); OPL, the outer plexiform layer; INL, the inner nuclear layer; IPL, the inner plexiform layer; G, the ganglion cell layer; NF, the nerve fiber layer. Scale bar 100 μm. From Adler's Physiology of the Eye, 8th ed., p. 463.

cells, amacrine cells, interplexiform cells, and ganglion cells. Special glial cells, Müller cells, separate the neurons.

Histologically, the retina is an extremely well-organized tissue in which 10 layers are classically recognized (Figure 4). The inner limiting membrane (ILM) is a basement membrane, secreted by the Müller cell endfeet processes. It forms the boundary to the vitreous. The retinal nerve fiber layer is just outside it and contains the ganglion cell axons *en route* to the optic nerve head (NF in Figure 4). The ganglion cell layer (GCL in Figure 4) is next, and contains the cell bodies of the ganglion cells. The inner plexiform layer (IPL in Figure 4) consists of the synapses between amacrine cells, ganglion cells, interplexiform cells, and bipolar cells. The inner nuclear layer (INL in Figure 4) contains the nuclei of amacrine cells, bipolar cells, interplexiform cells, and glial Müller cells. The outer plexiform layer (OPL in Figure 4) is composed of the synapses between the photoreceptor cells, horizontal cells, bipolar cells, and interplexiform cells. Photoreceptor cell bodies form the outer nuclear layer (ONL in Figure 4). The external limiting membrane is not a real membrane, but an optical phenomenon caused by the junctions between the photoreceptor inner segments and the Müller cells. The photoreceptor layer (Ph in Figure 4) consists of inner and outer segments of the photoreceptors. The outermost layer of the retina is the retinal pigment epithelium (RPE in Figure 4).

Most of our knowledge about the morphology of the different retinal cell types and their synaptic connections has been gained from studies with metal impregnations like the Golgi method. They give excellent information about the morphology of individual neurons, but it was understood already in the first studies more than 100 years ago that they do not give much information on the prevalence of different cell types. Only in the last decade has it been possible to get good estimates of the relative prevalences of the many different neuron types in the inner part of the retina (MacNeil et al., 1999; Marc and Jones, 2002).

The retina receives its photosensory input from its light-sensitive cells, the photoreceptors. The retinal output is produced by the ganglion cells, which send their axons to the brain via the optic nerve. The intercalated neurons (bipolar cells, horizontal cells, amacrine cells, and interplexiform cells) mediate and modulate the flow of information between the photoreceptors and the ganglion cells.

There are two major types of photoreceptor cells: rod cells and cone cells, and the two are connected to the brain by different pathways. The cone pathway is a three-neuron link. The rod pathway does not have ganglion cells of its own. Instead, the rod cells connect to the ganglion cells via specialized amacrine cells, making this pathway a four-neuron link (Figure 5). The cells forming the pathway will be discussed in more detail in connection with the inner nuclear and plexiform layers. Phylogenetically, the cone pathway is currently believed to be the oldest, with the rod pathway having been secondarily patched in.

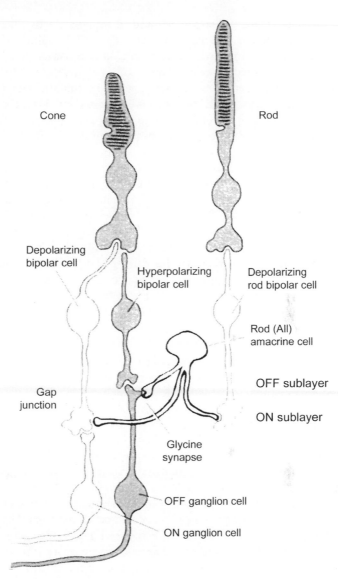

Figure 5. The pathway from rods to ganglion cells is complex, as shown, with the special AII glycinergic amacrine cell intercalated in the pathway. Cells that give depolarizing responses to light are shown as white, while those that give "off" responses and hyperpolarize are shaded. Rods connect to specialized rod bipolar cells. When depolarized, such a cell excites an AII rod amacrine cell. Through gap junctions this produces excitation of the depolarizing cone bipolar cell shown on the left, which influences the "ON" ganglion cell. At the same time the depolarization of the rod (AII) amacrine cells causes it to release inhibitory transmitter (glycine) onto the "OFF"-center ganglion cell. The glycinergic AII amacrine cell thus relays rod signals to both "ON" and "OFF" ganglion cells. Adapted from Daw et al. (1990).

In the human retina, there are around 1 million nerve fibers in the optic nerve, and around 100 million rod cells and 5 million cone cells (Curcio et al., 1990b). From these numbers, it is obvious that both rod and cone pathways involve convergence of cell contacts. The convergence is dependent on retinal eccentricity, with the highest degree of convergence in the periphery. In the human foveal center, the wiring is actually divergent, since each cone is connected to three ganglion cells (Sjöstrand et al., 1999). This divergence at the center allows the visual system to achieve the maximum spatial resolution allowed by the cone mosaic itself (i.e., no loss of central resolution is incurred by the wiring, Sjöstrand et al., 1999). In mammalian species without a fovea, the wiring is also convergent at the center, and the spatial resolution of the retina is less than what could theoretically be achieved by the cone mosaic (Wässle and Boycott, 1991).

The light-induced electrical responses from individual retinal neurons have been studied with microelectrodes during the last five decades. It is apparent that only the ganglion cells produce classical action potentials. The other cell types produce graded potentials that spread electrotonically. Both rod cells and cone cells hyperpolarize in response to light; cone cells have the fastest response kinetics. Also the horizontal cells hyperpolarize in response to light. For the bipolar cells, there is no universal response to light stimulation; some hyperpolarize, others depolarize (Wässle and Boycott, 1991). This dichotomy in the response of the bipolar cells is transferred to the ganglion cells, some of which respond to light with an increased frequency of action potentials, the ON center ganglion cells, while others respond with a decreased firing rate, the OFF center ganglion cells (Figure 6).

A key concept in visual physiology is the receptive field of a ganglion cell or any other neuron in the visual system. The receptive field for a given cell is defined as the area in the visual field, or the region of the retinal surface, where light stimulation elicits a change in the electrical activity of the cell. The sizes and shapes of receptive fields vary for retinal neurons. The smallest receptive fields are found in the foveal center, where midget ganglion cells have receptive fields of a few microns, whereas large parasol ganglion cells in the periphery can have receptive fields of 1 mm or more (Sjöstrand et al., 1999).

For most retinal ganglion cells, the response to light stimulation is not uniform throughout their receptive fields. ON center ganglion cells respond with increased activity to illumination in the center of their receptive field, whereas stimulation in the periphery of their receptive field results in decreased activity. Uniform stimulation throughout the receptive field results in largely unchanged activity. For OFF center ganglion cells, the response pattern is the opposite (i.e., light stimulation in the center results

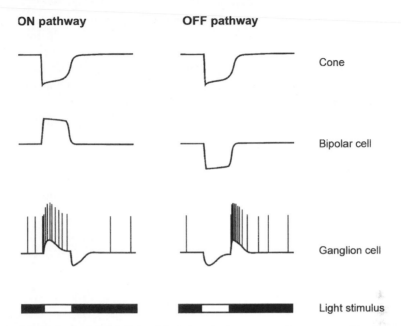

ON pathway **OFF pathway**

Cone

Bipolar cell

Ganglion cell

Light stimulus

Figure 6. Schematic drawing of the electrical responses from cones, bipolar cells, and ganglion cells to a light stimulus. The cones hyperpolarize in response to light, but two different responses are apparent in the bipolar cells. This dichotomy at the synapse between photoreceptors and bipolar cells is reflected in the existence of two functional classes of ganglion cells. ON-ganglion cells respond to light with an increased firing pattern. OFF-ganglion cells respond with a decreased firing pattern during the light stimulus.

in decreased activity, Figure 7). These responses are caused by lateral inhibition mediated by horizontal cells, whereby the periphery of the receptive field inhibits the center and vice versa. This so-called center-surround inhibition maximizes retinal contrast sensitivity throughout a wide operating range of background illumination intensities.

In the following paragraphs, the individual cells and synaptic layers of the retina will be described in more detail.

II. PHOTORECEPTOR CELLS

The photoreceptor cells are found in the outer part of the retina, external to the OPL. Based on morphological and physiological criteria, two types can be discerned: rod cells and cone cells. Both have a light-sensitive outer segment, a modified cilium loaded with stacks of membranous discs that

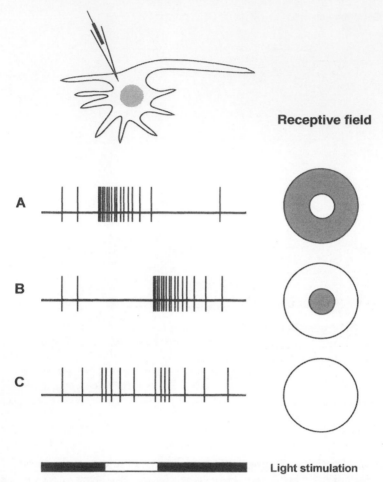

Receptive field

Light stimulation

Figure 7. Diagrammatic representation of the center-surround inhibition in the receptive field of large ganglion cells. Panels (A)–(C) represent intracellular recordings from an ON-center retinal ganglion cell. To the right of each panel is shown an illustration of the receptive field for the ganglion cell, and the localization of the white light stimulus within it. In panel (A), the center of the receptive field is stimulated, and the ganglion cell responds with an increased firing rate. In panel (B), the periphery of the receptive field is stimulated, and the cell responds with a decreased firing rate. In panel (C), the entire receptive field is stimulated, and only little change in the firing pattern results.

incorporate the light-sensitive pigments, denoted photopigments (Figure 8). The inner segments of the photoreceptors contain mitochondriae and the machinery for protein synthesis. The outer and inner segments of the

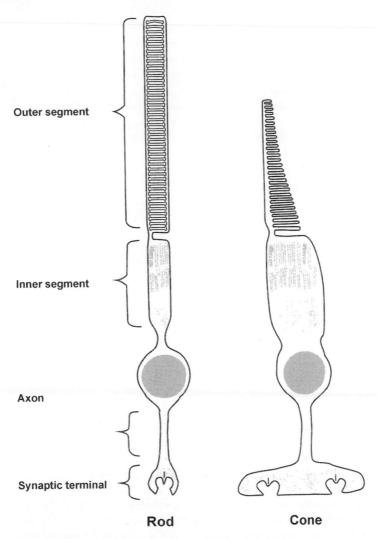

Outer segment

Inner segment

Axon

Synaptic terminal

Rod   **Cone**

Figure 8. Schematic drawing of rod and cone photoreceptors. Note the difference in the disc morphologies and in the synaptic terminals of rods and cones. Foveal cones have longer slender outer and inner segments than shown here, and their axons are more elongated, forming the so-called Henle fibers visible as oblique structures in the OPL of Figure 4.

photoreceptors constitute the photoreceptor layer seen in histological sections of the retina (Ph in Figure 4). The photoreceptor nuclei form the ONL. The photoreceptor inner fiber, or axon, connects the photoreceptor cell body with the synaptic terminals that form parts of the OPL. The major morphological distinctions between rod cells and cone cells are the differences in their synaptic terminals and their discs. Cone synaptic terminals, cone pedicles, are wider than the spherules of the rod cells (Ahnelt and Kolb, 2000). Cone cell discs are continuous with the outer segment cell membrane, whereas rod cell discs reside completely within the outer segment cytoplasm, without continuity with the cell membrane. In both rod cells and cone cells, the photopigments float relatively freely in the central flat part of the disc membrane, but they are somehow excluded from the edges, or rims, of the discs. The disc rims, as well as the confinement of photopigments to the central part of the discs, are thought to be maintained by structural disc rim associated proteins, such as the peripherin–ROM1 complex (Sears et al., 2000).

Functionally, rod cells are specialized to yield high sensitivity. Cone cells provide high temporal and spatial resolution, but are less sensitive than rod cells. In nonmammalian retinas, there are several types of both rod cells and cone cells. In the mammalian retina, there is only one type of rod cells, and two major types of cone cells: short-wavelength sensitive S-cone cells (or blue sensitive cone cells), and the medium- and long-wavelength sensitive M–L-cone cells. These cone cells can be discerned structurally and immunohistochemically. Old-world monkeys enjoy trichromatic vision due to the presence of two types of M–L-cone cells: medium wavelength, M-cone cells (green sensitive cone cells), and long wavelength L-cone cells (red sensitive cone cells). The visual pigments in M- and L-cone cells are closely related with only a few amino acids difference in their structure. The genes for these pigments are localized in tandem array on the X-chromosome. Apart from the pigments, primate M- and L-cone cells are identical and cannot be discerned ultrastructurally, or immunohistochemically (Ahnelt and Kolb, 2000).

A. Distribution of Photoreceptors

The photoreceptor mosaic varies tremendously between different species because it reflects the environmental demands facing the animal. Nocturnal animals have rod-dominated retinas, whereas diurnal animals have cone-dominated retinas (Ahnelt and Kolb, 2000). Like most mammalian retinas, the human retina is rod dominated. However, unlike other mammals, humans and other higher primates have a cone dominated fovea centralis, from the center of which rod cells are excluded. In the human fovea, the diameter of the central rod-free zone is 0.35 mm, and the peak cone density in the central fovea is 199,000 cone cells/mm^2; approximately the same density as that of pixels in CCD chips used in current state-of-the-art digital photography (fall 2003). This peak density is found at the bottom of the foveola, an area less than 100 μm in diameter. Outside the foveola, the cone density declines steeply and has fallen by one order of magnitude at an eccentricity of 1 mm (Figure 9). Among the cone cells, L- and M-cone cells are by far the more numerous. S-cone cells (blue sensitive cone cells) represent only 5–10% of the total cone population, and they are excluded from the central fovea (Calkins, 2001). Outside the fovea, rod cells outnumber cone cells (Figure 9). The peak density of rod cells is reached at an eccentricity of 3–6 mm, the so-called rod ring (Curcio et al., 1990b).

B. Photoreceptor Electrophysiology

The primary function of the photoreceptors is to convert light into electrical signals. This process is called phototransduction and is a cascade of events, initiated by the absorption of photons by the photopigments in the photoreceptor outer segment disc membranes. The phototransduction cascade ends with a hyperpolarization of the photoreceptor cell membrane potential. This membrane potential is determined by a light-insensitive K^+ conductance and a light-sensitive unspecific cation conductance, which is carried by cGMP-gated cation channels. In the dark, there is a high cytoplasmic cGMP concentration in the outer segment that keeps the cGMP-gated cation channels open, allowing influx of a current carried by Na^+ and Ca^{2+} into the

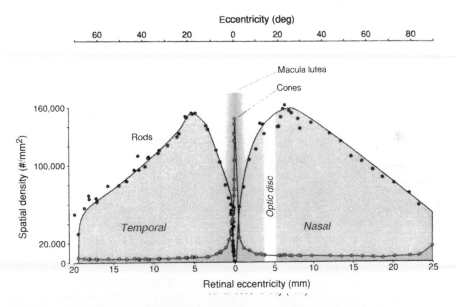

Figure 9. The photoreceptor cell distribution is not symmetrical across the retina, as exemplified by the accumulation of cone cells in the fovea (except for S-cones which are absent from its very center) and that of rods in the rod ring. After Østerberg (1935) as modified in Rodieck (1998).

outer segment. In rod cells, Ca^{2+} carries 10–15% of this so-called dark-current; in cone cells, this proportion is higher. The influx of sodium through the cGMP-gated cation channels is balanced by sodium efflux through Na^+/K^+ pumps, located in the inner segment. The combined action of the Na^+ influx through the outer segment and the Na^+ extrusion via the inner segment Na^+/K^+ pumps creates a current loop, carried by Na^+, between the inner and outer segment, the so-called dark current (Figure 10). The phototransduction cascade interrupts this current loop by closing the outer segment cGMP-gated cation channels. In the outer segment, there is a much shorter current loop carried by Ca^{2+}, which flows into the outer segment through the cGMP-gated cation channels, and is extruded via $4Na^+/Ca^{2+}/K^+$ exchangers, localized in pairs in close proximity to the cGMP-gated cation channels (Burns and Baylor, 2001).

The electrical response to light has been measured in single photoreceptors sucked into small glass pipettes (Baylor, 1987). Figure 11 shows the changes in the dark current in single mammalian photoreceptors. Rod cells are more sensitive than cone cells, but have slower photoresponses (Panels A and B in Figure 11).

C. Photopigments

The vertebrate photopigments consist of a membrane protein (the opsin) and the chromophore, which in mammals is 11-cis-retinal. The opsin consists of a single amino acid chain and contains seven membrane-spanning helices. The chromophore is buried in a pocket deep within the opsin, and it is covalently bound by means of a protonated Schiff-base linkage to the ε-amino group of a lysine residue in helix VII. The chromophore is oriented approximately parallel to the disc membrane and interacts noncovalently with a number of amino acid residues, including a glycine residue on helix III. The absorption spectrum of free 11-cis-retinal has a maximum at 380 nm (i.e., in the ultraviolet range). Protonation of the Schiff-base linkage and the interactions between the chromophore and the opsin shifts the absorption maximum of the visual pigments toward longer wavelengths and endows each photopigment with its unique absorption spectrum (Pepe, 2001).

Based on their amino acid composition, the vertebrate retinal photopigments can be divided into five evolutionarily distinct clusters (Yokoyama, 2000). All five photopigment clusters are found in retinas from species of fish, amphibians, reptiles, and birds. Mammals, on the other hand, probably descend from nocturnal ancestors, and only three of the photopigment clusters are preserved in mammalian retinas (Ahnelt and Kolb, 2000). Rhodopsin in mammalian rod cells belongs to the rhodopsin 1 (RH1) cluster. Most mammals are dichromats with only two cone pigments. One type of mammalian cone cell, the short wavelength (blue) sensitive S-cone cell contains a

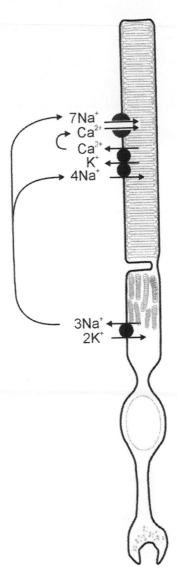

Figure 10. The photoreceptor dark current runs between the outer and inner segments of vertebrate photoreceptors. It is primarily carried by Na$^+$.

short wavelength sensitive-1 (SWS1) cluster pigment. The middle or long wavelength (green or red) sensitive cone cells contain M–LWS pigments (M–LWS = middle- and long-wavelength sensitive cluster). Some primate ancestors duplicated their M–LWS pigment, evolving the long wavelength

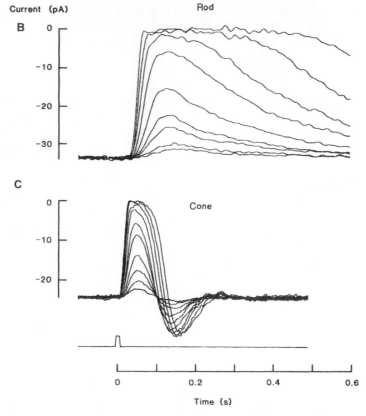

Figure 11. Recording of photocurrent families from single rod and cone cells in a monkey (*Macaca fascicularis*). Panel (A) shows the recording pipette into which the outer segments are sucked. An electrical seal forms between the outer segment

sensitive L-cone cells and middle wavelength sensitive M-cone cells of higher primate retinas (Ebrey and Koutalos, 2001). As a general rule, a single type of photoreceptor cell contains only one photopigment. However, there are reports of photoreceptors that contain a second pigment (Ebrey et al., 2001). Although rhodopsins are generally found in rod cells, and SWS and LWS pigments are primarily found in cone cells; it appears that it is not the visual pigment that defines whether a photoreceptor cell is a rod or a cone (Ma et al., 2001).

The photopigments, particularly rhodopsin, have a remarkable thermal stability. The noise derived from random thermal activation of rhodopsin can therefore be kept to a minimum despite the packing of many rhodopsin molecules into a single rod cell (Burns et al., 2001).

It is a characteristic property of the vertebrate photopigments that they bleach upon illumination. Hence the color of these pigments can only be appreciated in a brief moment after light onset. It took until 1878 before their existence within the neuroretina was discovered (Marmor and Martin, 1978).

D. Activation of the Phototransduction Cascade

The phototransduction cascade is an example of a G-protein signaling pathway (Arshavsky et al., 2002). G-proteins are heterotrimeric guanine nucleotide-binding proteins which become activated when GTP is bound to them. The G-protein in phototransduction is transducin, as described below (Figure 12).

Phototransduction begins with the absorption of a photon by the chromophore of a photopigment. Within less than a picosecond, this results in the isomerization of 11-cis-retinal to all-trans retinal, and a destabilization of the photopigment (Burns et al., 2001; Pepe, 2001). A series of conformational changes follows, and after a few milliseconds the activated, bleached, photopigment is formed, which in rod cells is denoted metarhodopsin II or R*. The photoactivated pigment, R*, activates the G protein, transducin, which is a disc membrane associated protein that consists of three subunits:

---

membrane and the pipette that allows changes in the dark current to be measured in responses to dim flashes of light that are projected through the pipette. Panel (B) shows a family of photocurrents from a single rod. Flash strengths were increased by a factor of two until saturation. The flashes were expected to cause between 2.9 and 860 photoisomerizations in the rod cell. Panel (C) shows similar photocurrents from cone cells. The flashes used in panel (C) were expected to cause between 190 and 36.000 photoisomerizations in the cone cell. Modified from Figures 4 and 11 in Baylor (1987).

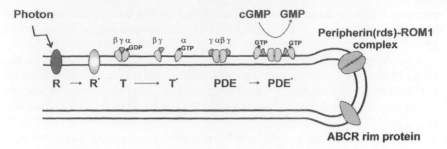

Figure 12. Diagram of the phototransduction cascade in a rod disc. R is rhodopsin, R* is activated rhodopsin (metarhodopsin II). T is transducin; T* is activated transducin. PDE is phosphodiesterase; PDE* is activated phosphodiesterase. Also shown are the rim proteins, ABCR, and the peripherin-ROM1 protein.

T_α, T_β, and T_γ with GDP bound to its $T\alpha$ subunit. The activation involves binding of the GDP–$T_\alpha T_\beta T_\gamma$ complex to R*, a replacement of GDP with GTP on the T_α subunit, and a subsequent dissociation of the now activated GTP–T_α subunit, and the $T_\beta T_\gamma$-dimer, from the still active R*. For simplicity, the activated transducin, GTP–T_α, will be denoted T* (Burns et al., 2001).

The next step in the cascade is the activation of the phosphodiesterase enzyme (PDE), which is a tetrameric disc membrane associated protein consisting of a catalytic $PDE_{\alpha\beta}$ dimer, and two inhibitory PDE_γ subunits. Activated transducin, T*, activates PDE by binding to the inhibitory PDE_γ subunits and thereby displacing them from the catalytic $PDE_{\alpha\beta}$ dimer that remains active as long as T* is bound to the PDE_γ subunits. Each dimer has two catalytic sites, and full activation requires binding of two T*, one for each inhibitory PDEγ subunit. The active $PDE_{\alpha\beta}$ dimer, denoted PDE*, hydrolyzes the outer segment cytoplasmic second messenger, cGMP, to $5'GMP$ at a very high rate, and hence decreases the outer segment cGMP concentration (Leskov et al., 2000).

The decease in outer segment cytoplasmic cGMP concentration closes the cGMP-gated cation channels in the outer segment cell membrane, thereby decreasing the unspecific cation conductance and hyperpolarizing the membrane potential across this membrane (Burns et al., 2001; Pepe, 2001). The cGMP-gated cation channels are voltage independent, unspecific cation channels with a very low unit conductance around 0.1 pS (Burns et al., 2001). Each channel incorporates four binding sites for cGMP, and three of these binding sites have to be occupied before significant channel opening occurs. The cGMP gated cation channels in the outer segment cell membrane ensure a steep, cubic dependence of the cell membrane cation conductance on the free cytoplasmic cGMP concentration (Karpen et al., 1988).

The remarkable sensitivity of rod cells is derived from two amplification steps in the phototransduction cascade. The first amplification step is the activation of transducin by R*, and the second is the hydrolysis of cGMP by the active PDE$\alpha\beta$ enzyme. The activation of PDE by T* does not amplify the signal (Arshavsky et al., 2002). In rod cells, the maximum rate of transducin activation by one R* is 120 T* per second, and the rate limiting factor is probably the diffusional encounter of R* and transducin at the disc membrane. Each activated PDE dimer hydrolyzes cGMP at a rate of 2200 molecules per second (Leskov et al., 2000).

E. Inactivation

Activation and amplification in the transduction cascade are necessary for sensing light. Timely inactivation, on the other hand, is essential for preservation of the sensitivity and time resolution of the visual system. Inactivation of the outer segment photocurrents requires restoration of the cytoplasmic cGMP concentration in the outer segment, and it involves inactivation of R*, T*, and PDE*, as well as cGMP synthesis.

R* is inactivated by phosphorylation at multiple sites near its C terminus, followed by the binding of the soluble protein, arrestin. R* phosphorylation is catalyzed by pigment kinases, collectively denoted G-protein-coupled receptor kinases, GRK's. Of these kinases, GRK7 is only expressed in cone cells, whereas GRK1 is expressed in both cone cells and rod cells in at least the primate retina. Sometimes, GRK1 is denoted rhodopsin kinase (RK) (Maeda et al., 2003).

Activated transducin is inactivated by the intrinsic GTPase activity of T_α that hydrolyzes bound GTP to GDP. The resulting Tα-GDP can no longer bind the inhibitory PDE$_\gamma$ subunit, which is released to shut off the activity of catalytic active sites on the active PDE$_{\alpha\beta}$ dimer. The intrinsic GTPase activity of T_α is stimulated by the proteins RGS9 and by the PDE$_\gamma$ subunit. Experiments with RGS9 gene knock-out mice have shown that this protein is essential for normal inactivation in both rod cells and cone cells (Arshavsky et al., 2002).

The outer segment cGMP level is restored by the action of the guanylate cyclases, GCs. GC1 is present in both in rod cells and cone cells, whereas CG2 is probably exclusive to rod cells (Koch et al., 2002). The activity of the guanylate cyclases is controlled by the Ca^{2+} binding guanylate cyclase activating proteins (GCAPs). In primate photoreceptors, there are currently three GCAPs known. When the cytoplasmic Ca^{2+} concentration in the outer segment is high, the GCAPs bind to and inhibit the guanylate cyclase (Koch et al., 2002).

F. Photoreceptor Adaptation

Fully dark-adapted rod cells are very sensitive and will produce highly reproducible electrical responses to single photons with a probability, or quantum-efficiency, of around 50% (Baylor et al., 1984). Despite this high sensitivity, the photoreceptor mosaic will remain responsive to changes in light energy through a dynamic range of 11 orders of magnitude (Fain et al., 2001). This remarkably wide operating range of the retina is accomplished in part by the duplicity of the retina, with rod cells responsible for the sensitivity at very low-energy levels and cone cells responsible for the preservation of sensitivity at higher energy levels. However, the photoreceptor cells themselves also adapt to increasing levels of background illumination by a reduction in sensitivity and a quickening of the photoresponses. Throughout the entire operating range of the photoreceptors, their sensitivity decreases in direct proportion to the level of the background illumination. This property, denoted Weber's law, is universal among biologic sensors (Pugh et al., 1999).

Adaptation in rod cells occurs at very low levels of illumination. In mammalian rod cells, sensitivity is reduced by one-half by light intensities that cause only 30–50 photoisomerizations per second per rod. Rod cells are saturated at light intensities that bleach only 1% of their rhodopsin. Cone cells can adapt to much higher levels of background illumination than rod cells. They remain responsive at intensities that cause 1 million photoisomerizations per second per cone, and bleach all but a small fraction of the cone pigments (Schnapf et al., 1990; Burkhardt, 1994).

Since adaptation affects the entire outer segment and is active at very low-light intensities, a cytoplasmic second messenger must be involved. This second messenger is now known to be Ca^{2+} in both rod cells and cone cells (Pugh et al., 1999; Fain et al., 2001). In constant light, activation of the phototransduction cascade closes a significant fraction of the cGMP-gated cation channels. Consequently, the rate of Ca^{2+} influx through these channels decreases, as does the outer segment cytoplasmic Ca^{2+} concentration. When a dark-adapted rod is subjected to saturating light, the outer segment Ca^{2+} concentration decreases 10- to 20-fold. In cone cells, this decrease is more pronounced, almost 100-fold (Pugh et al., 1999; Fain et al., 2001). The lowered Ca^{2+} concentration activates guanylate cyclase through the Ca^{2+}-dependence of the GCAPs (Fain et al., 2001; Koch et al., 2002). The combined effects of steady activation of the phototransduction cascade and increased activity of the guanylate cyclase are an increased rate of cGMP turnover (Burns and Baylor, 2001; Fain et al., 2001). This results in decreased sensitivity and accelerated photoresponses (Fain et al., 2001).

G. Regeneration of the Visual Pigments

Once the activated visual pigments have become inactivated by phosphorylation and arrestin binding, all-trans-retinal is released from the opsin. It is not known to what extent the chromophore is released into the rod disc lumen, but whatever all-trans-retinal that appears in the disc lumen is apparently flipped back into the outer segment cytoplasm by the rod disc rim protein, ABCR (Weng et al., 1999). Once in the cytoplasm, the all-trans-retinal is reduced to all-trans-retinol by a NADPH dependent process. Neither rod cells nor cone cells can efficiently reisomerize all-trans-retinol back to the 11-cis-isomer. Instead, all-trans-retinol has to be transported to other cell types that possess the enzymatic machinery necessary for reisomerization (Figure 13). Finally, the 11-cis-chromophore must be transported back to the photoreceptors and incorporated into the visual pigments. This process is denoted the visual cycle, or the retinoid cycle (McBee et al., 2001). Recently, it has become apparent that the visual cycle might be different in rod cells and cone cells (Mata et al., 2002). However, for both rod cells and cone cells, the visual cycle involves the transport of retinoids through the subretinal space (i.e., the extracellular space between the photoreceptors and the RPE). This transport might

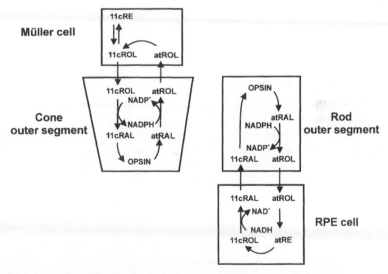

Figure 13. The rod and cone visual cycle. 11cRE = 11-cis-retinyl ester; atRE = all-trans retinyl ester; 11cROL = 11-cis retinol; atROL = all-trans retinol; 11cRAL = 11-cis retinal; atRAL = all-trans retinal; NAD = nicotinamide adenine dinucleotide; and NADP = nicotinamide adenine dinucleotide phosphate.

involve the interphotoreceptor matrix retinoid binding protein (IRBP), which is the most abundant extracellular protein in the subretinal space, and which has retinoid binding properties. Possibly, this is valid only for cone cells, since the rod visual cycle seems to proceed normally in mice that lack the IRBP gene (McBee et al., 2001).

The visual cycle is well studied in rod cells, and in this case the reisomerization takes place in the RPE, where many of the involved enzymes have been characterized. Within the RPE, retinoids are stored as retinyl esters, which probably form the substrate for the isomerization enzyme. The energy necessary for reisomerization is thought to be derived from the hydrolysis of the ester bond. The yet elusive critical enzyme is therefore called an isomerohydrolase (Saari, 2000). Rod cells need 11-cis-retinal to regenerate their bleached rhodopsin, and the oxidation from 11-cis-retinol to 11-cis-retinal takes place within the RPE with NAD^+ as cofactor (McBee et al., 2001).

In cone cells, the visual pigment regeneration is less well understood, but there are at least three important differences from the rod system (Mata et al., 2002). First of all, contrary to rod cells, cone cells are not dependent on the RPE for the regeneration of their bleached pigments. Secondly, contrary to rod cells, cone cells can oxidize 11-cis-retinol to 11-cis-retinal by an $NADP^+$ dependent process. They are therefore not dependent on exogenous 11-cis-retinal, but can regenerate their pigments from 11-cis-retinol. Finally, in order to maintain sensitivity during bright daylight, the cone system needs the 11-cis-chromophore to be supplied at a high rate; each individual cone might need 1 million molecules of 11-cis-chromophore per second. The output of 11-cis-retinal from the RPE is limited and insufficient to meet the demand of the cone system during bright daylight (Mata et al., 2002).

Recently, Mata et al. (2002) proposed a scheme for a cone-specific visual cycle. This scheme provides the cone cells with a private source of 11-cis chromophore in the form of 11-cis-retinol, which the rod cells cannot utilize. All-trans-retinol liberated from cone cells is transported to Müller cells, where it is reisomerized to 11-cis-retinol. It is then transported back to the cone cells, where it is oxidized, and incorporated into the cone pigments (see Figure 13). Mata's cone-specific visual cycle is attractive because it provides the cone system with a private supply of 11-cis-chromophore that will meet its needs even during bright daylight. With the scheme proposed by Mata, the limited capacity of the RPE for output of 11-cis-retinal makes sense, because it will limit the amount of energy wasted on futile reisomerization of the rod chromophore at light intensities, where rods do not contribute to vision, The key enzymes necessary in Mata's cone-specific visual cycle are yet uncharacterized, but it has been shown that the cone dominated chick retina can produce 11-cis-retinol at (almost) sufficient rates to regenerate cone pigments during daylight (Mata et al., 2002).

Table 1. Photoreceptor Gene Defects and Retinal Disease

Protein	Disease
Rhodopsin	Dominant RP, recessive RP, dominant CSNB
SWS opsin	Dominant tritanopia
MWS opsin	Deuteranopia
LWS opsin	Protanopia
Transducin	Dominant CSNB, Nougaret type
Cone transducin	Recessive achromatopsia
Phosphodiesterase	Dominant CSNB, recessive RP
cGMP gated cation channel	Recessive RP
Cone cGMP gated cation channel	Recessive achromatopsia
Rhodopsin kinase	Recessive CSNB, Oguchi type
Arrestin	CSNB Oguchi type, recessive RP
Guanylate cyclase	Leber's congenital amaurosis, dominant cone–rod dystrophy
Guanylate cyclase activating protein (GCAP)	Dominant cone dystrophy, dominant cone-rod dystrophy
ABCR rim protein	Recessive Stargardt's disease. Recessive cone–rod dystrophy
IRBP	Retinal dystrophy in mice, no human equivalent
Peripherin (rds protein)	Dominant RP, dominant macular degeneration (several types), digenic RP with ROM1
ROM1	Dominant RP, digenic RP with peripherin

Note: The table lumps proteins together and does not distinguish between defects in different subunits. The emphasis is on proteins involved in retinoid metabolism, but the table includes the outer segment rim proteins, peripherin and ROM1, necessary for disc morphogenesis and for the stabilization and compaction of outer segment disks, or for the maintenance of the curvature of the rim. References can be found on the RETNET Web site (Daiger et al., 2003).

Abbreviations: RP, retinitis pigmentosa; CSNB, congenital stationary night blindness; GCAP, guanylate cyclase activating protein; IRBP, interphotoreceptor matrix retinoid binding protein; ROM1, retinal outer segment membrane protein 1.

H. The Visual Cycle and Retinal Disease

The activation and inactivation of phototransduction as well as the regeneration of the visual pigments involve a number of specific proteins. In the last decade, an increasing number of retinal diseases have been shown to be caused by genetic defects in these proteins. Current information has been summarized in Table 1, and more is likely to come in the future. The RetNet internet site maintains updated tables (Daiger et al., 2003).

III. THE OUTER PLEXIFORM LAYER

This layer (OPL in Figure 4) is predominantly formed by the synapses between the photoreceptor, bipolar, and horizontal cells. In the fovea and parafovea, the long Henle fibers make the OPL thick (about 50 μm), but otherwise it is only a few microns thick, thinning toward the periphery. The most external part of the OPL is predominantly formed by the inner fibers axons of the photoreceptor cells. Their synaptic terminals occupy the middle of the layer, and dendrites of bipolar and horizontal cells together with Müller cell processes form the innermost part. The photoreceptor cell synapses involve at least two horizontal cells and one bipolar cell, and are called triads (Figure 14). Rod spherules usually feature one deeply invaginated triad; the cone pedicles contain several, up to 25.

There are two types of synapses between cone cells and bipolar cells. One belongs to the ON pathway and depolarizes the postsynaptic membrane when the cone cell hyperpolarizes in response to light. Morphologically, this "ON synapse" is a triad with an invaginated bipolar cell dendrite. The "OFF synapse" is a flat contact between a bipolar dendrite and the cone pedicle. In both cases, the transmitter is glutamate, and the dichotomy in their electrical responses is caused by different glutamate receptors in the postsynaptic membrane (Wässle and Boycott, 1991).

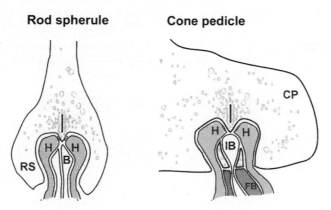

Rod spherule   **Cone pedicle**

Figure 14. The triads in the OPL constitute the synapses between photoreceptor, bipolar, and horizontal cells. In rod cells, the synaptic terminal is called a spherule (RS). The rod cell synapse has two deeply invaginated rod bipolar cell dendrites (B) at its center, and two horizontal cell processes (H) in its periphery. The cone pedicle has an invaginated bipolar dendrite (IB) at its center, flanked by two horizontal cell dendrites. Also shown is the flat contact between the cone pedicle and flat bipolar cells (FB). In both rod and cone cells, a synaptic ribbon is apparent opposite to the invaginated bipolar cell dendrite.

IV. HORIZONTAL CELLS

Horizontal cells occupy the outermost few rows of cells in the INL. They have long processes that arborize exclusively in the OPL. Their dendritic fields increase with eccentricity, but the shapes of the fields remain the same. The mammalian retina has two types of horizontal cells (Figure 15). The HII–H2 subtype (in nonprimates called the A type) contacts exclusively cone cells, predominantly S-cone cells (Ahnelt and Kolb, 1994). The other subtype, HI–H1, (in nonprimates called the B type) features a dendritic tree as well as an axon with an arborizing terminal. The dendrites contact cone cells, mostly M- and L-cone cells, and the axon contact rod cells.

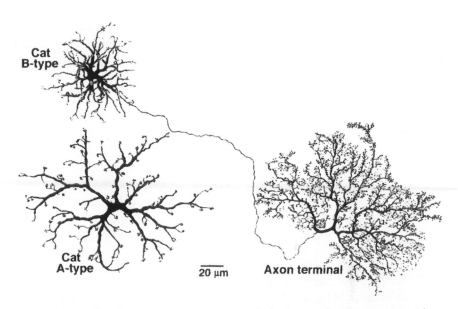

Figure 15. Golgi-stained horizontal cells in whole mounted cat retina, one with an axon and one axonless. The axon-bearing H1 type cell (in nonprimates called the B type) has a dendritic end which is cone connected, and an axon terminal portion contacting rod cells. The axonless A type horizontal cells are believed to be cone driven. From Fisher and Boycott (1974). The axon bearing type is found in all mammalian retinas and has identical connectivity and function, regardless of whether the animal has dichromatic or trichromatic vision. The axon in these cells arises from a dendrite and is very thin at the origin (0.5 μm in diameter), but increases in thickness as the axon branches out. The axon branches end with clusters of telodendrons, which enter rod triad synapses (Boycott and Kolb, 1973). The dendrites of the HI horizontal cells radiate in all directions, branch little and at their terminations give out clusters of small processes with tiny end swellings, which form the lateral processes of cone pedicle triads. (Scale bar: 20 μm.)

The narrow single axon of the horizontal cells does not transmit electrical signals; and the two parts of the cell, the dendritic and axonal arborizations, are electrically insulated from one another.

Neighboring horizontal cells are electrically connected with gap junctions, with a pore size of around 16 Å. These pores allow diffusion of ions, second-messenger molecules, and peptides smaller than about 1 kDa. Gap junctions are formed by proteins called connexins, and it has recently been shown that connexin36 dominates in the retina (Mills et al., 2001).

Horizontal cells form inhibitory synapses with cone cells and may use GABA as their neurotransmitter. The release mechanism is not the conventional one with synaptic vesicles but some other form of facilitated transport. Horizontal cells provide inhibitory feedback to photoreceptor cells and inhibitory feed-forward to bipolar cells. The classical lateral- or center-surround inhibition of ganglion cell receptive fields is caused by the lateral spread of feed-back and feed-forward inhibition by horizontal cells (Mangel, 1991). In human retinas, conventional synaptic contacts from horizontal cells onto rod cells and cone cells have been observed, but such structures are lacking in other species (Linberg and Fisher, 1988).

V. BIPOLAR CELLS

Photoreceptor cells pass their signals to bipolar cells, which form the next link in the chain of neurons leading to the brain. All bipolar cells share the same general morphology: a single dendrite branching in the OPL contacting one or more photoreceptor cells, a perikaryon in the middle of the INL, and an axon, branching in the IPL, contacting amacrine cells or ganglion cells there.

Bipolar cells contact exclusively either rod cells or cone cells, and the class of bipolar cells can therefore be subdivided into rod and cone bipolar cells. In the human retina, there is only one type of rod bipolar cell, but 9 or 10 different types of cone bipolar cells.

Electrophysiologically, cone bipolar cells can be classified as ON bipolar cells that respond with a depolarization to light stimulation in the center of their receptive fields, and OFF bipolar cells that respond to light with a hyperpolarization. Rod bipolar cells are always of the ON type. As already mentioned, there is a difference in the morphology of the synapse between the photoreceptor cells and ON and OFF cone bipolar cells (Figure 14). The dendrites of OFF bipolar cells form flat contacts with cone cells, whereas ON bipolar cells form triads with a central invaginated bipolar cell dendrite (Kolb et al., 2003). The distinction between ON and OFF bipolar cells is also reflected in their contacts with amacrine cells and ganglion cells in the IPL. Functionally, this layer is subdivided into the outer sublamina a, closest to the amacrine cells and the inner sublamina b, closest to the ganglion cells. OFF bipolar cell axons terminate in sublamina a (the OFF sublayer in Figure 5), whereas ON bipolar cells send their axons into sublamina b (the ON sublayer in Figure 5).

All bipolar cell axons end in synaptic bulbs, which characteristically contain ribbon synapses and contact a pair of processes, either two from amacrine cells or one from a ganglion cell and one from an amacrine cell. These bipolar cell synapses are called dyads, because there are two postsynaptic parts.

A. Rod Bipolar Cells

In mammals, primates included, only one variety of rod bipolar cells is known, and it is easily identified by its high content of the enzyme, protein kinase Ca (PKCa) (Figure 16). Rod bipolar cell dendrites enter rod triad synapses. In the parafoveal region, each cell may contact as many as 18–70 rod cells and have a dendritic field measuring 15–30 µm. Their axons descend

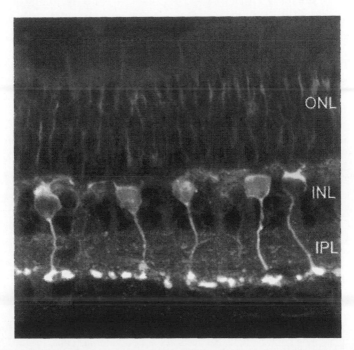

Figure 16. Fluorescence micrograph of rod bipolar cells labeled with an antibody against PKC, an enzyme which specifically occurs in these cells. Rabbit retina. ONL, outer nuclear layer; INL, inner nuclear layer; IPL, inner plexiform layer.

without much branching into sublamina b (the ON sublayer) of the IPL, where they give rise to small numbers of large synaptic terminals. These characteristically contain ribbon synapses and contact a pair of cell processes in dyad synapses. Usually, one of the processes is from a glycinergic AII amacrine cell and the other from a type A17 amacrine cell. Only rarely do rod bipolar cells contact ganglion cells (of the diffuse type), instead the AII amacrine cell links the rod bipolar cells to the ganglion cells (Wässle and Boycott, 1991). The rod signal must therefore pass through a chain of at least four neurons on its way to the brain (Figure 3).

B. Cone Bipolar Cells

Primate cone bipolar cells can morphologically be divided into 9 or 10 different types according to their dendritic branching pattern, the number of cone cells contacted and the shape and stratification of their processes in the IPL (Figure 17, Wässle, 1999; Kolb et al., 2003). Six or seven of the 10 cone bipolar cell types collect information from many cone cells and are called diffuse bipolar cells (DB1–DB6 in Figure 17). They make basal contacts or triad synapses with a number of cone cells, and the axons of ON cells terminate in the IPL, in sublamina b, and the axons of OFF cells terminate in sublamina a (Kolb et al., 2003). Otherwise, it is not well known what the morphological differences mean in terms of function.

Giant bipolar cells make flat contacts with large numbers of cone cells and their axons terminate in both sublaminae of the IPL (Mariani, 1984).

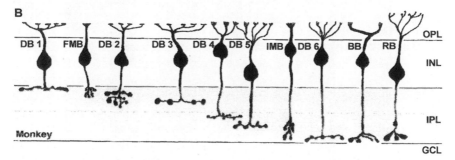

Figure 17. Summary diagram of the bipolar cells found in Macaque monkey retina with Golgi staining. DB1–DB6 are diffuse bipolar cells contacting several neighboring cone pedicles. FMB and IMB cells are flat and invaginating midget bipolar cells contacting a single cone pedicle. BB cells are S cone bipolar cells, and RB cells are rod bipolar cells. OPL, the outer plexiform layer; INL, the inner nuclear layer; IPL, the inner plexiform layer; GCL, the ganglion cell layer; NF, the nerve fiber layer. Adapted from Wässle (1999).

Midget bipolar cells (FMB and IMB in Figure 17) are at the other end of the spectrum, being comparatively small and contacting only a few cone cells each. In the fovea, midget bipolar cells contact only a single cone, and their axons contact a single midget ganglion cell. This midget pathway is found in humans and higher primates, and it ensures that each cone in the central fovea has a "private" pathway to the brain, thus allowing the maximum spatial resolution possible, given the cone mosaic and the optics of the eye (Kolb et al., 2003). Each cone in the central fovea is connected to both an ON and an OFF midget bipolar cell. Midget bipolar cells only contact M- and L-cone cells. S cone cells have a separate pathway with bipolar cells that are immunoreactive to cholecystokinin. They form triad synapses with 1–3 S-cone cells and reach sublamina b of the IPL, where they contact a nonmidget type of bistratified blue-yellow responsive ganglion cell. The resolution of the pathway they form is less than that of the midget cell system for L- and M-cone cells.

VI.   AMACRINE CELLS

Most of the cells in the innermost cell rows of the INL are amacrine cells. Many can also be found in the GCL, forming 20% or more of the neurons seen there, and are then usually referred to as displaced amacrine cells. These cells are not at all displaced, and the term is a misnomer, but it is so squarely rooted that it will be used here too. Occasionally, amacrine cells can also be found in the IPL, and are then called interstitial or A1 amacrine cells (Figure 18).

Amacrine cells all modulate signals in the IPL and are quite diverse in both morphology and neurochemistry (Kolb et al., 2003). There are at least 22 types in rabbits (MacNeil et al., 1999), and others have found as many as 40 (Dacey, 1999), but only few have been well enough characterized to be described in this text. Amacrine cells typically do not have axons, which are reflected in their name that Cajal coined in 1892; see Rodieck (1973), but certain large-field amacrine cells of the vertebrate retina can have long axon-like output processes, projecting exclusively within the retina. According to the distribution of the dendrites in the IPL, amacrine cells can broadly be classified as stratified or diffuse. They have also been classified according to their dendritic field diameters as narrow-field (30–150 μm), small-field (150–300 μm), medium-field (300–500 μm), or wide-field (300–500 μm) cells.

Amacrine cells all have in common that they form conventional chemical synapses with presynaptic clusters of small (300–400 nm) synaptic vesicles at a part of the cell membrane, which shows some increased electron density. Gap junctions are also present.

Most neurotransmitters or neuromodulators described in the brain have also been found in amacrine cells, with epinephrine and 5-hydroxytryptamine

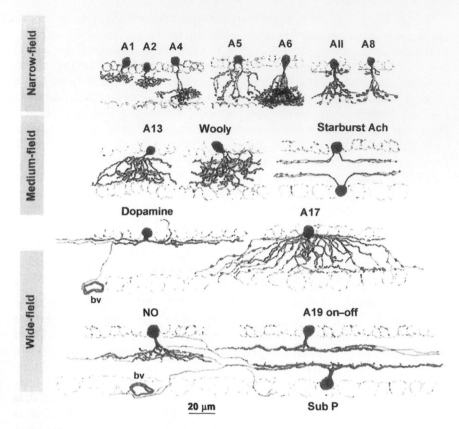

Narrow-field

A1 A2 A4   A5   A6   AII A8

Medium-field

A13   Wooly   Starburst Ach

Dopamine   A17

bv

Wide-field

NO   A19 on–off

bv

20 μm   Sub P

Figure 18. Drawings of morphologies of representative amacrine cells of the mammalian retina as seen with Golgi-type stainings. From Kolb et al. (2001).

as exceptions in mammalian retinas. The neurochemicals mediate both classical fast neurotransmission and more remanent signal modulation. Most amacrine cells contain an inhibitory neurotransmitter, either glycine or GABA.

A. The AII Amacrine Cell

This is a very well-characterized type, with a narrow receptive field and characteristic lobular appendages (Figure 19). Its name derives from a classification of amacrine cell types with Roman numerals, and should be pronounced "A two." This classification has been replaced by a system with

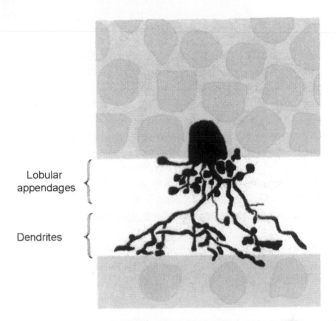

Lobular appendages

Dendrites

Figure 19. An AII amacrine cell at 3 mm eccentricity. The cell type is glycinergic and relays rod bipolar cell signals into the cone pathways. It spreads its two kinds of processes to two different parts of the IPL. Modified from Rodieck (1998).

Arabic digits, except for the AII cells. To increase confusion further, there is an amacrine cell in the cone pathway, which is denoted A2, and which is distinct from the AII cell described here. As already discussed, the AII amacrine cells form an important part of the rod pathway, relaying signals from rod cells to ON cone bipolar cells through gap junctions in sublamina b, and to OFF cone bipolar cells through a glycinergic chemical synapse in sublamina a (OFF in Figure 5). In the cone pathway, the dichotomy between the ON and OFF channels begins in the postsynaptic membrane of cone bipolar cells. In the rod pathway, this dichotomy arises in the AII amacrine cell with its excitatory gap junction synapse to ON ganglion cells and its inhibitory chemical synapse to OFF ganglion cells (Figure 5).

In at least murine and rabbit retinas, the gap junctions at AII cells contain connexin 36 (Feigenspan et al., 2001). AII cells typically contain the calcium binding protein, calretinin (Kolb et al., 2003), which is a good but not exclusive immunohistochemical marker for them. In cats and rabbits, the AII amacrine cells form about 10% of the total number of amacrine cells, and each cell may receive input from as many as 30 rod bipolar cells. The prevalence is likely to be similar in humans.

B. The Starburst Amacrine Cells

These cells are the only neurons in the retina definitely known to be cholinergic. They have a characteristic wide-field morphology with radiating dendrites resembling a starburst firework cracker (Figure 20). They are found both in the INL and in the GCL (Masland, 1994). In humans and in cats, there are more of these cells in the GCL than in the INL. The cholinergic starburst amacrine cells form two distinct layers of branching processes in the IPL, one in sublamina a and one in b (Figure 21). In addition to acetylcholine, cholinergic amacrine cells also contain and release GABA, suggesting that they are both excitatory and inhibitory in function. Starburst amacrine cells are involved in the motion direction detection system of the retina, but their exact role in it remains to be established.

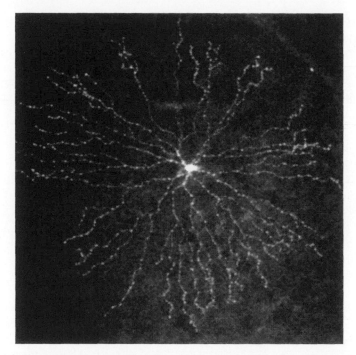

Figure 20. A cholinergic starburst amacrine cell with its characteristically radiating and dichotomously dividing processes, injected with Lucifer Yellow. Confocal micrograph, Zucker and Ehinger (1998).

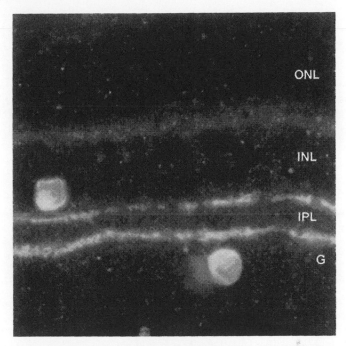

Figure 21. Two cholinergic amacrine cell bodies, one in the position of classic amacrine cells in the inner nuclear layer (INL) and the other, a so-called displaced amacrine cell, in the ganglion cell layer (G). The inner plexiform layer (IPL) contains two narrow bands of cholinergic processes. Choline acetyl transferase (ChAT) immunolabeling.

C. Dopaminergic Amacrine Cells

There are at least two distinct types of dopaminergic neurons. Both have very long processes and can thus cover the entire retina, despite being relatively few in number, approximately 8000. Type 1 cells synapse in sublayer b of the IPL on the AII amacrine cells, which form the connecting link between rod bipolar cells and ganglion cells. Type 1 cells thus seem to modulate the flow of information about scotopic vision, perhaps by regulating the gap junctions of the AII amacrine cells. Type 2 cells arborize in the inner layers of the IPL (Figure 22).

D. DAPI-3 Amacrine Cells

The DAPI-3 amacrine cells were first described as a third type of cells identified with the special nuclear dye DAPI in rabbit retina (Wright et al., 1997). They are glycinergic, bistratified, medium-field cells, and are morphologically similar

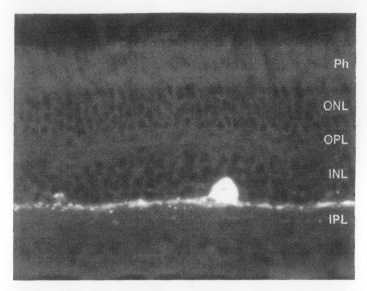

Figure 22. Dopaminergic amacrine cell and processes in the macaque retina, demonstrated by tyrosine hydroxylase immunohistochemistry. Layers: Ph, photoreceptor cells; ONL, outer nuclear layer, OPL, outer plexiform layer; INL, inner nuclear layer; IPL, inner plexiform layer. After Marshak (2001).

to the type A8 amacrine cell (Figure 18). They form less than 2% of the amacrine cells in rabbits, and appear to have synaptic associations with the cholinergic starburst amacrine cells, thought to be involved in the motion detection system of the retina (Figures 23 and 24).

E. Nitrergic Amacrine Cells

NADPH-diaphorase activity is found in amacrine cells that contain nitric oxide synthase (NOS) immunoreactivity, or produce NO (Eldred, 2001). They are therefore presumed to use NO as a neurotransmitter or neuromodulator. Two types of such cells have been identified, but the precise functions of NO in retinal neurotransmission remain to be established.

VII.   INTERPLEXIFORM CELLS

These cells have their perikarya among the amacrine cells in the INL, and they send their processes both to the OPLs and to the IPLs (Figure 25). In the IPL, these processes are both pre- and postsynaptic to amacrine cells and

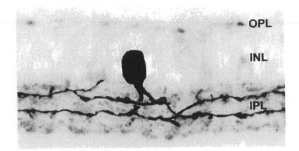

Figure 23. Schematic drawing of the cell body and distribution of the processes of a glycinergic DAPI-3 cell in the rabbit retina.

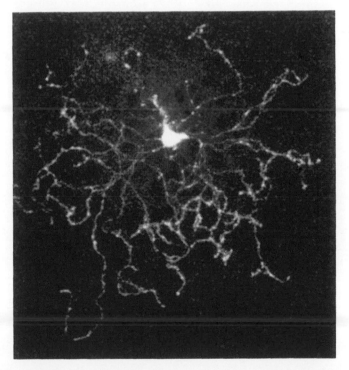

Figure 24. A DAPI-3 cell in a whole-mounted rabbit retina, injected with Lucifer yellow, showing the characteristic distribution of its undulating and entangled processes.

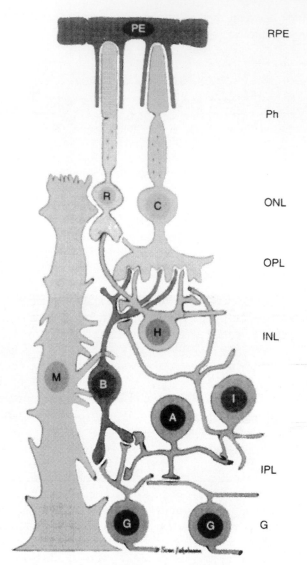

Figure 25. Cartoon showing an interplexiform cell (IC) in fish retina, receiving input from amacrine cells (AC) and transmitting information to cells in the OPL. RC, rod cells; CC, cone cell; HC, horizontal cells; BC, bipolar cell; GC, ganglion cell.

presynaptic to both rod and cone bipolar cells (Figure 25). In the OPL, the cells are presynaptic to rod or cone bipolar cells (Kolb et al., 2003). They use GABA or dopamine as their neurotransmitter. The interplexiform cells carry feedback signals between the IPL and the OPL, but further details about their physiology remain to be established.

VIII.   INNER PLEXIFORM LAYER

This layer (IPL in Figure 4) is the site of synapses between bipolar cells, amacrine cells, interplexiform cells, and ganglion cells, and it is the layer where the visual information is extensively processed so that the ganglion cells can send appropriately coded signals to the brain. However, the code is not known, and we have only begun to understand the most basic features of the processing in the IPL. Detection of movement and of changes in brightness, contrast, or hue take place in this layer, whereas static information is predominantly handled in the OPL. However, there is no consensus on the details of how these processes work in the IPL.

Morphologically, Cajal (1892) somewhat arbitrarily subdivided the IPL into five sublayers, and this system is still in use (Kolb et al., 2003). From a functional point of view, the IPL can be divided into two: the outer a layer (corresponding to Cajal's sublamina 1 and 2) and the inner b layer (corresponding to Cajal's sublamina 3, 4, and 5). OFF signals are processed in the a layer, whereas ON signals are handled in the b layer, see Figure 28 and Kolb et al. (2003). Synapses between cone bipolar cells and ganglion cells are essentially sign conserving, so that the OFF and ON center characteristics of ganglion cells are usually determined by the type of bipolar cells which dominate them. Bipolar cells and ganglion cells are in most cases restricted to either the a (OFF) or the b (ON) layers, whereas amacrine cells often form bridges between them.

In the electron microscope, three types of processes and synapses occur in the IPL.

Bipolar cell processes are rounded and contain an abundance of neurotubules, neurofilaments, and only few mitochondria. A synaptic ribbon characterizes their synapses, and sometimes also an arcuate density similar to the one seen in photoreceptor triad synapses. The synaptic ribbons are smaller and are surrounded by denser synaptic vesicles than in photoreceptor triad synapses. No other synapses are known to contain ribbon synapses in the IPL. The bipolar ribbon synapses usually have two postsynaptic members, either two amacrine cell processes or one ganglion cell and one amacrine cell process (Figure 26). These synapses are called dyads. Rod bipolar cell terminals occur in sublamina b of the IPL, are comparatively voluminous, and make many ribbon synapses. Occasionally, the amacrine cell processes form reciprocal contacts back onto the bipolar cell terminals within a few micrometers of its postsynaptic position, and this is particularly common with the indole-amine accumulating A17 amacrine cells with a

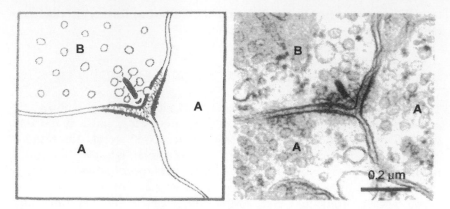

Figure 26. Drawing and electron micrograph of the characteristic synaptic ribbon in a bipolar cell dyad (B) in the IPL of a frog retina. Both postsynaptic processes belong to amacrine cells (A), and one is making a reciprocal synapse back onto the bipolar cell process. Redrawn from Dowling (1987). (Scale bar: 0.2 μm.)

characteristic morphology in rabbits. Bipolar cells occasionally form synapses with a single postsynaptic process, which by analogy is called a monad. Occasionally, bipolar cell processes also make ribbon synapses on amacrine cell somata.

Amacrine cell processes possess electron microscopic properties of both axons and dendrites, but can nevertheless usually be distinguished from bipolar cell axons. Synaptic vesicles aggregate at the site of synaptic contacts, forming so-called conventional synapses (Figure 27). Amacrine cell processes also frequently make synapses of the conventional type with bipolar cell processes and their terminals as well as ganglion cell processes or somata, and these synapses are similar to the ones found in most other parts of the central nervous system.

Ganglion cell dendrites lack specific features and are therefore difficult to identify.

IX. GANGLION CELLS

Ganglion cells occupy the innermost layer of cell bodies in the retina and comprise two-thirds to three-quarters of the cells in this layer (Marc and Jones, 2002), which is called the GCL. The term is slightly misleading, since 20% of the cell bodies in this layer are amacrine cells. Not all ganglion cell bodies are located in the GCL. They can be found also among amacrine cells in the INL, in the IPL, or, at least in rabbits, in a sparse layer of cells in the

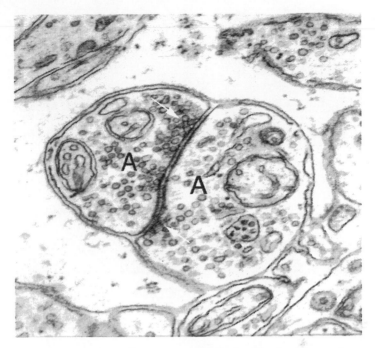

Figure 27. Two amacrine cell processes (A) in the IPL of a Cynomolgus monkey retina, making synaptic contacts with each other (reciprocal synapses; arrows). The synapses are of the conventional type, with an accumulation of synaptic vesicles and membrane thickenings at the synaptic sites.

nerve fiber layer of the central retina (Ehinger and Zucker, 1996). Ganglion cells have their dendrites in the IPL where they collect all visual information processed in the retina. They collect this information and send it to the brain by means of the retinal nerve fiber layer and the optic nerve, both of which are composed of ganglion cell axons. Contrary to other retinal neurons, ganglion cell axons readily produce action potentials, which are necessary to forward information over distances longer than a millimeter. Like many long axon cells in the CNS, the ganglion cells use glutamate for their neurotransmission, but some of these cells also contain substance P (Kolb et al., 2003).

Ganglion cells typically project to the lateral geniculate body, or its equivalents in nonprimate vertebrates, but about 10% of the fibers in the optic nerve project to supraoptic and subthalamic structures, participating in nonvisual processes like the pupillary reflexes and circadian rhythm regulation.

In the human retina, the most common ganglion cell types are the midget and the parasol ganglion cells (Figure 28). Midget ganglion cells are also

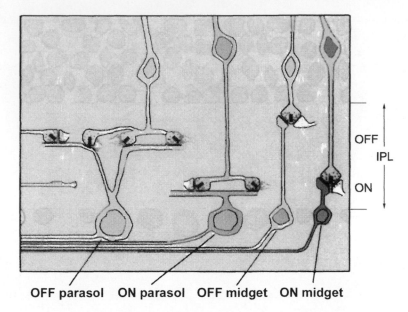

OFF parasol   ON parasol   OFF midget   ON midget

Figure 28. Four types of ganglion cells and their synapses with bipolar cells in the outer sublamina a (OFF layer), and the inner sublamina b (ON layer) of the inner plexiform layer (IPL). After Rodieck (1998).

known as P cells because of their projection to the parvocellular layers of the lateral geniculate nucleus. Parasol ganglion cells are also known as M cells because of their projection to the magnocellular layers of the lateral geniculate nucleus. The midget ganglion cells are concentrated in the macular area, and they constitute approximately 80% of the ganglion cells in the primate retina; parasol cells constitute 10% and other types the remaining 10% (Dacey and Petersen, 1992).

There are no ganglion cells in the foveola. The axons of the central cone cells, the Henle fibers, project centrifugally toward the ganglion cells, which reach their maximum densities of approximately 40.000 cells/mm^2 at the upward slope of the fovea, at an eccentricity of 0.5 mm (Curcio and Allen, 1990a; Sjöstrand et al., 1999). For any given eccentricity, the ganglion cell density is highest in the superior and nasal retina (Curcio and Allen, 1990a).

A. Midget Ganglion Cells

These cells are small, and they send a single dendritic process to the IPL, where they branch in either sublamina a or b. Through careful counting of midget ganglion cells and examination of their connections, it has become apparent that each central cone has two "private" midget ganglion cells: one ON and one OFF cell (Sjöstrand et al., 1999). More peripherally, they collect information from more cone cells, and in the midperipheral retina their dendritic field may reach 100 μm (Dacey and Petersen, 1992). Midget ganglion cells probably carry information regarding form and color.

B. Parasol Ganglion Cells

These cells have a large cell body. One or more dendrites arise from them, and form large horizontally spread arbors in either the proximal or distal parts of the IPL (Rodieck, 1998). Their receptive fields are wide, approximately 50 μm, centrally and up to several hundred microns in the periphery (Dacey and Petersen, 1992). Physiologically, they comprise more than one type of cells. Some of them probably selectively carry blue-yellow color information, whereas others receive input from all chromatic types of cone cells. Their comparatively wide dendritic fields allow them to cover a wide area of the retina. This makes them able to respond to moving or changing stimuli, and they are presumably responsible for carrying such information to the brain.

C. Light-Sensitive Ganglion Cells

A recently detected and very special population of ganglion cells contains a photopigment, melanopsin, and the neuropeptide PACAP. They respond to light with a characteristic sluggishness, send their axons to the supraoptic nucleus, and drive the circadian rhythm of at least rodents (Hannibal et al., 2002; Hattar et al., 2002; Provencio et al., 2002). The observation heralds hitherto unsuspected retinal functions.

X.   NERVE FIBER LAYER

Ganglion cell axons form the nerve fiber layer, converging from all parts of the retina towards the optic disc and forming progressively thicker bundles as they approach the optic nerve head, where the layer can be 20- to 30-μm thick. Peripherally, the layer becomes thinner, and at the very periphery it is indistinguishable. Glial cell processes belonging to Müller cells or astrocytes surround the bundles. Fibers coming from the temporal side arch above or below the fovea. Fibers coming from the fovea do not arch, but runs straight to the optic disc forming the straight maculopapillar bundle. The axons are normally not myelinated in mammals, but aberrant myelination is at times seen in patches at the optic disc margins or in the central fundus.

XI. CENTRIFUGAL FIBERS

Centrifugal fibers are nerve fibers in the retina whcih originate in cell somata in the brain. They are well known in cold-blooded vertebrates and birds, but they were not convincingly demonstrated in mammalian retinas. However, a modest number of histamine containing centrifugal fibers were recently described in macaque retinas. These fibers appear to originate in the posterior hypothalamus and reach the innermost layers of the retina (Gastinger et al., 1999).

XII. NEUROTRANSMITTERS IN THE RETINA

Basically, all neurosignalling systems found in the CNS have been found also in the retina, epinephrine, and 5-hydroxytryptamine being exceptions in mammals. Both conventional neurotransmitter or neuromodulatory release from presynaptic vesicles onto specific chemical receptors, and ion channel signaling through gap junctions are important in retinal neurotransmission; the latter have only begun to be known in the last few years. Nonvesicular facilitated neurotransmitter transport has also been postulated for at least horizontal cells. Dopaminergic neurons are thought partly to act by paracrine transmission. The cannabinoid synthesis-on-demand feedback system has also been described in retinal neurons, as further discussed below.

One excitatory and two inhibitory neurotransmitters dominate among the neurons of the mammalian retina: glutamate, GABA, and glycine. Glutamate is the neurotransmitter of photoreceptor, bipolar, and ganglion cells, and probably also some amacrine cells (Thoreson and Witkovsky, 1999). The cone signals are thus sent from cone cells to the brain by three glutamatergic neurons: cone, bipolar, and ganglion cells. In the rod pathway, the glycinergic AII amacrine cell is intercalated between rod bipolar cells and the final ganglion cells (Figure 6, see also Kolb et al., 2003). Glycine is a likely neurotransmitter in perhaps 40% of the amacrine cells, and GABA in another approximately 40% of the amacrine cells (Pourcho, 1996; Marc and Liu, 2000).

Glutamate released from photoreceptor cells act upon their second-order neurons in two different ways depending on the properties of the postsynaptic receptors (see Kolb et al., 2003). One type of receptor is the metabotropic glutamate receptor that is found in ON-bipolar cells. Signalling through this receptor involves cGMP as second messenger. Activation of the metabotropic glutamate receptors in ON-bipolar cells causes a decrease in the intracellular cGMP concentration, a closure of cGMP-gated cation channels in the cell membrane, and consequently a hyperpolarization of the cell membrane potential. The other receptor type is the ionotropic glutamate receptor found in OFF-bipolar cells and horizontal cells. This receptor incorporates an unspecific cation channel, which is rapidly activated when glutamate is bound to the receptor. Activation of the ionotropic glutamate receptor causes influx of current, carried by Na^+, and depolarization of the membrane potential. There are several subtypes of both metabotropic and the ionotropic glutamate receptors (see Kolb et al., 2003).

Acetylcholine is present in the starburst amacrine cells, as already described. Other locations have been described but not verified.

In mammals, dopamine is the likely neurotransmitter in certain amacrine cell types, including the A18 type. In the retinae of teleost fish and some New World monkeys, but not in Old World monkeys and humans, interplexiform cells also contain dopamine. Dopamine is known to affect gap junctions in the retina (Weiler et al., 1999; Kolb et al., 2003).

Dopamine receptors have been described on sites well away from dopaminergic synapses and cell processes. It has been argued that in the CNS, dopamine may act by paracrine transmission (i.e., extrasynaptic action by diffusion over much longer distances than the synaptic cleft, and possibly, this is then the case also in the retina).

Serotonin (5-hydroxytryptamine) is a likely neurotransmitter in some of amacrine cells of cold-blooded vertebrates and birds, but it has not been found in mammals. Nevertheless, some mammalian amacrine cells, called indoleamine accumulating cells (among them type A17 in rabbits) contain much of the biochemical machinery needed to produce and handle serotonin. The indoleamine accumulating cells are likely to have GABA, and not serotonin, as their neurotransmitter, so apparently the serotonin system of these cells has been downregulated, but left in place for some reason (Fletcher and Wässle, 1999). Humans and cynomolgus monkeys do not have any indoleamine accumulating neurons in the retina (Ehinger and Floren, 1979).

Histamine has in Macaque monkeys recently been demonstrated in a set of centrifugal fibers (Gastinger et al., 1999), reopening the old question of to what extent such fibers exist in primates.

Nitric oxide is regarded as a neurotransmitter candidate in many amacrine cells (Perez et al., 1995; Eldred, 2001). However, its physiological roles in the retina are largely unknown. It has been suggested that NO produced by amacrine and ganglion cells is a paracrine modulator of cell death within the retinal tissue (Guimaraes et al., 2001).

The effects of endogenous cannabinoids have been explored only recently in the retina. They are not stored in neurons or other cells, but synthesized on demand. Thus neurons using cannabinoids cannot be detected by their transmitter content but only by localizing appropriate receptors and enzymes involved. Cannabinoids appear mostly to act by presynaptic modulation of neurotransmitter release, and may well be active at many retinal synapses, including those of rod cells and cone cells (Straiker et al., 1999; Yazulla et al., 2000).

Numerous neuroactive small peptides have been described in retinal neurons of experimental animals, for instance neurotensin, galanin, TRH, LHRH, PHI, FMRFamid, NPY, MSH, substance P, other tachykinins, glucagon, somatostatin, met- and leu-enkephalins, VIP, GRP, bombesin, cholecystokinin (CCK), and PACAP. Most of them appear in small populations of amacrine cells. They have so far always been found in neurons that also are likely to have a conventional small molecule neurotransmission system, like the monoamine or amino acid systems. Some small other molecules have also been suggested, for instance epinephrine, norepinephrine, ATP, and adenosine, but, like with the small peptides, their possible functions in the retina are unclear.

XIII. GLIAL CELLS

The retina contains four kinds of glial cells: Müller cells, which are by far the most numerous; astrocytes, which mainly appear in the innermost parts of the retina; microglial cells, which are phagocytic, vary considerably in number, and appear whenever and wherever they are needed; and finally, oligodendroglial cells, which surround ganglion cell axons when they are myelinated. In primates, the ganglion cell axons are normally myelinated only in the optic nerve.

The retinal Müller cells have a unique and characteristic morphology, extending through the whole thickness of the neural retina, and with their perikarya in the INL (Figure 29). They issue many fine processes, ensheathing all retinal neurons from the outer (or external) limiting membrane (OLM) and inwards. In the plexiform layers, they cover the dendritic processes of the neurons up to the synaptic clefts, insulating them both electrically and chemically. In the nerve fiber layer the Müller cell processes cover most ganglion cell axons. Similarly, Müller cells cover blood vessels within the retina.

Retinal Müller cells, astroglial cells, and oligodendroglial cells are extensively coupled by gap junctions of different kinds (Zahs et al., 2003). The biologic significance of these contacts remains to be established.

The Müller cells form both the OLMs and ILMs of the retina. The OLM is not a true membrane, but a series of junctional complexes of the zonulae adherens type between Müller cells and photoreceptor cells (Figure 30). In nonprimate vertebrates, gap junctions can also be found at the OLM. Müller cells extend microvilli beyond the OLM, into the subretinal space. At the inner surface of the retina, Müller cells widen into coalescing endfeet. These

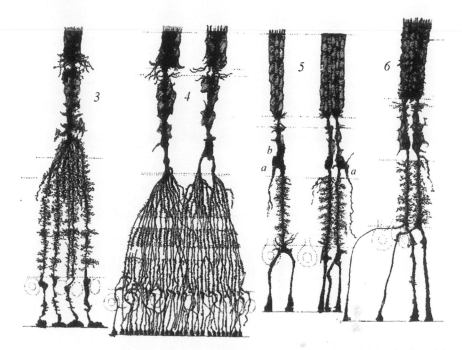

Figure 29. A collection of Golgi-stained Müller cells from lizard (3), chicken (4), and ox (5 and 6). a: descending collateral branches, b: region of the nucleus. After Ramón y Cajal (1972).

produce the 1- to 2-μm thick ILM, which is a regular basal lamina, covering the entire inner surface of the retina.

Müller cells not only provide a scaffold for retinal neurons, but also perform many tasks, which in the CNS are handled by astrocytes, oligodendrocyte, or ependymal cells. They regulate the extracellular environment of the retina by buffering the light-evoked variations of particularly the K^+ concentrations in the extracellular space, and they remove glutamate from the extracellular space by active uptake (Newman and Reichenbach, 1996; Thoreson and Witkovsky, 1999).

Müller cells can respond to injury in several different ways. For instance, they can proliferate or express glial fibrillary acidic protein, heat shock proteins, major histocompatability molecules, or intracellular adhesion molecules. The significance of these reactions is not known.

Star-shaped astrocytes occur sparsely in the GCL and the IPLs. Their processes contact ganglion cells and capillary surfaces. Most of the astrocytes in the nerve fiber layer send out two types of processes: one around nerve fibers and the other to blood vessel walls (Ramirez et al., 1996).

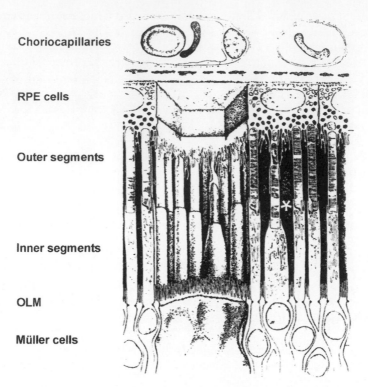

Figure 30. Diagram of the subretinal space, showing the relationship between the retinal pigment epithelium (RPE), the outer segment of the photoreceptors, the outer limiting membrane (OLM), and the Müller cells. The asterisk denotes the subretinal space. The loose organization of the subretinal space with abundant space between the cellular elements contrasts the tightly organized extracellular space in the inner retina (i.e., internal to the outer limiting membrane), From Steinberg et al. (1983).

Reticuloendothelial microglial cells are normally only found in small numbers in the nerve fiber layer. However, under pathologic conditions these mobile phagocytic cells can be found anywhere in the retina.

Oligodendroglia is normally not present in primate retina, and usually not in other mammals either. However, in rabbits, they abound in the myelinated streak (Newman et al., 1996).

XIV. THE RETINAL EXTRACELLULAR SPACE

The OLM divides the retinal extracellular space into two different compartments: the subretinal space and the inner retinal extracellular space. The subretinal space is the extracellular space between the apical membrane of the RPE and the external limiting membrane. It is a loosely organized space with several microns between the cells, and it contains glucosaminoglycans and proteins, including IRBP (Adler and Martin, 1982). The apical membrane of the RPE and the tight junctions between the epithelial cells is the outer border of the subretinal space, and constitutes a diffusion barrier (a part of the blood-retinal barrier) that hinders free diffusion of water soluble substances between the blood and the subretinal space. The OLM is the inner border of the subretinal space. The junctional complexes that form this membrane allow free diffusion of small molecules. However, the pores in the OLM retain molecules with a Stokes' radius of more than 36 Å. The outer limiting membrane protects the extracellular proteins, such as IRBP, from dilution by hindering their diffusion towards the vitreous (Bunt-Milam et al., 1985). The extracellular space of the inner retina is a tight labyrinth of narrow intercellular clefts, only a few nanometer wide. This extracellular space resembles that of the brain (Figure 30).

The different anatomy of these two parts of the retinal extracellular space is reflected in the morphology of pathologic extracellular fluid accumulation in the retina. Subretinal fluid tends to cause serous detachment of the retina, whereas inner retinal fluid tends to dissect pathologic extracellular spaces, and form what is clinically known as cystic macular edema.

XV. RETINAL CIRCULATION AND METABOLISM

In most vertebrates, primates included, the retina is supplied from two different vascular beds: the (inner) retinal circulation and the choroid. The retinal circulation is supplied from the central retinal artery, and its capillaries branch in inner part of the retina from the nerve fiber layer to the OPL. In some mammalian species, such as bats and rabbits, the retinal circulation is rudimentary or missing. The capillaries of the retinal circulation are of the same type as brain capillaries. Their endothelial cells are nonfenestrated and bound together with tight junctions. The capillaries of the choroid, the so-called choriocapillaries, are polarized towards Bruch's membrane and the RPE, and they are fenestrated, and leaky. The outer retina, where the photoreceptors are located, is avascular. The capillary blood flow in the retinal circulation is approximately 50 ml/min/100 g tissue, and resembles that in the brain. In the choriocapillaries, the capillary blood flow is among the highest measured in the body, approximately 2000 ml/min/100 g tissue (Gioffi et al., 2003). The tight junctions in the retinal capillary endothelial cells and the junctions between the retinal pigment epithelial cells constitute the blood-retinal barrier, which hinders the free diffusion of water soluble molecules between the blood and the extracellular space in the retina. The extracellular environment in the retina is controlled via specialized transport systems in the endothelial cells and in the retinal pigment epithelial cells (Tornquist et al., 1990).

The retina has high energy requirements and high-oxygen consumption. The photoreceptors in particular are highly metabolically active cells, especially in darkness, where the dark current requires Na^+ to be pumped out of the photoreceptor cells at a high rate. Since the outer retina is avascular, the oxygen and glucose that fuel photoreceptor metabolism must diffuse a hundred microns to reach the inner segments, where the photoreceptor mitochondriae are located. In darkness, the oxygen consumption by the photoreceptors is so high that the oxygen tension in the vicinity of the photoreceptor inner segments approaches zero. Thus, even under physiological conditions, the photoreceptors operate under near ischemic conditions. The retina has a high capacity for anaerobic glycolysis, and produces lactate, even under normal physiological conditions (Picaud, 2003).

XVI. DEVELOPMENT

Photosensitive patches are known in the simplest unicellular organisms, and it is therefore believed that light sensitivity is a very ancient property of living organisms. From there, photoreceptor cells and eyes have developed along many different lines, which nevertheless have all tended to give superficially similar morphologies; vertebrate and cephalopod squid and octopus eyes being points in this case. Both are spheroids with a cornea, lens, and retina, but the human retina is inverted, with photoreceptor cells being on the scleral side of the retina, close to the choriocapillary layer, whereas the octopus retina is not; a consequence of differing but converging evolution lines (Land and Nilsson, 2002). Vertebrate photoreceptor cells apparently need the heat sink provided by the very fast choriocapillary blood flow, but the octopus does not, living in deep waters with only little sunlight.

The ectodermal neural plate is the embryonic source of the entire nervous system, including the retina. This plate is first rolled up in the midline of the embryo to form the neural tube. In humans, the first signs of the developing eyes appear as two pits in transverse neural folds. This occurs at the third week of gestation (the 2.6-mm stage), when the brain is still in its three-vesicle stage. Within a few days (the 3.2-mm stage), these pits develop into optic vesicles, connected to the brain by the optic stalks (Figure 31).

The optic vesicles enlarge and start to invaginate their inferolateral parts during the 4th–6th week (the 4–5-mm stage and 15–18-mm stage). The concurrently developing lens is important for this, probably exerting its effect through extracellular matrix components. The invaginating vesicle forms a two-layered cup open laterally, and with a fissure appearing in its lower nasal quadrant (Figure 31B and C). This fissure eventually closes, but

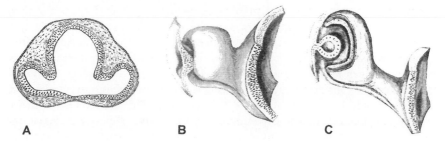

A   B   C

Figure 31. Panel (A): Transverse section through the anterior part of the forebrain and optic vesicles of a 4-mm human embryo. a, cavity of forebrain; b, surface ectoderm; c, wall of the optic vesicle; d, cavity of the optic vesicle. Panel (B): Model of optic the cup and lens thickening of a 5.5-mm human embryo showing the invagination of the optic vesicle to form the optic cup, and the deepening of the lens pit. Panel (C): Model of the optic cup of a 7.5-mm human embryo. The wall of the cup has been cut away to show the two layers. The lens vesicle has been opened to show the cavity and the narrow passage running from this through the lens stalk. From Mann (1923).

not before it has provided a path for the developing hyaloid artery that supplies the developing retina, vitreous, and lens. The retinal circulation develops from the hyaloid artery, but in humans the peripheral retina remains avascular until birth. If the optic fissure fails to close, an inferonasal coloboma or an even more malformed eye is formed.

The invaginating portion lining the inner wall of the cup will give rise to the neural retina along with most of its glial elements. The outer wall develops into the RPE. The optic stalk gives rise to the glia of the optic nerve.

By the time optic cup is formed (the 10-mm stage, around the 5th week), retinal differentiation is already in progress. Cell division in the outer wall of the optic cup occurs only in one plane, giving rise to the single layer of RPE cells. In these, pigment granules begin to appear by the 4th week (the 5–6-mm stage), and are well formed by the 5th week (the 10-mm stage). By the end of the 8th week (the 30-mm stage), a single-layered pigmented epithelium can readily be identified. It continues to grow, but there are no further significant changes in it. On the choroidal side, the RPE is firmly attached to Bruch's membrane. Both RPE and the choroid contribute to elements of this membrane, which thus has both ectodermal and mesodermal origins. It starts to develop by the 14–18-mm stage (the 6th week) and is well demarcated by the 6th month.

By the 12-mm stage (about 5 weeks), two nuclear layers are established at the posterior pole of the retina in the inner wall of the optic cup. They are called the inner and the outer neuroblastic layers. A narrow acellular strip called the Chievitz's transient fiber layer separates them (Figure 32). The development spreads radially through the optic cup, and by the 26-mm stage, the two neuroblastic layers are developed out to its equator. Proliferating cells synthesize DNA with their nuclei on the vitreal side of the neuroblastic cell mass. The nucleus then move towards the sclera to divide. If the daughter cells do not leave the mitotic cycle, the nucleus moves back to the vitreal side to synthesize more DNA again. The movement is called interkinetic nuclear migration (Linden et al., 1999). Cells that leave the cell cycle migrate to locations determined by their birthday and start to differentiate.

Only few of the factors that control the differentiation and proliferation are understood, but they appear to be both stimulatory and inhibitory. A number of transcription factors have been suggested to control the fate of developing retinal cells (Figure 33). Retinoic acid is an important early differentiation factor, followed by ath, BMP4, hedgehog, ephrin, Tbx5, and other factors (Peters, 2002; Marquardt, 2003). The ephrin system appears to

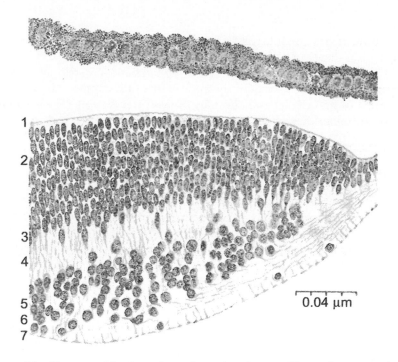

Figure 32. The neuroblastic region of a retina from a 17-mm human embryo. 1, basement membrane; 2, outer neuroblastic layer; 3, Müller fibers; 4, Chievitz' layer; 5, inner neuroblastic layer; 6, nerve fiber layer; 7, internal limiting membrane. Adapted from Mann (1923).

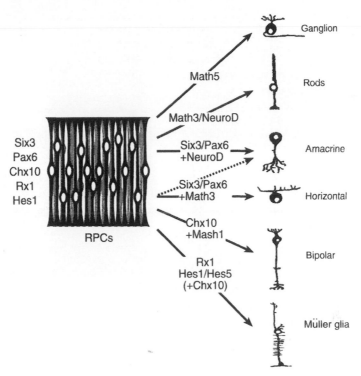

Figure 33. Transcription factors currently believed to direct retinal progenitor cells (RPCs) to different retinal cell fates. From Marquardt (2003).

be important for guiding ganglion cell axons into the optic nerve (Feldheim et al., 2000). Transforming growth factor alpha, acidic and basic fibro-blast growth factors, and epidermal growth factors are also among the agents currently believed to be involved because they stimulate the prolifer-ation of retinal progenitor cells (RPCs) in cultures, and they use intracellular signaling cascades ultimately influencing the cell-cycle control system.

Proliferation eventually ceases, first in the central retina and subsequently in more peripheral parts. The first cells to leave the cell cycle are ganglion cells along with cone cells, horizontal cells, and certain amacrine cells. Rod cells develop later and bipolar cells are the last (Figure 34). In humans, the ganglion cells, horizontal cells, cone cells, and amacrine cells are present already at birth. Some Müller cells, and some peripheral rod cells, may continue to be produced up to the 3rd postnatal month. The birthday sequence of the retinal neurons is very similar in most vertebrates, suggesting they have followed an important common line during the evolution.

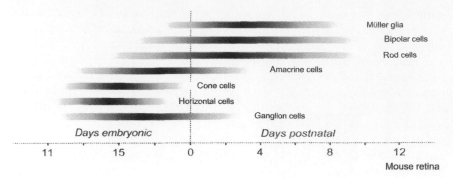

Figure 34. Birthday periods for different cell types in the mouse retina. Other species develop similarly, although their time scales are different. After Young (1985).

It is well established that in adult birds and cold-blooded vertebrates, new retina cells can be formed by progenitor cells, mainly present at the ora serrata. Mammalian retinas were not known to contain such cells, but very recently, a small number of progenitor cells have been described in murine ciliary processes. The observation has spurred hopes that these normally deeply dormant cells can be activated to replace lost or damaged retinal cells (Ahmad, 2001).

The development of the retinal layers as well as the differentiation of various cell types begin in the central retina and spread peripherally. The GCL and the IPL are the first to appear, starting in humans at 12th week, and they are well established by the 5th month. The OPL forms at the 4 month, and cells located between the OPLs and the IPLs then form the INL, erasing the Chievitz's transient fiber layer, which leaves no identifiable mark in the adult retina.

The cell distribution is not symmetrical across the retina, as exemplified by the accumulation of cone cells in the fovea except for S-cone cells which are absent from its very center and that of rod cells in the rod ring (Figure 6). Already from before birth, the cone and ganglion cell densities are also higher in the nasal than the temporal retina. This suggests the presence of important positional information, both in the fetal and in the young developing retina, and details of differences along the dorsoventral and the anteroposterior axes of the retina have recently been presented, but much remains to be established (Ross et al., 2000; Peters, 2002).

Programmed cell death (apoptosis) is an important event during the development of most tissues, retina included (Linden et al., 1999). Apoptosis is likely to regulate the number of cells of the different types at different locations. During the early development of the eye, it is responsible for the sculpting of the optic primodium and the formation of the optic fissure. It may also be responsible for opening extracellular channels, thereby creating passages where the optic fibers could grow (Silver and Hughes, 1973). The process appears most prominent in ganglion cells. Approximately 70% of these die in a developing human retina, but different subclasses of ganglion cells die at different rates (O'Leary et al., 1986). Factors responsible for the control of apoptosis include the availability of trophic factors with different kinds of dendritic interactions, as well as intrinsic or induced activity in the neurons. Most recently, it has been shown that gap junctions can mediate apoptosis triggering signals, and there is evidence that they may be directly involved in regulating apoptosis during development (Cusato et al., 2003).

During development, the first sign of the human fovea is present already at embryonic week 11–12 on the temporal side of the optic disc. In the early stages, the macular region is actually elevated from the surface of the retina due to the accumulation of ganglion cells there, with as many as up to nine rows of ganglion cells in the macular region at the 6th month of gestation. The region then starts thinning, with ganglion cells and other cells of the inner retina becoming displaced centrifugally. The displacement results in a depression in the center of the macula, the foveal pit, which is fully formed at about 11–15 months postnatally. The cells maintain their synaptic contacts made at an earlier stage (Hendrickson and Yuodelis, 1984).

There are no rod cells in the fovea, and S-cone cells are also lacking at its very center. It is not clear how this is achieved. There is no evidence of selective apoptosis in the fovea. It is possible that the progenitor cells produce only L- and M-cone cells in this region.

At the time of birth, the cone density of the human fovea is only 20% of the adult value, which it reaches only 4–5 years after birth, probably by centripetal migration of cone cells. As the fovea matures, the cone cells become thinner so that they are more compactly packed. Foveal cone cells continue to change their shape and develop the outer segments well into adulthood.

REFERENCES

Adler, A.J. and Martin, K.J. (1982). Retinol-binding proteins in bovine interphotoreceptor matrix. Biochem. Biophys. Res. Comm. 108, 1601–1608.

Ahmad, I. (2001). Stem cells: New opportunities to treat eye diseases. Invest. Ophthalmol. Vis. Sci. 42, 2743–2748.

Ahnelt, P. and Kolb, H. (1994). Horizontal cells and cone photoreceptors in human retina: A Golgi-electron microscopic study of spectral connectivity. J. Comp. Neurol. 343, 406–427.

Ahnelt, P.K. and Kolb, H. (2000). The mammalian photoreceptor mosaic-adaptive design. Prog. Retin. Eye Res. 19, 711–777.

Arshavsky, V.Y., Lamb, T.D. and Pugh, E.N., Jr. (2002). G proteins and phototransduction. Annu. Rev. Physiol. 64, 153–187.

Baylor, D.A. (1987). Photoreceptor signals and vision. Proctor lecture. Invest. Ophthalmol. Vis. Sci. 28, 34–49.

Baylor, D.A., Nunn, B.J. and Schnapf, J.L. (1984). The photocurrent, noise and spectral sensitivity of rods of the monkey Macaca fascicularis. J. Physiol. 357, 575–607.

Boycott, B.B. and Kolb, H. (1973). The horizontal cells of the Rhesus monkey retina. J. Comp. Neurol. 148, 115–139.

Bunt-Milam, A.H., Saari, J.C., Klock, I.B. and Garwin, G.G. (1985). Zonulae adherentes pore size in the external limiting membrane of the rabbit retina. Invest. Ophthalmol. Vis. Sci. 26, 1377–1380.

Burkhardt, D.A. (1994). Light adaptation and photopigment bleaching in cone photoreceptors in situ in the retina of the turtle. J. Neurosci. 14, 1091–1105.

Burns, M.E. and Baylor, D.A. (2001). Activation, deactivation, and adaptation in vertebrate photoreceptor cells. Annu. Rev. Neurosci. 24, 779–805.

Cajal, S.R.Y. (1892). The Structure of the Retina (English translation by S.A. Thorpe and M. Glickstein, 1972). Charles C Thomas, Springfield; IL.

Calkins, D.J. (2001). Seeing with S cones. Prog. Retin. Eye Res. 20, 255–287.

Cohen, A.I. (1987). The Retina. In: Adler's Physiology of the Eye. (Mosses, R.A. and Hart, W.M., Eds.), 8th ed. Mosby, St.Louis.

Curcio, C.A. and Allen, K.A. (1990a). Topography of ganglion cells in human retina. J. Comp. Neurol. 300, 5–25.

Curcio, C.A., Sloan, K.R., Kalina, R.E. and Hendrickson, A.E. (1990b). Human photoreceptor topography. J. Comp. Neurol. 292, 497–523.

Cusato, K., Bosco, A., Rozental, R., Guimaraes, C.A., Reese, B.E., Linden, R. and Spray, D.C. (2003). Gap junctions mediate bystander cell death in developing retina. J. Neurosci. 23, 6413–6422.

Dacey, D.M. (1999). Primate retina: Cell types, circuits and color opponency. Prog. Retin. Eye Res. 18, 737–763.

Dacey, D.M. and Petersen, M.R. (1992). Dendritic field size and morphology of midget and parasol ganglion cells of the human retina. Proc. Natl. Acad. Sci. USA 89, 9666–9670.

Daiger, S.P., Sullivan, L.S. and Rossiter, B.J.F. (2003). The RetNet. Laboratory for the Molecular Diagnosis of Inherited Eye Diseases. http://www.sph.uth.tmc.edu/Retnet/home.htm

Daw, N.W., Jensen, R.J. and Brunken, W.J. (1990). Rod pathways in mammalian retinae. Trends Neurosci. 13, 110–115.

Dowling, J.E. (1987). The Retina. An Approachable Part of the Brain, pp. 1–282. The Belknap Press of the Harvard University Press, Cambridge and London.

Dowling, J.E. (1998). Creating Mind. How the Brain Works. W. W. Norton & Company, New York and London.

Dowling, J.E. and Boycott, B.B. (1966). Organization of the primate retina: Electron microscopy. Proc. R. Soc. Lond. B. Biol. Sci. 166, 80–111.

Ebrey, T. and Koutalos, Y. (2001). Vertebrate photoreceptors. Prog. Retin. Eye Res. 20, 49–94.

Ehinger, B. and Floren, I. (1979). Absence of indoleamine-accumulating neurons in the retina of humans and cynomolgus monkeys. Albrecht. Von. Graefes. Arch. Klin. Exp. Ophthalmol. 209, 145–153.

Ehinger, B. and Zucker, C.L. (1996). GABAA receptors in neurons of the nerve fiber layer in rabbit retina. Vis. Neurosci. 13, 991–994.

Eldred, W.D. (2001). Real time imaging of the production and movement of nitric oxide in the retina. Prog. Brain. Res. 131, 109–122.

Fain, G.L., Matthews, H.R., Cornwall, M.C. and Koutalos, Y. (2001). Adaptation in vertebrate photoreceptors. Physiol. Rev. 81, 117–151.

Feigenspan, A., Teubner, B., Willecke, K. and Weiler, R. (2001). Expression of neuronal connexin36 in AII amacrine cells of the mammalian retina. J. Neurosci. 21, 230–239.

Feldheim, D.A., Kim, Y.I., Bergemann, A.D., Frisen, J., Barbacid, M. and Flanagan, J.G. (2000). Genetic analysis of ephrin-A2 and ephrin-A5 shows their requirement in multiple aspects of retinocollicular mapping. Neuron 25, 563–574.

Fisher, S.K. and Boycott, B.B. (1974). Synaptic connections made by horizontal cells within the outer plexiform layer of the retina of the cat and the rabbit. Proc. R. Soc. London. Series B. 186, 317–331.

Fletcher, E.L. and Wässle, H. (1999). Indoleamine-accumulating amacrine cells are presynaptic to rod bipolar cells through GABA(C) receptors. J. Comp. Neurol. 413, 155–167.

Gastinger, M.J., O'Brien, J.J., Larsen, N.B. and Marshak, D.W. (1999). Histamine immuno-reactive axons in the macaque retina. Invest. Ophthalmol. Vis. Sci. 40, 487–495.

Gioffi, G.A., Grandtam, E. and Alm, A. (2003). Ocular circulation. In: Adler's Physiology of the Eye. (Kaufman, P.L. and Alm, A., Eds.), 10 ed., pp. 747–784. Mosby, St. Louis.

Guimaraes, C., Assreuy, J. and Linden, R. (2001). Paracrine neuroprotective effect of nitric oxide in the developing retina. J. Neurochem. 76, 1233–1241.

Hannibal, J., Hindersson, P., Knudsen, S.M., Georg, B. and Fahrenkrug, J. (2002). The photopigment melanopsin is exclusively present in pituitary adenylate cyclase-activating polypeptide-containing retinal ganglion cells of the retinohypothalamic tract. J. Neurosci. 22, RC191.

Hattar, S., Liao, H.-W., Takao, M., Berson, D.M. and Yau, K.-W. (2002). Melanopsin-containing retinal ganglion cells: Architecture, projections, and intrinsic photosensitivity. Science 295, 1065–1070.

Hendrickson, A.E. and Yuodelis, C. (1984). The morphological development of the human fovea. Ophthalmology 91, 603–612.

Karpen, J.W., Zimmerman, A.L., Stryer, L. and Baylor, D.A. (1988). Gating kinetics of the cyclic-GMP-activated channel of retinal rods: Flash photolysis and voltage-jump studies. Proc. Natl. Acad. Sci. USA 85, 1287–1291.

Koch, K.W., Duda, T. and Sharma, R.K. (2002). Photoreceptor specific guanylate cyclases in vertebrate phototransduction. Mol. Cell. Biochem. 230, 97–106.

Kolb, H., Fernandez, E. and Nelson, R. (2003). Webvision. The organization of the vertebrate retina. http://webvision.med.utah.edu/

Kolb, H., Nelson, R., Ahnelt, P. and Cuenca, N. (2001). Cellular organization of the vertebrate retina. Prog. Brain Res. 131, 3–26.

Land, M.F. and Nilsson, D.E. (2002). Animal Eyes, pp. 1–221. Oxford University Press, Oxford.

Last, R.J. (1968). Eugene Wolff's Anatomy of the Eye and Orbit, 6th ed. Lewis, London.

Leskov, I.B., Klenchin, V.A., Handy, J.W., Whitlock, G.G., Govardovskii, V.I., Bownds, M.D., Lamb, T.D., Pugh, E.N., Jr. and Arshavsky, V.Y. (2000). The gain of rod phototrans-duction: Reconciliation of biochemical and electrophysiological measurements. Neuron 27, 525–537.

Linberg, K.A. and Fisher, S.K. (1988). Ultrastructural evidence that horizontal cell axon terminals are presynaptic in the human retina. J. Comp. Neurol. 268, 281–297.

Linden, R., Rehen, S.K. and Chiarini, L.B. (1999). Apoptosis in developing retinal tissue. Prog. Retin. Eye Res. 18, 133–165.

Ma, J., Znoiko, S., Othersen, K.L., Ryan, J.C., Das, J., Isayama, T., Kono, M., Oprian, D.D., Corson, D.W., Cornwall, M.C., Cameron, D.A., Harosi, F.I., Makino, C.L. and Crouch, R.K. (2001). A visual pigment expressed in both rod and cone photo-receptors. Neuron 32, 451–461.

MacNeil, M.A., Heussy, J.K., Dacheux, R.F., Raviola, E. and Masland, R.H. (1999). The shapes and numbers of amacrine cells: Matching of photofilled with Golgi-stained cells in the rabbit retina and comparison with other mammalian species. J. Comp. Neurol. 413, 305–326.

Maeda, T., Imanishi, Y. and Palczewski, K. (2003). Rhodopsin phosphorylation: 30 years later. Prog. Retin. Eye Res. 22, 417–434.

Mangel, S.C. (1991). Analysis of the horizontal cell contribution to the receptive field surround of ganglion cells in the rabbit retina. J. Physiol. 442, 211–234.

Mann, I.C. (1923). The Development of the Human Eye, pp. 1–306. Cambridge University Press, Cambridge.

Marc, R.E. and Jones, B.W. (2002). Molecular phenotyping of retinal ganglion cells. J. Neurosci. 22, 413–427.

Marc, R.E. and Liu, W. (2000). Fundamental GABAergic amacrine cell circuitries in the retina: Nested feedback, concatenated inhibition, and axosomatic synapses. J. Comp. Neurol. 425, 560–582.

Mariani, A.P. (1984). Bipolar cells in monkey retina selective for the cones likely to be blue-sensitive. Nature 308, 184–186.

Marmor, M.F. and Martin, L.J. (1978). 100 years of the visual cycle. Surv. Ophthalmol. 22, 279–285.

Marquardt, T. (2003). Transcriptional control of neuronal diversification in the retina. Prog. Retin. Eye Res. 22, 567–577.

Marshak, D.W. (2001). Synaptic inputs to dopaminergic neurons in mammalian retinas. Prog. Brain Res. 131, 83–91.

Masland, R.H. (1994). Cell mosaics and neurotransmitters. In: Principles and Practice of Ophthalmology; Basic Sciences. (Albert, D.M. and Jakobiec, F.A., Eds.), pp. 384–398. W. B. Saunders Company, Philadelphia-London-Toronto-Montreal-Sydney-Tokyo.

Mata, N.L., Radu, R.A., Clemmons, R.C. and Travis, G.H. (2002). Isomerization and oxidation of vitamin A in cone-dominant retinas: A novel pathway for visual-pigment regeneration in daylight. Neuron 36, 69–80.

McBee, J.K., Palczewski, K., Baehr, W. and Pepperberg, D.R. (2001). Confronting complexity: The interlink of phototransduction and retinoid metabolism in the vertebrate retina. Prog. Retin. Eye Res. 20, 469–529.

Mills, S.L., O'Brien, J.J., Li, W., O'Brien, J. and Massey, S.C. (2001). Rod pathways in the mammalian retina use connexin 36. J. Comp. Neurol. 436, 336–350.

Newman, E. and Reichenbach, A. (1996). The Müller cell: A functional element of the retina. Trends Neurosci. 19, 307–312.

O'Leary, D.D., Fawcett, J.W. and Cowan, W.M. (1986). Topographic targeting errors in the retinocollicular projection and their elimination by selective ganglion cell death. J. Neurosci. 6, 3692–3705.

Pepe, I.M. (2001). Recent advances in our understanding of rhodopsin and phototransduction. Prog. Retin. Eye Res. 20, 733–759.

Perez, M.T., Larsson, B., Alm, P., Andersson, K.E. and Ehinger, B. (1995). Localisation of neuronal nitric oxide synthase-immunoreactivity in rat and rabbit retinas. Exp. Brain Res. 104, 207–217.

Peters, M.A. (2002). Patterning the neural retina. Curr. Opin. Neurobiol. 12, 43–48.

Picaud, S. (2003). Retinal biochemistry. In: Adler's Physiology of the Eye. (Kaufman, P.L. and Alm, A., Eds.), 10 ed., pp. 382–408. Mosby, St. Louis.

Pourcho, R.G. (1996). Neurotransmitters in the retina. Curr. Eye Res. 15, 797–803.

Provencio, I., Rollag, M.D. and Castrucci, A.M. (2002). Photoreceptive net in the mammalian retina. Nature 415, 493.

Pugh, E.N., Jr., Nikonov, S. and Lamb, T.D. (1999). Molecular mechanisms of vertebrate photoreceptor light adaptation. Curr. Opin. Neurobiol. 9, 410–418.

Ramirez, J.M., Trivino, A., Ramirez, A.I., Salazar, J.J. and Garcia Sanchez, J. (1996). Structural specializations of human retinal glial cells. Vision Res. 36, 2029–2036.

Ramón and y Cajal, S. (1972). The structure of the retina. Compiled and translated by Sylvia A Thorpe and Mitchell Glickstein.

Rodieck, R.W. (1973). The Vertebrate Retina. W.H. Freeman & Co., San Francisco.

Rodieck, R.W. (1998). The First Steps in Seeing, pp. 1–562. Sinauer Associates, Inc., Sunderland, MA.

Ross, S.A., McCaffery, P.J., Drager, U.C. and De Luca, L.M. (2000). Retinoids in embryonal development. Physiol. Rev. 80, 1021–1054.

Saari, J.C. (2000). Biochemistry of visual pigment regeneration: The Friedenwald lecture. Invest. Ophthalmol. Vis. Sci. 41, 337–348.

Schnapf, J.L., Nunn, B.J., Meister, M. and Baylor, D.A. (1990). Visual transduction in cones of the monkey Macaca fascicularis. J. Physiol. 427, 681–713.

Sears, S., Erickson, A. and Hendrickson, A. (2000). The spatial and temporal expression of outer segment proteins during development of Macaca monkey cones. Invest. Ophthalmol. Vis. Sci. 41, 971–979.

Silver, J. and Hughes, A.F. (1973). The role of cell death during morphogenesis of the mammalian eye. J. Morphol. 140, 159–170.

Sjöstrand, J., Olsson, V., Popovic, Z. and Conradi, N. (1999). Quantitative estimations of foveal and extra-foveal retinal circuitry in humans. Vision Res. 39, 2987–2998.

Steinberg, R.H., Linsenmeier, R.A. and Griff, E.R. (1983). Three light-evoked responses of the retinal pigment epithelium. Vision Res. 23, 1315, 1983.

Straiker, A., Stella, N., Piomelli, D., Mackie, K., Karten, H.J. and Maguire, G. (1999). Cannabinoid CB1 receptors and ligands in vertebrate retina: Localization and function of an endogenous signaling system. Proc. Natl. Acad. Sci. USA 96, 14565–14570.

Thoreson, W.B. and Witkovsky, P. (1999). Glutamate receptors and circuits in the vertebrate retina. Prog. Retin. Eye Res. 18, 765–810.

Tornquist, P., Alm, A. and Bill, A. (1990). Permeability of ocular vessels and transport across the blood-retinal-barrier. Eye 4 (Pt 2), 303–309.

Wässle, H. (1999). Mammlian rod and cone pathways. In: The Retinal Basis of Vision. (Toyoda, J.-I., Murakami, M., Kaneko, A. and Saito, T., Eds.), pp. 185–195. Elsevier, Amsterdam-Lausanne-New York-Oxford-Shannon-Singapore-Tokyo.

Wässle, H. and Boycott, B.B. (1991). Functional architecture of the mammalian retina. Physiol. Rev. 71, 447–480.

Weiler, R., He, S. and Vaney, D.I. (1999). Retinoic acid modulates gap junctional permeability between horizontal cells of the mammalian retina. Eur. J. Neurosci. 11, 3346–3350.

Weng, J., Mata, N.L., Azarian, S.M., Tzekov, R.T., Birch, D.G. and Travis, G.H. (1999). Insights into the function of Rim protein in photoreceptors and etiology of Stargardt's disease from the phenotype in abcr knockout mice. Cell 98, 13–23.

Wright, L.L., Macqueen, C.L., Elston, G.N., Young, H.M., Pow, D.V. and Vaney, D.I. (1997). The DAPI-3 amacrine cells of the rabbit retina. Visual Neurosci. 14, 473–492.

Yazulla, S., Studholme, K.M., Mcintosh, H.H. and Fan, S.F. (2000). Cannabinoid receptors on goldfish retinal bipolar cells: Electron-microscope immunocytochemistry and whole-cell recordings. Visual Neurosci. 17, 391–401.

Yokoyama, S. (2000). Molecular evolution of vertebrate visual pigments. Prog. Retin. Eye Res. 19, 385–419.

Young, R.W. (1985). Cell differentiation in the retina of the mouse. Anat. Rec. 212, 199–205.

Zahs, K.R., Kofuji, P., Meier, C. and Dermietzel, R. (2003). Connexin immunoreactivity in glial cells of the rat retina. J. Comp. Neurol. 455, 531–546.

Zucker, C.L. and Ehinger, B. (1998). Gamma-aminobutyric acid A receptors on a bistratified amacrine cell type in the rabbit retina. J. Comp. Neurol. 393, 309–319.

THE RETINAL PIGMENT EPITHELIUM

Morten la Cour and Tongalp Tezel

ABSTRACT

The retinal pigment epithelium (RPE) is a monolayer of cuboidal epithelial cells intercalated between the photoreceptors and the choriocapillaries. The human RPE incorporates some 3.5 million epithelial cells arranged in a regular hexagonal pattern. The density of RPE cells is relatively uniform throughout the retina, approximately 4000 cells/mm^2. With age, the cell density decreases

Advances in Organ Biology
Volume 10, pages 253–272.
© 2006 Elsevier Inc. All rights reserved.
ISBN: 0-444-50925-9
DOI: 10.1016/S1569-2590(05)10009-3

particularly in the periphery, where it is reduced to approximately 2000 cells/ mm^2 in individuals over 40 years. The peripheral RPE cells are larger and more pleomorphic than central cells (Harman et al., 1997; del Priore et al., 2002). In the primate retina, each RPE cell faces 30–40 photoreceptors, a number that is rather constant throughout the retina, although perhaps somewhat lower in the fovea (Robinson and Hendrickson, 1995). In fully developed primate retinas, no mitoses are seen in the RPE, and the epithelium is currently believed to consist of a stable, nondividing, pool of cells (Tso and Friedman, 1967).

The retinal membrane of the RPE faces the subretinal space, which is the extracellular space surrounding the photoreceptor outer segments (Figure 1). Between the optic disc and the ora serrata, there are no anatomical contacts between the photoreceptors and the RPE. The RPE forms numerous long microvilli that interdigitate with the rod outer segments. In mammals, the cone outer segments are ensheathed by multilamellar specializations of the RPE, the so-called cone sheaths. The epithelial cells are bound together by junctional complexes with tight junctions that separate the cells into an apical half that faces the retina and a basal half that faces the choroid. The nucleus and mitochondriae are located in the basal half of the cell. Numerous pigment granules, located predominantly in the apical cytoplasm, give the epithelium its macroscopic black appearance, from which it derives its name.

The choroidal side of the RPE directly apposes Bruch's membrane, a pentalaminar, approximately 2 μm thick, elastic membrane. The innermost part of Bruch's membrane is the basement membrane of the RPE. The outer part of Bruch's membrane is the basement membrane of the choriocapillaries. In between are two collagenous layers and a central elastic layer.

I. RPE FUNCTION

The retinal pigment epithelial cells act as supportive cells for the photoreceptors. The best studied of these functions are the participation of the RPE in photoreceptor outer segment renewal, the storage and metabolism of vitamin A, the production of cytokines necessary for retinal development and survival, and the transport and barrier functions of the epithelium. These four aspects of RPE physiology will be discussed in separate sections later.

Other, less well-characterized functions of the RPE are the absorption of stray light by the melanin pigment in the epithelium, the scavenging of free radicals by the melanin pigment, and the drug detoxification by the smooth endoplasmic reticulum cytochrome p-450 system (Shichi and Nebert, 1980).

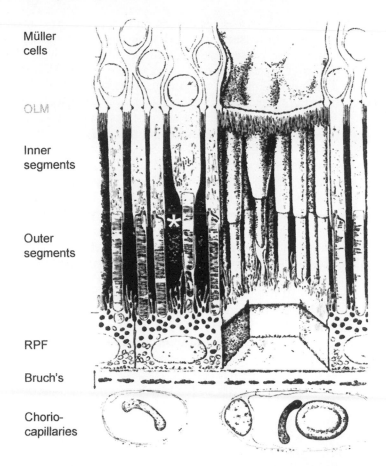

Müller
cells

OLM

Inner
segments

Outer
segments

RPF

Bruch's

Chorio-
capillaries

Figure 1. Diagram of the subretinal space, showing the relationship between the retinal pigment epithelium (RPE), the outer segment of the photoreceptors, the outer limiting membrane (OLM), and the Müller cells. The asterisk denotes the subretinal space. (From Steinberg R.H., Linsenmeier R.A. and Griff E.R. (1983). Vision Res. 23, 1315.)

II. PHOTORECEPTOR OUTER SEGMENT TURNOVER

It has been known for more than 20 years that the outer segments of both rod and cone photoreceptors undergo continuous renewal. New membrane material is added at the base of the outer segments, and old membrane material is removed from the tip of the outer segments by RPE phagocytosis (Besharse and Defoe, 1998). In the Rhesus monkey, the renewal time for rod outer segments is 13 days in the parafoveal region and 9 days in the peripheral retina. These estimates are based on the classical experiments by Young, where he used pulse labeling of rod outer segment discs with radioactive amino acids and subsequent autoradiography (Young, 1971). In cones, a similar outer segment renewal process takes place. However, because of the continuity of the disc membranes with the outer segment cell membrane, the rate of cone outer segment membrane renewal cannot be assessed by pulse labeling. Count of phagosomes, containing outer segment material, within the RPE cells suggests that the rate of outer segment renewal is slower in cones than in rods (Anderson et al., 1980). The outer segment discs are shed in a circadian rhythm. In rods there is a burst of disc shedding and phagocytosis in the morning, immediately after light onset. In cones, there seem to be some species variability in the pattern of disc shedding (Besharse and Defoe, 1998). In rhesus monkeys there is a burst of disc shedding associated with light onset, although discs are also shed during the dark period in this species (Anderson et al., 1980).

Retinal pigment epithelium phagocytosis is important for photoreceptor survival. This was first discovered in RCS rats, where a retinal dystrophy is caused by defective RPE phagocytosis of outer segment material (Chaitin and Hall, 1983). More recently, this dystrophy was linked to the MERKT gene (D'Cruz et al., 2000). Transfer of the MERKT gene has been shown to rescue photoreceptors in RCS rats (Vollrath et al., 2001). The MERKT gene product is a receptor tyrosinase kinase, necessary for normal RPE phagocytosis (Feng et al., 2002). MERKT gene defects have subsequently been shown to cause retinitis pigmentosa in humans, and a murine retinal dystrophy, similar to the one seen in RCS rats (Gal et al., 2000; Duncan et al., 2003). In Usher's syndrome type 1b, a major deaf–blind disorder in humans, RPE phagocytosis is affected; with the causative defect in the myosine 7A gene (Gibbs et al., 2003).

The amount of membrane material ingested and degraded by the RPE cells is impressive. Based on a quantitative study in the Rhesus monkey (Young, 1971), it can be calculated that each extrafoveal RPE cell, every day, must ingest and degrade a volume of rod outer segment material that corresponds to 7% of the volume of the RPE cell itself. Since the RPE cells do normally not divide, the burden of membrane material that these cells must ingest and degrade in their lifetime, far exceeds that of any other phagocytic cell. Probably as a consequence of this massive phagocytic load on the RPE cells, lipofuscin granules accumulate with age in these cells, and makes them autofluorescent (Nilsson et al., 2003). Exocytosis through the RPE choroidal cell membrane of lipofuscin and other waste products from phagocytosis may lead to the accumulation of hydrophobic material in Bruch's membrane, and reduced water permeability of this membrane (Marshall et al., 1998). The major fluorophore in RPE lipofuscin is A2E, which is a product of retinal and ethanolamine (Sakai et al., 1996). Early accumulation of lipofuscin is seen in Stargardt's disease, which is a macular dystrophy caused by a defective ABCR gene (Allikmets, 1997). The ABCR gene product seems to be involved in the transfer of all-trans-retinal from the disc lumen and into the cytosol of outer segments in rods (Weng et al., 1999). Accumulation of A2E and other indigestible retinoid metabolites in the rods might be cause of the excess RPE lipofuscin accumulation in Stargardt's disease and possibly in age-related macular degeneration as well (Mata et al., 2001).

III. RETINOID METABOLISM AND THE VISUAL CYCLE

The retinal pigment epithelium plays an important role in the uptake, storage and metabolism of vitamin A and related compounds. It has been known for more than 100 years that photoreception involves bleaching of the visual pigments, and that the RPE is required for the regeneration of these pigments, at least in rods (Marmor and Martin, 1978). The underlying mechanisms are now known in some detail (McBee et al., 2001). The retinoid 11-cis-retinaldehyde is the chromophore of the visual pigments in mammals. When light is absorbed by the visual pigment, the phototransduction cascade is initiated, and the visual pigment bleaches. During the bleaching process, the chromophore is released from the visual pigment and converted within the photoreceptor to all-trans-retinol, which is then released from the outer segment. Some of this all-trans-retinol finds its way to the retinal pigment epithelium, where it is reisomerized to 11-cis-retinol, oxidized to 11-cis-retinaldehyde, and subsequently transported back to the photoreceptor outer segments. The light induced movement of retinoids between the photoreceptors and the retinal pigment epithelium, and the involved transformations between the different retinoids, is denoted the *visual cycle* (Thompson and Gal, 2003).

Retinoids enter the RPE cell via three different mechanisms: Firstly, all-trans-retinol is taken up from the blood through the choroidal membrane of the RPE cells. All-trans-retinol circulates in the blood bound to a small, 21 kD, carrier protein, serum retinol binding protein, RBP, which is complexed to another and larger, protein, transthyretin. Autoradiographic studies have demonstrated specific membrane binding sites for RBP on the RPR choroidal membrane (Thompson and Gal, 2003, 1976). However, the receptor has not been identified with certainty, and the retinol-uptake mechanism in the RPE choroidal membrane remains to be elucidated (Chader et al., 1998). Secondly, retinoid bound to the visual pigments enter the RPE cells via the phagocytosis of outer segment material (see previous section). Finally, retinoid is shuttled back and forth between the RPE and the outer segments as a part of the visual cycle. The all-trans-retinol that is liberated from the outer segments is shuttled to the RPE through the subretinal space. The inter-photoreceptor matrix retinoid-binding protein, IRBP, has been implicated as a carrier protein in retinoid transport through the subretinal space, since this protein is the most abundant protein in the subretinal space, and capable of binding both retinol and retinal in their all-trans- as well as their 11-cis configurations (Chader et al., 1998). However, the role of IRBP as the obligate carrier of retinoids in the subretinal space has been recently been challenged. In mice with a targeted disruption of the IRBP gene (IRBP-/- gene knockout mice), it was found that the kinetics of 11-cis-retinal and rhodopsin recovery after a flash of light was only marginally slower in IRBP-/- mice than in normal, wild-type, mice (Palczewski et al., 1999). IRBP may play a role in retinoid transport between Müller cells and cones, and it may play a role in retinal development (Gonzalez-Fernandez, 2003).

The fate of retinol inside the RPE cell has recently been reviewed (Thompson and Gal, 2003), and is illustrated in Figure 2. Once inside the RPE cell, the retinol is bound to a small, 16 kDa, carrier protein, cellular retinol-binding protein (CRBP), which is a relative ubiquitous protein, not specific for the RPE. Retinol might then be esterified to a palmitate group derived from the membrane phospholipids lecithin. This reaction is cata-lyzed by lecithin-retinol acyltransferase (LRAT). The retinyl ester is a stable, nontoxic form of the retinoid, and both the all-trans- and the 11-cis-isomer of retinol is stored as retinyl esters within the RPE cell. The formation of retinyl esters is an important step in the synthesis of the 11-cis-isomer, since all-trans-retinyl ester is the substrate for the isomerohydrolase enzyme that catalyzes the combined hydrolysis of the ester bond and isomerization of the all-trans-retinol to 11-cis-retinol (Bernstein et al., 1987). Isomerization of all-trans-retinol to the 11-cis-isomer is an endothermic reaction, and the energy required is probably derived from the energy-rich ester bond in the retinyl-lester (Dreigner et al., 1989). The isomerohydrolase enzyme has not been purified or cloned. Another RPE protein, RPE65, is necessary for the isomerization. RPE65 is distinct from the isomerohydrolase, although it is probably associated with it (Thompson and Gal, 2003). The 11-cis-retinol that is the product of the isomerization is bound to cellular retinaldehyde binding protein, CRALBP. This is a 36 kDa protein localized mainly in RPE cells and Müller cells. It binds retinol and retinal, but is specific to the 11-cis isomer. CRALBP might facilitate the isomerization reaction by trapping the 11-cis product. The 11-cis-retinol can be reesterified by LRAT and stored in the RPE cell as 11-cis-retinylester, or it can be oxidized to 11-cis-retinal by 11-cis-retinol dehydrogenase (11-cis-RDH). This reaction uses NAD^+ as cofactor. While bound to CRALBP, 11-cis-retinal travels to the retinal membrane of the RPE cell where it is released. The mechanism for this is not clear.

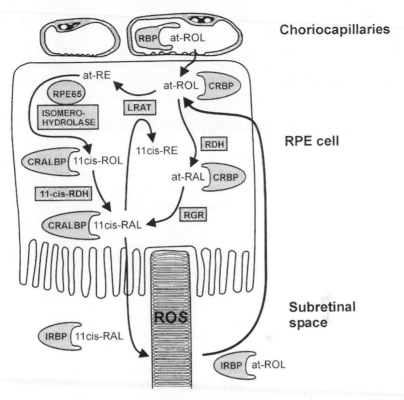

Figure 2. Diagram of retinoid trafficking in and around the retinal pigment epithelium (RPE) cell. ROS= rod outer segment; 11cis-ROL = 11–cis-retinol; 11cis-RAL = 11-cis-retinal; at-ROL = all-trans-retinol; at-RAL = all-trans-retinal; RGR = retinal G-protein coupled receptor; RBP = serum retinol binding protein, CRBP = cellular retinol binding protein, CRALBP = cellular retinal binding protein; RDH = all-trans-retinol dehydrogenase; 11-cis-RDH = retinol dehydrogenase specific for 11-cis-retinol; IRBP = interphotoreceptor matrix retinoid binding protein.

An alternative pathway for the isomerization of all-trans-retinol to 11-cis-retinol within the RPE cell is light dependent and resembles the mechanism used in eyes of cephalopods (e.g., octopus and squid). In this pathway the substrate for the isomerization reaction is all-trans-retinal. All-trans-retinol must therefore be oxidized to all-trans-retinal. This is accomplished by an all-trans-retinol dehydrogenase (all-trans-RDH). The isomerization is catalyzed by the enzyme retinal G-protein coupled receptor (RGR), which binds all-trans-retinal covalently with a Schiff base linkage resembling the attachment in rhodopsin. Illumination of the RGR-all-trans-retinal compound results in the formation of 11-cis-retinal. The RGR protein corresponds to retinochrome found in the myeloid bodies of cephalopod retinula cells (Thompson and Gal, 2003).

Proteins involved in retinoid metabolism seem to be important for photoreceptor survival. Table 1 shows gene products, expressed in the RPE, in which genetic defects have been shown to cause retinal disease.

IV. PRODUCTION OF CYTOKINES

Both photoreceptors and the choriocapillaries depend on the RPE for their survival. If the RPE is destroyed by chemical or mechanical means, the photoreceptors and the choriocapillaries atrophy (Peyman and Bok, 1972). As will be discussed further below, the RPE is necessary for the normal development of both the neuroretina and choroid (Raymond and Jackson, 1995; Zhao and Overbeek, 2001).

The RPE cell has been shown to produce many growth factors, and the roles of some of these have begun to be elucidated. Vascular endothelial growth factor, VEGF or VEGF-A, is a potent angiogenic factor, which is secreted by RPE cells in a polarized fashion, with 2–7 fold more being secreted towards the choroidal side of the epithelium (Blaauwgeers et al., 1999). The polarized secretion targets the vascular endothelium of the choriocapillaries, which expresses two different VEGF receptors, VEGFR-1 and VEGFR-2, preferentially towards its RPE facing side. During embryogenesis, a peak of VEGF production by the RPE is necessary for the development of the fetal choroidal vasculature (Witmer et al., 2003). VEGF also increases the permeability of vessel walls, and may contribute to the maintenance of the fenestrated phenotype of the choriocapillaries (Blaauwgeers et al., 1999). Ischemia, mechanical distortion, advanced glycation end products all upregulate VEGF production by RPE cells, and may result in

Table 1. RPE Gene Defects and Retinal Disease

Protein	Disease
RPE65	Leber congenital amaurosis
LRAT	Leber congenital amaurosis
RGR	Retinitis pigmentosa
CRALBP	Recessive retinitis pigmentosa
11-cis-RDH	Fundus albipunctatus
IRBP	Retinal dystrophy in mice

Proteins involved in retinoid metabolism in the retinal pigment epithelium (RPE), where genetic defects cause retinal disease. LRAT = lechitin retinol acyl transferase; RGR= retinal G-protein coupled receptor; CRALBP = cellular retinaldehyde binding protein; 11-cis-RDH = retinol dehydrogenase, specific for 11-cis-retinol; IRBP = Interphotoreceptor matrix retinoid binding protein. References in (Gonzalez-Fernandez, 2003; Thompson and Gal, 2003).

choroidal and/or retinal neovascularization (Seko et al., 1999; Ohno-Matsui et al., 2001; Uhlmann et al., 2002; Witmer et al., 2003). On the other hand, blockage of VEGF signalling inhibits choroidal neovascularization in a murine model (Kwak et al., 2000).

The angiogenic stimulation by VEGF is balanced by an RPE-derived antiangiogenic factor, pigment epithelium derived factor, PEDF (Tombran and Barnstable, 2003). PEDF provides autocrine negative feedback of the angiogenic stimulation caused by VEGF, partly through upregulation of PEDF expression though VEGFR-1 (Ohno-Matsui et al., 2003). Apart from its antiangiogenic properties, PEDF has neurotropic and neuroprotective properties (Tombran and Barnstable, 2003). The balance between PEDF and VEGF has been proposed as a major mechanism keeping the ocular vasculature patent and avoiding pathological angiogenesis (Ohno-Matsui et al., 2001).

Several members of the FGF family are involved in autocrine and paracrine control of RPE growth, resistance to apoptotic cell death, and wound repair (Schweigerer et al., 1987; Bost et al., 1992; Yamamoto et al., 1996; Mascarelli et al., 2001). FGF-2, previously named basic FGF, stimulates RPE growth, but also has neurotrophic and angiogenic properties. Failure of RPE to generate FGF signaling in transgenic mice results in disrupted photoreceptors and thinned choroid (Rousseau et al., 2000).

The RPE cells have been shown to produce a number of other cytokines, with less well-characterized roles (Campochiaro, 1998; Holtkamp et al., 2001).

V. TRANSPORT

The outer retina is avascular, and the choriocapillaries are the main source of oxygen and nutrients for the photoreceptors. The retinal pigment epithelium is interposed between the choriocapillaries and the photoreceptors, and the epithelial cells are bound together by tight junctions that effectively hinder transfer of water soluble compounds in-between the cells (i.e., via the paracellular route). Hence, the RPE cells control the exchange of water-soluble nutrients and metabolites between the choroid and the subretinal space.

In vitro experiments have shown that the retinal pigment epithelium has a retina positive transepithelial potential between 2 and 15 mV. This potential is responsible for the cornea positive standing potential of the eye, which can be recorded by DC electroretinography (Gallemore et al., 1998). The transepithelial electrical resistance across isolated RPE preparations has been measured to be between 79 and 350 Ωcm^2 (Hughes et al., 1998). The isolated epithelium *in vitro* absorbs Na^+, Cl^-, HCO_3^-, and K^+. The transepithelial transport of these ions has been extensively studied in isolated preparations of frog and bovine RPE (Figure 3). In both preparations, a number of transport mechanisms have been identified in the RPE membranes, and models have been proposed that explain quantitatively the transepithelial transport of the major ions in these species (la Cour, 1993; Hughes et al., 1998). The retinal membrane of the RPE incorporates an electrogenic Na^+/K^+ pump that pumps Na^+ out of the cell and K^+ into the cell at the expenditure of metabolic energy. The Na^+/K^+ pump lowers the intracellular concentration of Na^+, which is below the electrochemical equilibrium for this ion. Secondary active transport systems use the energy invested in the inwardly directed Na^+ gradient to drive transport of other ions. The retinal membrane incorporates three such secondary active transport systems: a $Na^+:K^+:2Cl^-$ cotransport system, a Na^+/H^+ exchange mechanism, and a $Na^+:2HCO_3^-$ cotransport system. These transport systems accumulate Cl^- and HCO_3^- intracellularly above their electrochemical equilibrium.

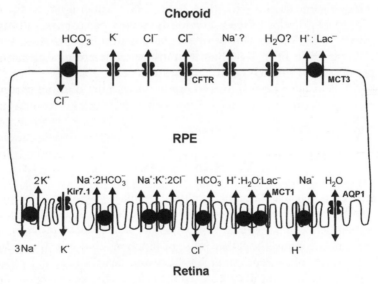

Figure 3. Model of mammalian RPE ion transport mechanisms. The retinal membrane incorporates an active Na^+/K^+ pump, an inward rectifying K^+ conductance of the Kir7.1 type, a rheogenic $Na^+:2HCO_3^-$ cotransport system, a $Na^+:K^+:2Cl^-$ cotransport system, a Cl^-/HCO_3^- exchange mechanism, a Na^+/H^+ exchange mechanism, and a H^+:water:lactate cotransport mechanism of the MCT1 type. The choroidal membrane incorporates a K^+ conductance, a Cl^- conductance, part of which is the CFTR channel, a Cl^-/HCO_3^- exchange mechanism, and a H^+: lactate transport system of the MCT3 type. There are yet uncharacterized efflux mechanisms for Na^+, and water across the choroidal membrane. The figure also shows the aquaporin water channel AQP1 is located in the retinal membrane.

The choroidal membrane incorporates a cAMP dependent Cl^- conductance that serves as exit mechanism for this ion (la Cour, 1992). Some of this cAMP dependent chloride conductance stems from the cystic fibrosis transmembrane conductance regulator (CFTR), which has recently been identified in human RPE choroidal membrane by reverse transcriptase PCR (Quinn et al., 2001; Blaug et al., 2003). The choroidal membrane also incorporates a cAMP independent Cl^- conductance, which is regulated by the intracellular Ca^{2+} concentration (Quinn et al., 2001). Bicarbonate exit is mediated by Cl^-/HCO_3^- exchangers in the RPE choroidal membrane. Most of the potassium that is pumped into RPE cells is recycled across the retinal membrane, through an unusual inward rectifying K^+ conductance, the major component of which is due to the Kir7.1 channel (la Cour et al., 1986; Segawa and Hughes, 1994; Hughes and Takahira, 1996; Yang et al., 2003). Some potassium exits the RPE cells via a smaller, yet uncharacterized K^+ conductance in the choroidal membrane (la Cour et al., 1986).

The RPE cells incorporate transport systems for a number of organic nutrients and metabolites. The GLUT1 transport protein, which mediates facilitated transport of glucose across brain capillaries, is found in both the retinal and choroidal membrane of the RPE (Sugasawa et al., 1994). In photoreceptors, glycolysis is active even under aerobic conditions, and lactate has to be cleared from the subretinal space (Wang et al., 1997). The RPE incorporates the lactate transport systems MCT1 (MCT = mono carboxylic acid transporter) in its retinal membrane and MCT3 in its choroidal membrane (la Cour et al., 1994; Philp et al., 2003). Other transport mechanisms, including transport systems for GABA, ascorbic acid, fluorescein, and amino acids, have also been described in the retinal pigment epithelium (Hughes et al., 1998).

Ion transport across the RPE has been shown to be influenced by receptors in the retinal membrane of the epithelium. Both alpha- and beta receptors, as well as a purine receptor $P2Y_2$ have been identified (Quinn et al., 2001; Yang et al., 2003). Stimulation of these receptors results in an increased chloride conductance in the choroidal membrane of the epithelium. The intracellular second messenger that conveys the signal from the retinal to the choroidal membrane is cAMP for the beta receptor, and Ca^{2+} for the $P2Y_2$ purine receptor (Quinn et al., 2001; Yang et al., 2003).

The retinal pigment epithelium from a number of species has been shown to absorb water (Hughes et al., 1998). This is consistent with the clinical observation that fluid under a rhegmatogenous retinal detachment absorbs quickly once the holes in the neurosensory retina are surgically closed. The rate of subretinal fluid resorption across the entire RPE in this setting has been estimated to be 2 ml per 24 hours. This corresponds to more than 50% of the aqueous secretion in that period (Marmor, 1998a). This high rate of water transport is probably not present under physiological conditions, since the retina with its tight extracellular spaces has a very low-hydraulic conductivity (Tsuboi, 1987). Nevertheless, the retinal pigment epithelium has a large reserve capacity for removal of excess fluid from the subretinal space (Marmor, 1998a). The RPE water transport may have a role in the maintenance of normal retinal adhesion by exerting suction on the neurosensory retina. However, other mechanisms also seem to contribute to retinal adhesion (Marmor, 1998b). Under the pathologic condition of retinal detachment, the RPE capacity for fluid absorption is important for the surgical treatment of this condition. The RPE fluid absorption can be enhanced by stimulation of the purinoceptor $P2Y_2$ as well as by beta adrenergic stimulation (Edelman and Miller, 1991; Peterson et al., 1997; Yang et al., 2003). Interestingly, the increased fluid absorption has been shown to increase the rate of resolution of experimental retinal detachments in rabbits (Yang et al., 2003).

The mechanisms underlying RPE water transport is poorly understood. It is a clinical observation, confirmed in laboratory experiments, that subretinal fluid can be absorbed despite a high-protein content (Marmor, 1998a). Proteinaceous retinal exudates can eventually be dehydrated to the extent that lipoprotein crystals precipitate in the retina as the so-called hard exudates. Proposed mechanisms for RPE fluid transport must therefore be able to account for the apparent ability of the RPE to transport fluid against an osmotic gradient. It is known that the passive water permeability in the retinal membrane is larger than that in the choroidal membrane (la Cour and Zeuthen, 1993). This water permeability may be due to the water channel aquaporin-1, AQP1, which recently was found to be expressed in the retinal membrane of human RPE (Stamer et al., 2003). Aquaporins are thought to be passive water channels. In order to explain vectorial water transport against an osmotic gradient, active transport systems are needed that can transport water against an electrochemical gradient (Stein and Zeuthen, 2002). The only such transport system so far known in the RPE is a H^+: lactate: water cotransport system in the retinal membrane of both frog and porcine RPE (Zeuthen et al., 1996; Hamann et al., 2003). However, the physiological significance of this transport system is unknown at the present time.

VI. THE RPE IN WOUND HEALING AND PROLIFERATIVE VITREORETINAL DISEASE

The RPE is often considered a glial element in the retina (Bok, 1993). Despite being a resting epithelium in adult individuals, the epithelium has a high-proliferative capacity, and after retinal trauma or rhegmatogenous retinal detachment, RPE cells can proliferate vigorously. The proliferating cells liberated from the epithelium can be seen clinically as free pigmented cells, so-called tobacco dust, in the vitreous (Fisher and Anderson, 1998;

Hiscott et al., 1999). The free RPE cells settle on all available surfaces, dedifferentiate, and acquire macrophage and fibroblast-like characteristics. Eventually they may participate in the formation of contractile membranes, a condition called proliferative vitreoretinopathy (PVR), which is a feared complication of rhegmatogenous retinal detachment (Kirchhof and Sorgente, 1989). Several factors may induce this dedifferentiation, including vitreous, albumin, serum, activated macrophages, and several growth factors, including hepatocyte growth factor (HGF), and connective tissue growth factor (CTGF) (Kirchhof and Sorgente, 1989; Hinton et al., 2002). The dedifferentiation goes along with down-regulation of proteins associated with the highly specialized functions and the cell shape of the RPE cells (Alge et al., 2003). Proliferative vitreoretinopathy might be characterized as an inappropriate and excessive wound-healing response of RPE, which results in the distortion and tractional detachment of the neurosensory retina.

Despite the high-proliferative capacity of RPE cells, it is a clinical observation that the RPE monolayer has a limited capacity to heal acquired *in situ* defects resulting from aging, inflammation, surgery, or trauma (Grierson et al., 1994). Recent experimental studies of RPE wound healing have shown that the RPE cells have the capacity to repopulate experimentally induced defects in the RPE monolayer, but that concurrent damage to Bruch's membrane may prevent healing of RPE defects *in situ* (Tezel and del Priore, 1999a; Wang et al., 2003). Healing of RPE defects requires attachment of RPE cells to Bruch's membrane at the base of the defect. It has become apparent that this attachment is critically dependent on age and on the general status of this membrane. If the basal laminar layer of Bruch's membrane is intact, RPE cells can attach to and repopulate both young and aged (>60 years) Bruch's membrane, whereas they fail to attach and survive on the deeper layers of aged human Bruch's membrane (Tezel and del Priore, 1999a). Age-related changes such as cross-linking and deposition of long-spacing collagen, drusen formation, calcifications, cracks, or loss of inner layers can make these areas uninhabitable for RPE. Such RPE defects eventually lead to the development of progressive choriocapillaries and photoreceptor atrophy, and eventually turn into areas of geographic atrophy (del Priore et al., 1996).

RPE wound healing does not necessarily involve cell proliferation. RPE cells are postmitotic and occasional cell loss does not stimulate cell proliferation, but is compensated for by the flattening and migration of the neighboring RPE cells (Grierson et al., 1994). This mechanism is an important role for the repopulation of small involutional RPE defects especially in the macula. Changes in the topographical density of aging human RPE suggest that there is a continuous proliferation and sliding of the peripheral RPE to cover midperipheral and macular RPE defects (del Priore et al., 2002). Mitosis, on the other hand, seems to be necessary to cover larger defects in the RPE monolayer (Hergott and Kalnins, 1991). Proliferating and migrating RPE cells become flat, display prominent stress fibers, and acquire the appearance of fibroblasts. This wound-healing process requires upregulation of various adhesion molecules, cytokines, and signaling genes (Singh et al., 2001). At later stages of wound healing, cultured RPE cells produce extracellular matrix molecules and remodel their matrix (Kamei et al., 1998).

VII. DEVELOPMENT

RPE develops from a layer of ciliated and pseudostratified neuroectodermal precursor cells that form the outer layer of the optic cup. At the 4th week of gestation, these cells start melanogenesis by the activation of the tyrosinase promoter and form simple cuboidal epithelia (Oguni et al., 1991). They are initially hexagonal in shape, and have short microvilli. A smooth RPE basal lamina can be identified in as early as the 5th week of gestation (Fu and Li, 1989). RPE basal lamina, elastic layer, and choriocapillaries basal lamina form the primordial Bruch's membrane (Marmorstein et al., 1998). At around the 11th week of gestation, the inner and outer collagen layers appear flanking the central elastic layer, thus completing the pentalaminar structure of the adult Bruch's membrane (Fu and Li, 1989). Further development of the embryonic RPE waits for the maturation of fetal photoreceptors. As photoreceptor's outer segments start to develop into the subretinal space, RPE extends long-apical microvilli to ensheath them. In order to adapt to photoreceptors' high need of nutrients, RPE cells develop basal infoldings that increase their surface area for uptake and transport purposes (Marmorstein et al., 1998). Maturation of photoreceptors also induce polarization of certain proteins, such as apically polarized Na^+, K^+-ATPase, N-CAM, EMMPRIN, $\alpha V\beta 5$ integrin, and basally polarized $\alpha 6\beta 1$ integrin (Marmorstein et al., 1998). Junctional complexes are observed as early as the 6th week of gestation (Oguni et al., 1991).

Survival and development of primordial RPE is controlled by a number of transcription and growth factors, such as brain derived neurotrophic factor, BDNF (Liu et al., 1997; Hackett et al., 1998). Interestingly, RPE precursors can be induced to transdifferentiate into neural retinal tissue by FGF stimulation (Galy et al., 2002).

Growth of RPE is required for the development of the choroid, neural retina, and vitreous. Neural retina and vitreous fail to develop in transgenic mice lacking RPE due to expression of diphtheria toxin-A within the developing epithelial cells (Raymond and Jackson, 1995). Similarly, choroid fails to develop in transgenic mice where RPE is transdifferentiated into neural retina, due to expression of FGF9 in the developing RPE (Zhao and Overbeek, 2001).

REFERENCES

Alge, C.S., Suppmann, S., Priglinger, S.G., Neubauer, A.S., May, C.A., Hauck, S., Welge-Lussen, U., Ueffing, M. and Kampik, A. (2003). Comparative proteome analysis of native differentiated and cultured dedifferentiated human RPE cells. Invest. Ophthalmol. Vis. Sci. 44, 3629–3641.

Allikmets, R. (1997). A photoreceptor cell-specific ATP-binding transporter gene (ABCR) is mutated in recessive Stargardt macular dystrophy. Nat. Genet. 17, 122.

Anderson, D.H., Fisher, S.K., Erickson, P.A. and Tabor, G.A. (1980). Rod and cone disc shedding in the rhesus monkey retina: A quantitative study. Exp. Eye Res. 30, 559–574.

Bernstein, P., Law, W. and Rando, R. (1987). Isomerization of all-trans-retinoids to 11-cis-retinoids *in vitro*. Proc. Natl. Acad. Sci. USA 84, 1849–1853.

Blaauwgeers, H.G., Holtkamp, G.M., Rutten, H., Witmer, A.N., Koolwijk, P., Partanen, T.A., Alitalo, K., Kroon, M.E., Kijlstra, A., van Hinsbergh, V.W. and Schlingemann, R.O. (1999). Polarized vascular endothelial growth factor secretion by human retinal pigment epithelium and localization of vascular endothelial growth factor receptors on the inner choriocapillaris. Evidence for a trophic paracrine relation. Am. J. Pathol. 155, 421–428.

Blaug, S., Quinn, R., Quong, J., Jalickee, S. and Miller, S.S. (2003). Retinal pigment epithelial function: A role for CFTR? Doc. Ophthalmol. 106, 43–50.

Bok, D. (1993). The retinal pigment epithelium: A versatile partner in vision. J. Cell. Sci. 17 (suppl.), 189–195.

Bost, L.M., Aotaki-Keen, A.E. and Hjelmeland, L.M. (1992). Coexpression of FGF-5 and bFGF by the retinal pigment epithelium *in vitro*. Exp. Eye Res. 55, 727–734.

Campochiaro, P.A. (1998). Growth factors in the retinal pigment epithelium and retina. In: The Retinal Pigment Epithelium. (Marmor, M.F. and Wolfensberger, T.J., Eds.), 2nd edn., pp. 459–477. Oxford University Press, New York.

Chader, G.J., Pepperberg, D.R. and Crouch, R. (1998). Retinoids and the retinal pigment epithelium. In: The Retinal Pigment Epithelium. (Marmor, M.F. and Wolfensberger, T.J., Eds.), 2nd edn., pp. 135–151. Oxford University Press, New York.

D'Cruz, P.M., Yasumura, D., Weir, J., Matthes, M.T., Abderrahim, H., LaVail, M.M. and Vollrath, D. (2000). Mutation of the receptor tyrosine kinase gene Mertk in the retinal dystrophic RCS rat. Hum. Mol. Genet. 9, 645–651.

del Priore, L.V., Kaplan, H.J., Hornbeck, R., Jones, Z. and Swinn, M. (1996). Retinal pigment epithelial debridement as a model for the pathogenesis and treatment of macular degeneration. Am. J. Ophthalmol. 122, 629–643.

del Priore, L.V., Kuo, Y.H. and Tezel, T.H. (2002). Age-related changes in human RPE cell density and apoptosis proportion *in situ*. Invest. Ophthalmol. Vis. Sci. 43, 3312–3318.

Dreigner, P., Law, W., Cañada, F. and Rando, R. (1989). Membranes as the energy source in the endogenic transformation of vitamin A to 11-cis-retinol. Sci. 244, 968–971.

Duncan, J.L., La Vail, M.M., Yasumura, D., Matthes, M.T., Yang, H., Trautmann, N., Chappelow, A.V., Feng, W., Earp, H.S., Matsushima, G.K. and Vollrath, D. (2003). An RCS-like retinal dystrophy phenotype in mer knockout mice. Invest. Ophthalmol. Vis. Sci. 44, 826–838.

Feng, W., Yasumura, D., Matthes, M.T., LaVail, M.M. and Vollrath, D. (2002). Mertk triggers uptake of photoreceptor outer segments during phagocytosis by cultured retinal pigment epithelial cells. J. Biol. Chem. 277, 17016–17022.

Gal, A., Li, Y., Thompson, D.A., Weir, J., Orth, U., Jacobson, S.G., Apfelstedt-Sylla, E. and Vollrath, D. (2000). Mutations in MERTK, the human orthologue of the RCS rat retinal dystrophy gene, cause retinitis pigmentosa. Nat. Genet. 26, 270–271.

Gallemore, R.P., Hughes, B.A. and Miller, S.S. (1998). Light-induced responses of the retinal pigment epithelium. In: The Retinal Pigment Epithelium. (Wolfensberger, T.J. and Marmor, M.F., Eds.), 2nd edn., pp. 175–198. Oxford University Press, New York.

Galy, A., Neron, B., Planque, N., Saule, S. and Eychene, A. (2002). Activated MAPK/ERK kinase (MEK-1) induces transdifferentiation of pigmented epithelium into neural retina. Dev. Biol. 248, 251–264.

Gibbs, D., Kitamoto, J. and Williams, D.S. (2003). Abnormal phagocytosis by retinal pigmented epithelium that lacks myosin VIIa, the Usher syndrome 1B protein. Proc. Natl. Acad. Sci. USA 100, 6481–6486.

Gonzalez-Fernandez, F. (2003). Interphotoreceptor retinoid-binding protein—An old gene for new eyes. Vis. Res. 43, 3021–3036.

Grierson, I., Hiscott, P., Hogg, P., Robey, H., Mazure, A. and Larkin, G. (1994). Development, repair, and regeneration of the retinal pigment epithelium. Eye 8(Pt 2), 255–262.

Hackett, S.F., Friedman, Z., Freund, J., Schoenfeld, C., Curtis, R., DiStefano, P.S. and Campochiaro, P.A. (1998). A splice variant of trkB and brain-derived neurotrophic factor are coexpressed in retinal pigmented epithelial cells and promote differentiated character-istics. Brain Res. 789, 201–212.

Hamann, S., Kiilgaard, J.F., la Cour, M., Prause, J.U. and Zeuthen, T. (2003). Cotransport of H+, lactate, and H_2O in porcine retinal pigment epithelial cells. Exp. Eye Res. 76, 493–504.

Hinton, D.R., He, S., Jin, M.L., Barron, E. and Ryan, S.J. (2002). Novel growth factors involved in the pathogenesis of proliferative vitreoretinopathy. Eye 16, 422–428.

Hiscott, P., Sheridan, C., Magee, R.M. and Grierson, I. (1999). Matrix and the retinal pigment epithelium in proliferative retinal disease. Prog. Retin. Eye Res. 18, 167–190.

Holtkamp, G.M., Kijlstra, A., Peek, R. and de Vos, A.F. (2001). Retinal pigment epithelium-immune system interactions: Cytokine production and cytokine-induced changes. Prog. Retin. Eye Res. 20, 29–48.

Hughes, B.A., Gallemore, R.P. and Miller, S.S. (1998). Transport mechanisms in the retinal pigment epithelium. In: The Retinal Pigment Epithelium. (Marmor, M.F. and Wolfensberger, T.J., Eds.), 2nd edn., pp. 103–134. Oxford University Press, New York.

Kamei, M., Kawasaki, A. and Tano, Y. (1998). Analysis of extracellular matrix synthesis during wound healing of retinal pigment epithelial cells. Microsc. Res. Tech. 42, 311–316.

Kwak, N., Okamoto, N., Wood, J.M. and Campochiaro, P.A. (2000). VEGF is major stimula-tor in model of choroidal neovascularization. Invest. Ophthalmol. Vis. Sci. 41, 3158–3164.

la Cour, M. (1992). Cl⁻ transport in frog retinal pigment epithelium. Exp. Eye Res. 54, 921–931.

la Cour, M. (1993). Ion transport in the retinal pigment epithelium. A study with double barrelled ion-selective microelectrodes. Acta Ophthalmol. 71 (suppl. 209), 1–32.

la Cour, M., Lin, H., Kenyon, E. and Miller, S.S. (1994). Lactate transport in freshly isolated human fetal retinal pigment epithelium [published erratum appears in Invest. Ophthalmol. Vis. Sci. 1995 Apr; 36(5):757]. Invest. Ophthalmol. Vis. Sci. 35, 434–442.

la Cour, M., Lund-Andersen, H. and Zeuthen, T. (1986). Potassium transport of the frog retinal pigment epithelium: Autoregulation of potassium activity in the subretinal space. J. Physiol. 375, 461–479.

Liu, Z.Z., Zhu, L.Q. and Eide, F.F. (1997). Critical role of TrkB and brain-derived neurotrophic factor in the differentiation and survival of retinal pigment epithelium. J. Neurosci. 17, 8749–8755.

Marmor, M.F. (1998a). Control of subretinal fluid and mechanisms of serous detachment. In: The Retinal Pigment Epithelium. (Wolfensberger, T.J. and Marmor, M.F., Eds.), 2nd edn., pp. 420–438. Oxford Universtiy Press, New York.

Marmor, M.F. (1998b). Mechanisms of retinal adhesion. In: The Retinal Pigment Epithelium. (Wolfensberger, T.J. and Marmor, M.F., Eds.), 2nd edn., pp. 392–405. Oxford University Press, New York.

Marmorstein, A.D., Finnemann, S.C., Bonilha, V.L. and Rodriguez-Boulan, E. (1998). Morphogenesis of the retinal pigment epithelium: Toward understanding retinal degenerative diseases. Ann. NY Acad. Sci. 857, 1–12.

Marshall, J., Hussain, A.A., Starita, C., Moore, D.J. and Patmore, A.L. (1998). Aging and Bruch's membrane. In: The Retinal Pigment Epithelium. (Marmor, M.F. and Wolfensberger, T.J., Eds.), 1st edn., pp. 669–692. Oxford University Press, New York.

Mascarelli, F., Hecquet, C., Guillonneau, X. and Courtois, Y. (2001). Control of the intracellular signaling induced by fibroblast growth factors (FGF) over the proliferation and survival of retinal pigment epithelium cells: Example of the signaling regulation of growth factors endogenous to the retina. J. Soc. Biol. 195, 101–106.

Mata, N.L., Tzekov, R.T., Liu, X., Weng, J., Birch, D.G. and Travis, G.H. (2001). Delayed dark-adaptation and lipofuscin accumulation in abcr+/− mice: Implications for involvement of ABCR in age-related macular degeneration. Invest. Ophthalmol. Vis. Sci. 42, 1685–1690.

McBee, J.K., Palczewski, K., Baehr, W. and Pepperberg, D.R. (2001). Confronting complexity: The interlink of phototransduction and retinoid metabolism in the vertebrate retina. Prog. Retin. Eye Res. 20, 469–529.

Nilsson, S.E., Sundelin, S.P., Wihlmark, U. and Brunk, U.T. (2003). Aging of cultured retinal pigment epithelial cells: Oxidative reactions, lipofuscin formation, and blue light damage. Doc. Ophthalmol. 106, 13–16.

Oguni, M., Tanaka, O., Shinohara, H., Yoshioka, T. and Setogawa, T. (1991). Ultrastructural study on the retinal pigment epithelium of human embryos, with special reference to quantitative study on the development of melanin granules. Acta Anat. (Basel) 140, 335–342.

Ohno-Matsui, K., Morita, I., Tombran-Tink, J., Mrazek, D., Onodera, M., Uetama, T., Hayano, M., Murota, S.I. and Mochizuki, M. (2001). Novel mechanism for age-related macular degeneration: An equilibrium shift between the angiogenesis factors VEGF and PEDF. J. Cell Physiol. 189, 323–333.

Ohno-Matsui, K., Yoshida, T., Uetama, T., Mochizuki, M. and Morita, I. (2003). Vascular endothelial growth factor upregulates pigment epithelium-derived factor expression via VEGFR-1 in human retinal pigment epithelial cells. Biochem. Biophys. Res. Commun. 303, 962–967.

Palczewski, K., Van, H.J., Garwin, G.G., Chen, J., Liou, G.I. and Saari, J.C. (1999). Kinetics of visual pigment regeneration in excised mouse eyes and in mice with a targeted disruption of the gene encoding interphotoreceptor retinoid-binding protein or arrestin. Biochem. 38, 12012–12019.

Peterson, W.M., Meggyesy, C., Yu, K. and Miller, S.S. (1997). Extracellular ATP activates calcium signaling, ion, and fluid transport in retinal pigment epithelium. J. Neurosci. 17, 2324–2337.

Philp, N.J., Wang, D., Yoon, H. and Hjelmeland, L.M. (2003). Polarized expression of monocarboxylate transporters in human retinal pigment epithelium and ARPE-19 cells. Invest. Ophthalmol. Vis. Sci. 44, 1716–1721.

Quinn, R.H., Quong, J.N. and Miller, S.S. (2001). Adrenergic receptor activated ion transport in human fetal retinal pigment epithelium. Invest. Ophthalmol. Vis. Sci. 42, 255–264.

Rousseau, B., Dubayle, D., Sennlaub, F., Jeanny, J.C., Costet, P., Bikfalvi, A. and Javerzat, S. (2000). Neural and angiogenic defects in eyes of transgenic mice expressing a dominant-negative FGF receptor in the pigmented cells. Exp. Eye Res. 71, 395–404.

Sakai, N., Decantur, J., Nakanishi, K. and Eldred, G.E. (1996). Ocular age pigment 'A2E': An unprecedented pyridinium bisretinoid. J. Chem. Soc. 118, 1559–1560.

Schweigerer, L., Malerstein, B., Neufeld, G. and Gospodarowicz, D. (1987). Basic fibroblast growth factor is synthesized in cultured retinal pigment epithelial cells. Biochem. Biophys. Res. Commun. 143, 934–940.

Seko, Y., Fujikura, H., Pang, J., Tokoro, T. and Shimokawa, H. (1999). Induction of vascular endothelial growth factor after application of mechanical stress to retinal pigment epithelium of the rat in vitro. Invest. Ophthalmol. Vis. Sci. 40, 3287–3291.

Singh, S., Zheng, J.J., Peiper, S.C. and McLaughlin, B.J. (2001). Gene expression profile of ARPE-19 during repair of the monolayer. Graefes Arch. Clin. Exp. Ophthalmol. 239, 946–951.

Stamer, W.D., Bok, D., Hu, J., Jaffe, G.J. and McKay, B.S. (2003). Aquaporin-1 channels in human retinal pigment epithelium: Role in transepithelial water movement. Invest. Ophthalmol. Vis. Sci. 44, 2803–2808.

Steinberg, R.H., Linsenmeier, R.A. and Griff, E.R. (1983). Three light-evoked responses of the retinal pigment epithelium. Vision Res. 23:1315, 1983.

Sugasawa, K., Deguchi, J., Okami, T., Yamamoto, A., Omori, K., Uyama, M. and Tashiro, Y. (1994). Immunocytochemical analyses of distributions of Na, K-ATPase and GLUT1, insulin and transferrin receptors in the developing retinal pigment epithelial cells. Cell Struct. Funct. 19, 21–28.

Tezel, T.H., Kaplan, H.J. and del Priore, L.V. (1999b). Fate of human retinal pigment epithelial cells seeded onto layers of human Bruch's membrane. Invest. Ophthalmol. Vis. Sci. 40, 467–476.

Tsuboi, S. (1987). Measurement of the volume flow and hydraulic conductivity across the isolated dog retinal pigment epithelium. Invest. Ophthalmol. Vis. Sci. 28, 1776–1792.

Uhlmann, S., Rezzoug, K., Friedrichs, U., Hoffmann, S. and Wiedemann, P. (2002). Advanced glycation end products quench nitric oxide in vitro. Graefes Arch. Clin. Exp. Ophthalmol. 240, 860–866.

Vollrath, D., Feng, W., Duncan, J.L., Yasumura, D., D'Cruz, P.M., Chappelow, A., Matthes, M. T., Kay, M.A. and La Vail, M.M. (2001). Correction of the retinal dystrophy phenotype of the RCS rat by viral gene transfer of Mertk. Proc. Natl. Acad. Sci. USA 98, 12584–12589.

Wang, H., Ninomiya, Y., Sugino, I.K. and Zarbin, M.A. (2003). Retinal pigment epithelium wound healing in human Bruch's membrane explants. Invest. Ophthalmol. Vis. Sci. 44, 2199–2210.

Wang, L., Tornquist, P. and Bill, A. (1997). Glucose metabolism in pig outer retina in light and darkness. Acta Physiol. Scand. 160, 75–81.

Weng, J., Mata, N.L., Azarian, S.M., Tzekov, R.T., Birch, D.G. and Travis, G.H. (1999). Insights into the function of Rim protein in photoreceptors and etiology of Stargardt's disease from the phenotype in abcr knockout mice. Cell 98, 13–23.

Witmer, A.N., Vrensen, G.F., Van Noorden, C.J. and Schlingemann, R.O. (2003). Vascular endothelial growth factors and angiogenesis in eye disease. Prog. Retin. Eye Res. 22, 1–29.

Yamamoto, C., Ogata, N., Yi, X., Takahashi, K., Miyashiro, M., Yamada, H., Uyama, M. and Matsuzaki, K. (1996). Immunolocalization of basic fibroblast growth factor during wound repair in rat retina after laser photocoagulation. Graefes Arch. Clin. Exp. Ophthalmol. 234, 695–702.

Yang, D., Pan, A., Swaminathan, A., Kumar, G. and Hughes, B.A. (2003). Expression and localization of the inwardly rectifying potassium channel Kir7.1 in native bovine retinal pigment epithelium. Invest. Ophthalmol. Vis. Sci. 44, 3178–3185.

Young, R.W. (1971). The renewal of rod and cone outer segments in the rhesus monkey. J. Cell Biol. 49, 303–318.

Zeuthen, T., Hamann, S. and la Cour, M. (1996). Cotransport of H^+, lactate, and H_2O by membrane proteins in retinal pigment epithelium of bullfrog. J. Physiol. (Lond.) 497, 3–17.

FURTHER READING

Besharse, J.C. and Defoe, D.M. (1998). Role of the retinal pigment epithelium in the photoreceptor membrane turnover. In: The Retinal Pigment Epithelium. (Marmor, M.F. and Wolfensberger, T.J., Eds.), 1st edn., pp. 152–172. Oxford University Press, New York.

Bok, D. and Heller, J. (1976). Transport of retinal from the blood to the retina: An autoradiographic study of the pigment epithelial cell surface receptor for plasma retinol-binding protein. Exp. Eye Res. 22, 395–402.

Chaitin, M.H. and Hall, M.O. (1983). Defective ingestion of rod outer segments by culutred dystrophic rat pigment epithelial cells. Invest. Ophthalmol. Vis. Sci. 24, 812–820.

Edelman, J.L. and Miller, S.S. (1991). Epinephrine stimulates fluid absorption across bovine retinal pigment epithelium. Invest. Ophthalmol. Vis. Sci. 32, 3033–3040.

Fisher, S.K. and Anderson, D.H. (1998). Cellular responses of the retinal pigment epithelium to retinal detachment and reattachment. In: The Retinal Pigment Epitelium. (Wolfensberger, T.J. and Marmor, M.F., Eds.), pp. 406–419. Oxford University Press, New York.

Fu, J. and Li, F.M. (1989). Embryonic development and structure of human Bruch's membrane. Zhonghua Yan Ke Za Zhi 25, 18–19.

Harman, A.M., Fleming, P.A., Hoskins, R.V. and Moore, S.R. (1997). Development and aging of cell topography in the human retinal pigment epithelium. Invest. Ophthalmol. Vis. Sci. 38, 2016–2026.

Hergott, G.J. and Kalnins, V.I. (1991). Expression of proliferating cell nuclear antigen in migrating retinal pigment epithelial cells during wound healing in organ culture. Exp. Cell Res. 195, 307–314.

Hughes, B.A. and Takahira, M. (1996). Inwardly rectifying K+ currents in isolated human retinal pigment epithelial cells. Invest. Ophthalmol. Vis. Sci. 37, 1125–1139.

Kirchhof, B. and Sorgente, N. (1989). Pathogenesis of proliferative vitreoretinopathy. Modulation of retinal pigment epithelial cell functions by vitreous and macrophages. Dev. Ophthalmol. 16, 1–53.

la Cour, M. and Zeuthen, T. (1993). Osmotic properties of the frog retinal pigment epithelium. Exp. Eye Res. 56, 521–530.

Marmor, M.F. and Martin, L.J. (1978). 100 years of the visual cycle. Surv. Ophthalmol. 22, 279–285.

Peyman, G.A. and Bok, D. (1972). Peroxidase diffusion in the normal and laser-coagulated primate retina. Invest. Ophthalmol. 11, 35–45.

Raymond, S.M. and Jackson, I.J. (1995). The retinal pigmented epithelium is required for development and maintenance of the mouse neural retina. Curr. Biol. 5, 1286–1295.

Robinson, S.R. and Hendrickson, A. (1995). Shifting relationships between photoreceptors and pigment epithelial cells in monkey retina: Implications for the development of retinal topography. Vis. Neurosci. 12, 767–778.

Segawa, Y. and Hughes, B.A. (1994). Properties of the inwardly rectifying K^+ conductance in the toad retinal pigment epithelium. J. Physiol. (Lond.) 476, 41–53.

Shichi, H. and Nebert, D.W. (1980). Drug metabolism in ocular tissues. In: Extrahepatic Metabolism of Drugs and Other Foreign Compounds. (Gram, T.E., Ed.), 1st edn., pp. 333–363. MTP press limited.

Snodderly, D.M., Sandstrom, M.M., Leung, I.Y., Zucker, C.L. and Neuringer, M. (2002). Retinal pigment epithelial cell distribution in central retina of rhesus monkeys. Invest. Ophthalmol. Vis. Sci. 43, 2815–2818.

Stein, W. and Zeuthen, T. (2002). In Molecular Mechanisms of Water Transport, 1st edn., pp. 1–475. Academic Press, London.

Tezel, T.H. and del Priore, L.V. (1999a). Repopulation of different layers of host human Bruch's membrane by retinal pigment epithelial cell grafts. Invest. Ophthalmol. Vis. Sci. 40, 767–774.

Thompson, D.A. and Gal, A. (2003). Genetic defects in vitamin A metabolism of the retinal pigment epithelium. Dev. Ophthalmol. 37, 141–154.

Tombran-Tink, J. and Barnstable, C.J. (2003). PEDF: A multifaceted neurotrophic factor. Nat. Rev. Neurosci. 4, 628–636.

Tso, M.O.M. and Friedman, E. (1967). The retinal pigment epithelium. I. Comparative histology. Arch. Ophthalmol. 78, 641–649.

Zhao, S. and Overbeek, P.A. (2001). Regulation of choroid development by the retinal pigment epithelium. Mol. Vis. 7, 277–282.

THE CHOROID AND OPTIC NERVE HEAD

Jens Folke Kiilgaard and Peter Koch Jensen

I. DEFINITION

The major vascular layer of the eye consists of the choroid, the iris, and the ciliary body. This vascular layer is also called the uvea, because of its resemblance with the inside of a purple grape (*uva* means grape in Greek). The Choroid is located in between sclera and retina (Duke-Elder and Wybar, 1961).

Advances in Organ Biology
Volume 10, pages 273–290.
© 2006 Elsevier Inc. All rights reserved.
ISBN: 0-444-50925-9
DOI: 10.1016/S1569-2590(05)10010-X

II. ANATOMY

External examination of the choroid reveals a rich vascularised tissue, which resembles the outer chorion of the fetus. It is a thin (0.1–0.15 mm) sheath located between retina and sclera. It is light to dark brown in color, depending on the pigmentation of the stroma. In albinos it is bluish. The choroid continues anteriorly into the ciliary body and iris, and ends posteriorly at the optic nerve. The consistency of the choroid is soft and spongy. The uvea is fixed to the sclera anteriorly at the scleral spur, at the exit points of the vortex veins, and posteriorly at the optic nerve (Figure 1). The choroid can easily be removed from the sclera by blunt dissection. The choroid is classically divided into four or five layers by light microscopy.

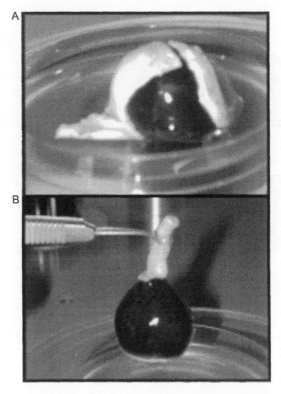

Figure 1. Dissected pig eye. (A) Shows the correlation between Sclera, which is partly removed and choroid. (B) Sclera is completely removed and only the optic nerve and choroid remains. Note the resemblance with the inside of a purple grape, which lead early anatomists to give the vascular layer its name uvea (uva means grape in Greek).

The epichoroid, or suprachoroidal layer is the outermost part of the choroid facing the sclera. This layer is possibly derived from both sclera and choroid. Suprachoroidea is approx. 30 μm thick. It consists of membranes of thin elastic and collagen fibers, which forms a fine network with several fibroblasts and melanocytes. Within the fibers and membranes of the suprachoroidal layer is a potential space, the suprachoroidal space, in which fluid can accumulate under pathological conditions forming a choroidal detachment.

The Vessel layer is the thickest layer of the choroid. The essential constituent of this layer is the blood vessels embedded in loose connective tissue with collagen, elastic and reticular fibers, but almost no capillaries are found in this layer. The vessel layer is also described as consisting of an outer Haller's layer, which consist of larger vessels, and inner Sattler's layer, which consist of smaller vessels. A distinct border between the two layers cannot be distinguished, but the larger vessels are seen nearest the suprachoroid zone and only small vessels are seen in the foveal region.

The choriocapillaris layer is a single layer of large, densely packed capillaries consisting of single endothelial tubes arranged in small lobules. Each lobule is supplied by a central arteriole (Hayreh, 1990) (Figure 2). The lumen of the choriocapillaries is 20–50 μm, which is larger than the lumen of the retinal capillaries (5 μm). Ultrastructurally, the choriocapillaris are of the fenestrated type, similar to those found in the renal glomerulus and other organs. The fenestrations are circular, and measure approximately 800 Å in diameter (Nagy and Ogden 1990) (Figure 4). In contrast to the renal glomerulus capillaries, the fenestrations in the choriocapillaris are covered by a diaphragm. The choriocapillaries are polarized towards the retina with the endothelial nuclei located externally, and the fenestrations located internally, towards Bruch's membrane, and the retinal pigment epithelium. This arrangement of capillaries is unique to the choriocapillaris and is believed to enhance the metabolic exchange between the choroidal blood and the retina.

The choriocapillaries lobules are drained by venoles situated in the periphery of the lobules (Hayreh, 1990) (Figure 2). These venoles unite into small venous whorls, or vortices leading off into the choroidal veins, which again combine into the larger vortex, or vorticose veins that penetrate the sclera.

The innermost layer of the choroid is Bruch's membrane. It consists of five layers (Figure 5). A central elastic layer is sandwiched in between the inner and outer collagenous layer. The outermost layer of Bruch's membrane is the basement membrane of the choriocapillaris; the innermost layer is the basement membrane of the retinal pigment epithelium. Bruch's membrane varies in thickness from 4 μm near the optic disk disc to 1–2 μm in the periphery, and its thickness increases with age (Hogan et al., 1971).

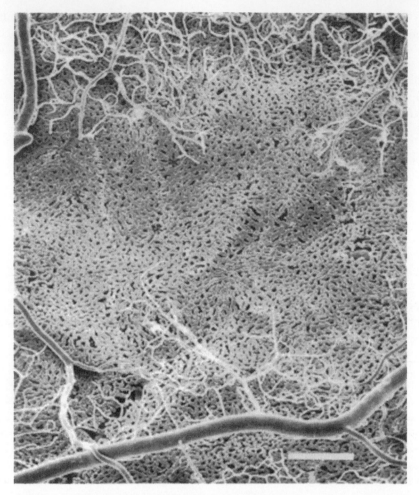

Figure 2. Posterior pole choriocapillaris viewed from the retinal aspect. Remnants of the inferior retinal vessel arcade and retinal capillaries are visible. Lobular appearance is difficult to distinguish but can be identified. * = Choroidal arteriole opening. Bar = 250 μm (From Olver, J.M. (1990). Eye 4, 255–261).

III. THE CHOROIDAL CIRCULATION

In humans, the choroidal circulation is derived from the ophthalmic artery, which branches into the central retinal artery and the main posterior ciliary arteries and several anterior ciliary arteries. One to five main posterior ciliary arteries usually supply the choroid. These main posterior ciliary arteries run anteriorly along the optic nerve. Just before reaching sclera they

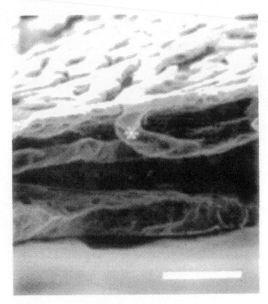

Figure 3. Cross section of equatorial choroid showing flattened, wide-diameter capillaries and draining venule (*) Bar = 60 μm (From Olver, J.M. (1990). Eye 4, 255–261).

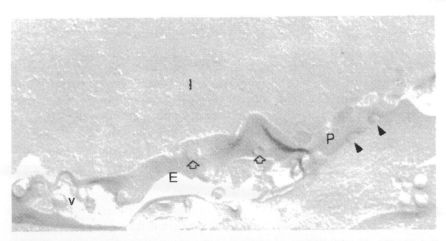

Figure 4. Freeze-fracture electron micrograph of an endothelial cell of the choriocapillaris. Fenestration (arrowheads) appear as papillae on the P face (P) or as craters (open arrowheads) on the E face (E). Cytoplasmic vesicles. v. Capillary lumen I. (x56280) (From Nagy, A.R. and Ogden, T.E. (1990). Eye 4, 290–302).

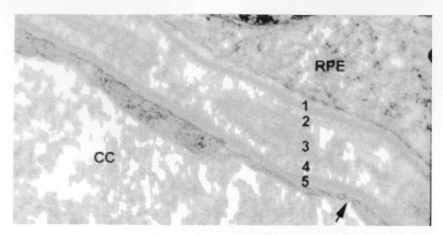

Figure 5. Transmission electron micrograph of Bruch's membrane from the domestic pig. RPE = cytoplasm of a retinal pigment epithelial cell, CC = lumen of choriocapillaris, 1 = basement membrane of the RPE, 2 = inner collagenous layer, 3 = lamina elastica, 4 = outer collagenous layer, 5 = basement membrane of the choriocapillaris, arrow marks a fenestration of the endothelial cell.

divide into the short ciliary arteries, 15–20 branches, which then penetrate sclera in a ring formation around the optic nerve (Amalric, 1983). The interscleral course is straight and very short and perpendicular to the sclera. Hereafter, the arteries branch extensively to supply the choriocapillaries. Anteriorly the choroidal circulation form anastomoses with recurrent branches from the long posterior ciliary arteries.

The long posterior ciliary arteries are normally two in number. They penetrate sclera obliquely 3–4 disc-diameters medial and temporal from the optic nerve, and have in contrast to the short ciliary arteries a long interscleral course. The long posterior ciliary arteries supply the ciliary body, iris, and the temporal and medial part of choroid.

Most of the choroidal vessels are venoles that drain the choriocapillaris. Small venoles collect the blood from the choriocapillaris lobules, and they join to form whorl-shaped systems in the outer part of the vascular layer of the choroid (Haller's layer). These venoles then join to form the vortex veins. There are a total of four to seven vortex veins, or, one or two for each quadrant of the eye. The vortex veins exit the globe at the equatorial region in oblique channels, or emissaries, after forming ampullae near the internal sclera (Figure 7, K). The vortex veins drains to the superior and inferior orbital veins, which drains into the cavernous sinus and pterygoid plexus, respectively. *In vivo* fluorescein-angiographic studies have

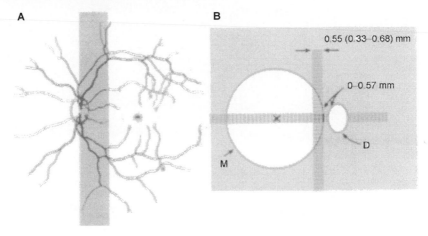

Figure 6. (A) Localization of the watershed between the medial and lateral main arteries and their supply areas (From Hayreh, S.S. (1980). Trans. Ophthalmol. Soc. UK. 100, 400–407). (B) The watersheds between the vortex veins. X = fovea, M = macular region, D = disc (From Hayreh, S.S. (1974). Albrecht Von Graefes Arch. Klin. Exp. Ophthalmol. 192, 181–196).

shown that the posterior ciliary arteries functionally works as end arteries in a segmental fashion (De Laey, 1977; Hayreh, 1990), however, several postmortem vascular cast studies have shown several anastomoses between the posterior arteries (Duke-Elder and Wybar 1961; Hogan et al., 1971; Olver, 1990).

Hayreh (1990) has described the presence of watershed zones (i.e., the territory between choriocapillaris lobules supplied by two different end arteries). These watershed zones are predilection sites for choroidal ischemia (Figure 6).

IV. THE CIRCULATION OF THE OPTIC DISC

Blood flow to the optic disc is obtained from three sources, classically described as supplying three distinct layers of optic nerve head tissue. The immediate retrobulbar part of the optic nerve is supplied from the central retinal artery. The major part of the optic disc is supplied from the circle of Zinn-Haller (Figures 8 and 9) that receives blood from choroidal arteries, from the short posterior ciliary arteries, and with a small contribution from pial arterial network (Hayreh, 2001). Small branches from the central arterial artery, supply the superficial nerve fiber layer in the optic

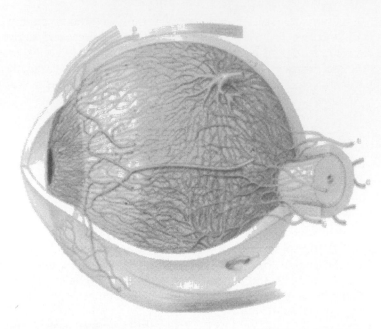

Figure 7.   The Uveal Blood vessels. The blood supply of the eye is derived from the
ophthalmic artery. Except for the central retinal artery that supplies the inner retina,
almost the entire blood supply of the eye comes from the uveal vessels. There are two
long posterior ciliary arteries: one entering the uvea nasally and one temporally
along the horizontal meridian of the eye near the optic nerve (A). These two arteries
give off three to five branches (B) at the ora serrata that pass directly back to form the
anterior choriocapillaris. These capillaries nourish the retina from the equator forward. The short posterior ciliary arteries enter the choroid around the optic nerve (C).
They divide rather rapidly to form the posterior choriocapillaris that nourishes the
retina as far as the equator (the choriocapillaris is not shown in this drawing). This
system of capillaries is continuous with those derived from the long posterior ciliary
arteries. The anterior ciliary arteries (D) pass forward with the rectus muscles, and
then pierce the sclera to enter the ciliary body. Before joining the major circle of the
iris they give off 8–12 branches (E) that pass back through the ciliary muscle to join
the anterior choriocapillaris. The major circle of the iris (F) lies in the corona ciliaris
and sends branches posteriorly into the ciliary body as well as forward into the iris
(G) and limbus (O). The circle of Zinn-Haller (H) is formed by pial branches (I) as
well as branches from the short posterior ciliary arteries. The circle lies in the sclera
and furnishes part of the blood supply to the optic nerve and disc. The vortex veins
exit from the eye through the posterior sclera (J) after forming an ampulla (K) near the
internal sclera. Venous branches that join the anterior and posterior part of the vortex
system are meridionally oriented and are fairly straight (L), while those joining the
vortices on their medial and lateral sides are oriented circularly about the eye (M).
The venous return from the iris and ciliary body is mainly posterior into the vortex
system, but some veins cross the anterior sclera and limbus (O) to enter the episcleral
system of veins. (From Hogan, et al. (1971). Histology of the Human Eye. Saunders.)

disc. Experimental studies have shown that the blood supply to the optic disk except for the most superficial part completely derives from the choroid circulation. (Hayreh, 2001). At the capillary level, there are no direct anastomoses between the choriocapillaries and the capillaries of the optic disc.

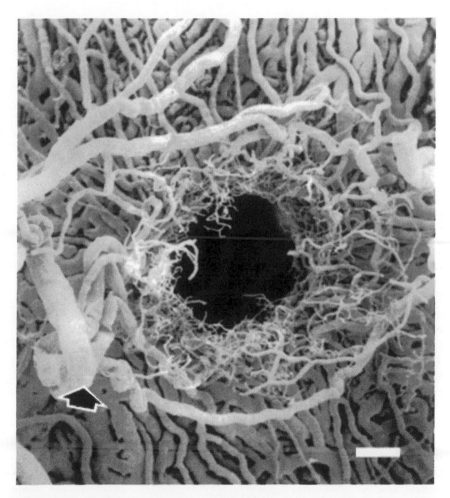

Figure 8. Anastomoses between medial and lateral short posterior ciliary arteries forming complete circle of Haller and Zinn. Note double supply temporally (solid arrow), which forms an incomplete anastomosis from which pial vessels also arise. (From a left eye). Bar = 400 μm (From Olver, J.M. (1990). Eye 4, 7–24).

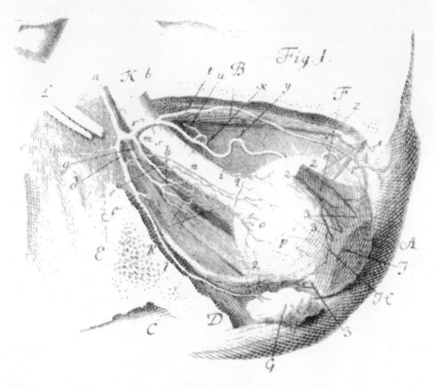

Figure 9. Zinn's original woodcut of the blood supply of the eye based on dissections showing the position of the "circle" of Haller and Zinn at (q) Made in 1755 (copied from Olver, J.M. (1990). Eye 4, 7–24).

V. INNERVATION OF THE CHOROID

The innervation of the choroid is mainly vasomotor. The choroid receives several small branches from the short posterior ciliary nerves, mainly containing sympathetic fibers, but parasympathetic and sensory fibers are also found. Branches from the long ciliary nerve run in the suprachoroidal layer, en route to the anterior uvea and cornea, which they innervate. Choroidal axons are observed in the suprachoroid and vessel layer but absent from the choriocapillary layer (Trivino et al., 2002). The axons form fine plexes in the vascular layer, called vasomotor, and innervate primarily the large vessels and their primary branches. Several ganglion cells are also found in the choroid vascular layer. The concentration of nerve fibers is highest in the posterior part of the human choroid, where they could be necessary to maintain control of blood flow in this zone. These neurons could induce vasodilatation reflexes in the vessels controlling the entrance of blood into the choroidal circulation.

VI. PHYSIOLOGY

The principal functions of the choroid are to nourish the outer retina and to provide a pathway for vessels that supply the anterior part of the eye. However, it also plays an important role in temperature regulation, maintenance of intraocular pressure, and it provides a smooth internal surface for the retina.

The choroid has some exceptional characteristics. The blood flow through the choriocapillaries is extremely high even when compared with tissues such as the kidney cortex (Figure 10).

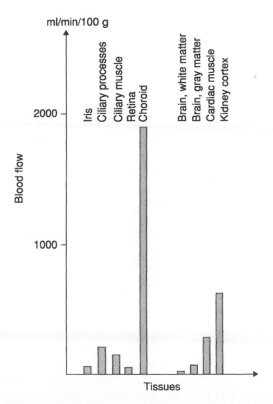

Figure 10. Blood flow through tissues of monkey eye. Flow through other tissues is included for comparison. (From Cioffi, et al. (2003). Adlers Physiology.)

Interestingly the oxygen extraction is low in the choroidal circulation. The arterio-venous difference for blood oxygenation in choroid has been measured in cats to be a mere 3% (Alm and Bill 1970). This difference is much lower than what is found in almost all other vascular beds; in the human retinal circulation it is approximately 38% (Cioffi et al., 2003). The low-oxygen extraction is a consequence of the high-capillary flow in the choriocapillaris, and it results in an extracellular oxygen tension at the level of the retinal pigment epithelium that approaches the oxygen tension in the arterial blood, approximately 100 mmHg (Figure 11). The outer retina is avascular, and the oxygen supply of photoreceptors is derived from the choriocapillaris. Since the mitochondria of the photoreceptors reside in the inner segments, approximately 100 μm from the choriocapillaris, and the oxygen has to be transported by diffusion over this distance, there is a steep gradient in the retinal oxygen tension from the choriocapillaris to the level of the photoreceptor inner segments (Figure 11). In darkness, where the photoreceptor dark current requires energy and oxygen, the oxygen tension at the level of the inner segments approaches zero, and the retina is in a borderline ischemic state (Figure 11). It seems that the high-choroidal blood flow and the steep oxygen gradient are needed to ensure sufficient diffusional oxygen flux into the photoreceptor inner segments

The regulation of choroidal blood flow is incompletely understood. There is ambiguity in the literature as to whether the choroidal circulation is autoregulated. By controlling the intraocular pressure, Alm changed the intraocular perfusion pressure in cats and monkeys, and found a linear correlation between the choroidal blood flow and the perfusion pressure (Figure 12), indicating that the choroidal blood flow is not autoregulated (Alm and Bill 1970). In contrast, Kiel manipulated the arterial blood pressure, and found that the choroidal blood flow indeed is autoregulated in rabbits (Kiel, 1994, 1995). The discrepancies may be species dependent since the rabbit has a nonvascular retina, whereas the cat (like humans) has a vascularised retina. The nonvascularised retina may need some kind of autoregulation of the oxygen supply to the retina.

The innervation of the choroidal circulation is mainly sympathetic (Delaey et al., 2000), and local factors as nitric oxide and carbon dioxide have been shown to play a role for the vascular tone in choroidal vessels (Geiser et al., 2000; Schmetterer and Polak 2001). Choroidal blood flow in pigeons is regulated by parasympathetic fibers derived from the medial part of the nucleus of Edinger-Westphal (EW) via the ipsilateral ciliary ganglion (Fitzgerald et al., 1990; Zagvazdin et al., 1996).

The high-choroidal blood flow act as a heat sink that drains away the heat generated from light absorbed by melanin pigments. Thus the choroid serves

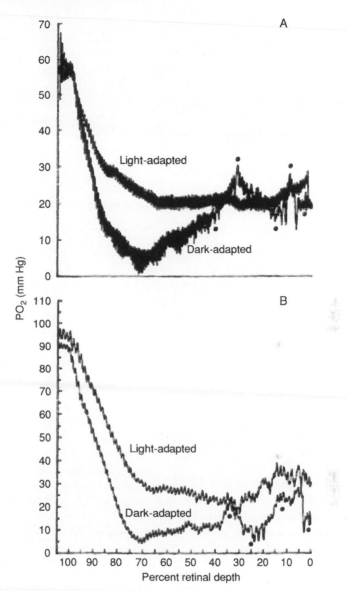

Figure 11. Profiles of oxygen tension as a function of retinal depth. Panel A is from light- and dark-adapted retina of M. fascicularis. Panel B from M. nemestrina. (From Ahmed, et al. (1993). IOVS 34, 517).

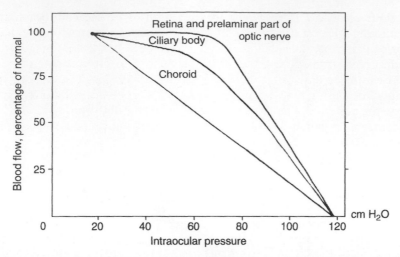

Figure 12. When intraocular pressure (IOP) is increased, there is no decrease in retinal blood flow and optic nerve head blood flow up to a certain level. Change in flow in ciliary body is also small. In the choroid, even moderate increments in eye pressure reduce blood flow. At high IOP, further increments in pressure reduce flow in all intraocular tissues (from Cioffi, et al. (2003). Adler's Physiology).

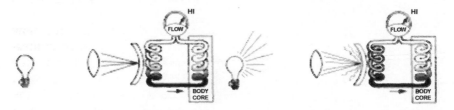

Figure 13. Model of function of the retinochoroidal circulation in normal animals and under conditions of thickened choroidal vessel walls. Left, focusing of light on the macula by the eye's optical system produces a local increase in tissue temperature. The high-choroidal circulation, shown as a coil just behind the retina, absorbs the heat and stabilizes local tissue temperature. The absorbed heat is dissipated by central body mechanisms (body core). Right, thickening of the blood vessel walls in the choroid decreases the efficiency of the choroidal vasculature in dissipating heat, allowing an increase in tissue temperature. (From Parver, L.M. (1991). Eye 5, 181–185.)

as a temperature regulator, and stabilizes retinal temperature especially in the foveal region, probably preventing heat damage to the photoreceptors (Parver, 1991; Tamai et al., 1999) (Figure 13).

There is a close relationship between the retinal pigment epithelium (RPE), and the choriocapillaris. It has been known for almost 20 years that the RPE somehow is necessary for the survival of the choriocapillaris. Atraumatic RPE removal causes atrophy of the underlying choriocapillaris (Korte et al., 1984; Del Priore et al., 1995; Nasir et al., 1997). On the other hand, survival of the RPE cell layer, and the photoreceptors, is dependent on intact choriocapillaris. If blood flow in choriocapillaris is obstructed experimentally by raising the IOP to very high values, or clinically by thrombosis of one the posterior ciliary arteries, the RPE layer is destroyed and RPE and photoreceptor atrophy result in what clinically are known as Elschnigs spots (Amalric, 1991).

VII. CHOROIDAL BLOOD FLOW MEASUREMENTS

Direct clinical measurement of choroidal blood is currently not feasible. Only indirect methods are available. These include Doppler measurements of velocity of blood flow either by ultrasound in the ciliary arteries (Quaranta et al., 1997), or diode-laser light in the submacular choroidal vasculature (Riva et al., 1994). In both instances, lacking knowledge of vascular caliber prohibits calculation of the absolute blood flow. Therefore, only relative measurements can be performed and the influences of physiologic or pharmacological perturbations are dependent upon the assumption that the cross-sectional diameter of the vasculature is unchanged.

Alternatively, it is possible to calculate the pulsatile ocular blood flow (pOBF) from the intraocular pulse pressure. The average intraocular pressure is proportional to the rate of aqueous humor production and the outflow resistance (R) of the eye; $R = 1.5$ mmHg min/μl (Brubaker, 1982). The time constant for this balance is $CR = 1.5$ minutes where C is the compliance of the eye about 1 μl/mmHg (Friedenwald, 1957). Hence, a perturbation in either determinant will cause the average pressure to settle at its new value within 10–20 minutes. At a much shorter time-scale, the intraocular pressure is pulsating synchronously with the heart beat due to pulsatile systolic inflow of blood exceeding the average outflow. Knowing the rigidity of the eye enables one to calculate pOBF (Krakau, 1995). The inherent assumptions in this calculation are a steady venous outflow and that the ocular elasticity *in vivo* is known. The pulsatile pressure in the ciliary arteries is the driving force that increases the blood volume of the choroids temporarily during systole, thereby changing axial length.

This can be measured by interferometry (Schmetterer et al., 1997). Both methods, however, merely quantitate the pulsatile component of the choroidal blood flow. Dye saturation curves obtained by indocyanine green angiography show that there is a substantial accumulation of dye during diastole (Figure 14).

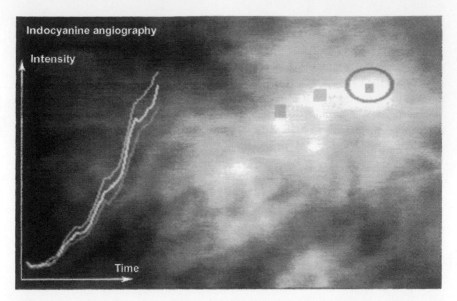

Figure 14. Angiography of a choroidal artery in a normal person after injection of indocyanine green intravenously. A laser scanning ophthalmoscope records successive frames that are digitized and aligned to extract the density in three windows. The intensity curves obtained from each window are shown versus time in the left panel. Pulsatile inflow with a substantial diastolic component is evident.

After the curves reach their maximum, the pulsatile component disappears as a result of insignificant pulsations of the vascular calibers. Hence, the optimal measurement of total choroidal blood flow would require measurement of ocular rigidity and the intraocular pulse pressure combined with an angiographic estimation of the ratio of systolic/diastolic inflow rates.

VIII.   DEVELOPMENT

The choriocapillaris starts to differentiate simultaneously with the development of the RPE during the 4th and 5th week (Sellheyer, 1990). Some controversy exists regarding whether these developing capillaries are fenestrated or not before the 12th week (Mund et al., 1972; Sellheyer, 1990). The basement membrane of choriocapillaris is starting to develop in the 7th week, but a continuous membrane is first seen in the 9th week. Four different layers in Bruch's membrane can be distinguished at this time-point; only the elastic layer of Bruch's membrane is not fully developed until the 13th week (Sellheyer, 1990).

Arterioles and venous are first observed outside the choriocapillary layer at week 15. Before this time-point, the only arterioles are at optic nerve entrance in sclera; and rudimentary vortex veins are seen starting their development in the 6th week (Hogan et al., 1971; Sellheyer, 1990). Arteries and veins are first observed in the 22nd week. Pigmentation of the choroid is first seen in the 5th month. It is believed that these pigmented cells are derived from the neural crest during early embryonic life (Hogan et al., 1971).

REFERENCES

Alm, A. and Bill, A. (1970). Blood flow and oxygen extraction in the cat uvea at normal and high-intraocular pressures. Acta Physiol. Scand. 80, 19–28.

Amalric, P. (1983). The choriocapillaris in the macular area. Clinical and angiographic study. Int. Ophthalmol. 6, 149–153.

Amalric, P. (1991). Choroidal vascular ischaemia. Eye 5 (Pt 5), 519–527.

Brubaker, R.F. (1982). The flow of aqueous humor in the human eye. Trans. Am. Ophthalmol. Soc. 80, 391–474.

Cioffi, G.A., Granstam, E. and Alm, A. (2003). Ocular circulation. In: Adler's Physiology of the Eye. (Kaufman, P.L. and Alm, A., Eds.), pp. 747–784, 10th edn., Mosby, St. Louis, MO.

De Laey, J.J. (1977). Fluoro-angiographic study of the choroid in man. Bull. Soc. Belge. Ophtalmol. 174, 1–217.

Del Priore, L.V., Hornbeck, R., Kaplan, H.J., Jones, Z., Valentino, T.L., Mosinger-Ogilvie, J. and Swinn, M. (1995). Debridement of the pig retinal pigment epithelium in vivo. Arch. Ophthalmol. 113, 939–944.

Delaey, C., Van, D. and V. (2000). Regulatory mechanisms in the retinal and choroidal circulation. Ophthalmic. Res. 32, 249–256.

Fitzgerald, M.E., Vana, B.A. and Reiner, A. (1990). Evidence for retinal pathology following interruption of neural regulation of choroidal blood flow: Muller cells express GFAP following lesions of the nucleus of Edinger-Westphal in pigeons. Curr. Eye Res. 9, 583–598.

Duke-Elder, s.S. and Wybar, K.C. (1961). The ocular tissues–The vascular tunic (the uvea). In: The Anatomy of the Visual System. (Duke-Elder, s.S., Ed.), pp. 131–146. Henry Kimpton, London.

Friedenwald, J.S. (1957). Tonometer calibration; An attempt to remove discrepancies found in the 1954 calibration scale for Schiotz tonometers. Trans. Am. Acad. Ophthalmol. Otolaryngol. 61, 108–122.

Geiser, M.H., Riva, C.E., Dorner, G.T., Diermann, U., Luksch, A. and Schmetterer, L. (2000). Response of choroidal blood flow in the foveal region to hyperoxia and hyperoxia-hypercapnia. Curr. Eye Res. 21, 669–676.

Hayreh, S.S. (1990). In vivo choroidal circulation and its watershed zones. Eye 4 (Pt 2), 273–289.

Hayreh, S.S. (2001). The blood supply of the optic nerve head and the evaluation of it—Myth and reality. Prog. Retin. Eye Res. 20, 563–593.

Hogan, M.J., Alvarado, J.A. and Weddell, J.E. (1971). Choroid. In: Histology of the Human Eye. (Hogan, M.J., Alvarado, J.A. and Weddell, J.E., Eds.), 1st edn., pp. 320–392. Saunders, London.

Kiel, J.W. (1994). Choroidal myogenic autoregulation and intraocular pressure. Exp. Eye Res. 58, 529–543.

Kiel, J.W. (1995). The effect of arterial pressure on the ocular pressure–volume relationship in the rabbit. Exp. Eye Res. 60, 267–278.

Korte, G.E., Reppucci, V. and Henkind, P. (1984). RPE destruction causes choriocapillary atrophy. Invest. Ophthalmol. Vis. Sci. 25, 1135–1145.

Krakau, C.E. (1995). A model for pulsatile and steady ocular blood flow. Graefes Arch. Clin. Exp. Ophthalmol. 233, 112–118.

Mund, M.L., Rodrigues, M.M. and Fine, B.S. (1972). Light and electron microscopic observations on the pigmented layers of the developing human eye. Am. J. Ophthalmol. 73, 167–182.

Nagy, A.R. and Ogden, T.E. (1990). Choroidal endothelial junctions in primates. Eye 4 (Pt 2), 290–302.

Nasir, M.A., Sugino, I. and Zarbin, M.A. (1997). Decreased choriocapillaris perfusion following surgical excision of choroidal neovascular membranes in age-related macular degeneration. Br. J. Ophthalmol. 81, 481–489.

Olver, J.M. (1990). Functional anatomy of the choroidal circulation: Methyl methacrylate casting of human choroid. Eye 4 (Pt 2), 262–272.

Parver, L.M. (1991). Temperature modulating action of choroidal blood flow. Eye 5 (Pt 2), 181–185.

Quaranta, L., Harris, A., Donato, F., Cassamali, M., Semeraro, F., Nascimbeni, G., Gandolfo, E. and Quaranta, C.A. (1997). Color Doppler imaging of ophthalmic artery blood flow velocity: A study of repeatability and agreement. Ophthalmol. 104, 653–658.

Riva, C.E., Cranstoun, S.D., Grunwald, J.E. and Petrig, B.L. (1994). Choroidal blood flow in the foveal region of the human ocular fundus. Invest. Ophthalmol. Vis. Sci. 35, 4273–4281.

Schmetterer, L., Krejcy, K., Kastner, J., Wolzt, M., Gouya, G., Findl, O., Lexer, F., Breiteneder, H., Fercher, A.F. and Eichler, H.G. (1997). The effect of systemic nitric oxide-synthase inhibition on ocular fundus pulsations in man. Exp. Eye Res. 64, 305–312.

Schmetterer, L. and Polak, K. (2001). Role of nitric oxide in the control of ocular blood flow. Prog. Retin. Eye Res. 20, 823–847.

Sellheyer, K. (1990). Development of the choroid and related structures. Eye 4 (Pt 2), 255–261.

Tamai, K., Toumoto, E., Yamada, K. and Ogura, Y. (1999). Effects of irrigation fluid temperature on choroidal circulation during vitrectomy. Curr. Eye Res. 18, 249–253.

Trivino, A., De Hoz, R., Salazar, J.J., Ramirez, A.I., Rojas, B. and Ramirez, J.M. (2002). Distribution and organization of the nerve fiber and ganglion cells of the human choroid. Anat. Embryol. (Berl.). 205, 417–430.

Zagvazdin, Y.S., Fitzgerald, M.E., Sancesario, G. and Reiner, A. (1996). Neural nitric oxide mediates Edinger-Westphal nucleus evoked increase in choroidal blood flow in the pigeon. Invest. Ophthalmol. Vis. Sci. 37, 666–672.

INNATE AND ADAPTIVE IMMUNITY
OF THE EYE

Mogens Holst Nissen and Carsten Röpke

Advances in Organ Biology
Volume 10, pages 291–305.
© 2006 Elsevier Inc. All rights reserved.
ISBN: 0–444-50925-9
DOI: 10.1016/S1569-2590(05)10011-1

I. INTRODUCTION

Vision is of fundamental importance for each individual to interact with and to survive in the world. Therefore, a clear visual axis focusing the images from the world outside on retina, as well as a proper function of retina itself, is of paramount importance. In order to achieve this, clear media (cornea, aqueous humor (AqH), lens, and vitreous body) for transmission and focusing of the light is needed. The perception of precise images by the retina requires maintaining the organization and structure of retina within narrow limits.

This organization can be disrupted by a potent inflammatory response that normally occurs in other sites in the body. Inflammation in the eye must therefore be very strictly regulated to avoid collateral damage to delicate structures lacking regenerative capacity.

It has for many years been known that the eye is an immune privileged site (i.e., that allogeneic grafts placed in the anterior chamber or subretinally can survive for an extended period of time without being rejected). Medawar proposed immune privilege of the anterior chamber in the 1950s, observing this phenomenon by placing allogeneic skin transplants in the anterior chamber of the eye. He believed that the survival of the transplant was due to immunologic ignorance based on "passive" mechanisms, such as the blood:ocular barrier and the lack of lymphatics. This "dogma" prevailed until investigations initiated by Kaplan and Streilein (1977, 1978) 20 years later demonstrated that immunization via the anterior chamber created a deviant immune response (i.e., that F1 cells or allogeneic cells placed in the anterior chamber created a systemic immune response). This response was deficient in cell-mediated immunity and devoid of complement fixing antibodies. In the eye, the vitreous cavity and the subretinal space have also been found to be immune privileged areas (Jiang and Streilein, 1991; Wenkel et al., 1999). From the last decades of research it is now known that the immune privilege of the eye relies not only on passive mechanisms, such as the blood:ocular barrier and the lack of lymphatics, but also on active processes involving a complex array of different mechanisms.

Other anatomic sites characterized as "immune privileged" include brain, testis, ovary, pregnant uterus, and adrenal cortex, where histoincompatible tissue grafts can be placed, and they survive for an extended or indefinite period of time (Streilein, 2003b).

Tissues derived from immune privileged sites, such as the cornea, lens, retina, or retinal pigment epithelium (RPE) from the eye or tissues from the brain and spinal cord can, when placed at an immune competent site, survive without rejection for a prolonged period in a histoincompatible individual— a feature designating these tissues as "immune privileged tissues." This finding further indicates that the structural organization including blood:

Table 1. Immune Privileged Tissues and Sites

Sites	Tissues
Eye	Cornea, lens, pigment epithelium, retina, anterior chamber, vitreous, subretinal space
Brain	Brain, spinal cord
Ovary	Ovary
Testis	Testis
Adrenal cortex	Liver
Pregnant uterus	Fetoplacental unit
Hair follicles	
Tumors	Tumors

Source: Adapted from Streilein (1999, 2003b).

ocular or blood:brain barriers of the "immune privileged sites" is not the full explanation of this phenomenon.

Today the "immune privileged" status of the eye is believed to be based on five different mechanisms including: (1) the blood:ocular barrier, (2) the absence of lymphatic drainage from the eye, (3) soluble factors with immune regulatory properties in ocular fluids, (4) the expression of immune regulatory molecules on the epithelial cells lining the interior of the eye, and (5) tolerance-inducing antigen presenting cells (APC). Some of these mechanisms will be described in the following sections (Table 1).

II. INNATE AND ADAPTIVE IMMUNITY

Apart from physical and chemical barriers to impede the invasion of pathogens, the immune system is capable of discriminating "self" from "nonself." This is done by two highly specified systems: the innate and adaptive immune system. The adaptive immune system is highly specific, capable of recognizing even subtle changes, and having the ability to generate specific antibodies and specific cytotoxic effector cells against these changes. The innate immune system recognizes and reacts to major structural differences by use of lactoferrin, lysozyme, defensins, complement, mannose-binding lectin, or by the use of Toll-like receptors and other receptors expressed on cell surfaces. The cellular network of the innate immune system includes neutrophils, macrophages, as well as NK-cells and $\gamma\delta$ T cells. In contrast to the adaptive immune system, the effector phase of the innate immune system is immediate (Table 2).

III. THE INNATE IMMUNE SYSTEM AND THE EYE

The innate immune system plays a central role in the immunity of the eye as the first responder to foreign pathogens. The responses elicited are not

Table 2.	Differences Between the Innate and the Adaptive Immune Response

Innate Immunity	Adaptive Immunity
Antigen-independent	Antigen-dependent
Immediate maximal response (minutes)	Delayed maximal response (days)
Not antigen-specific	Antigen specific
No immunological memory	Immunological memory

based on specific recognition of the individual pathogens, but on general differences displayed by pathogens compared to the individual they invade.

## IV.	LACTOFERRIN, LYSOZYME, AND DEFENSINS

The iron-binding protein lactoferrin constitutes around 25% of the protein content in tear fluid. The main biologic properties of lactoferrin can be ascribed to its very strong binding of iron cations. It is suggested that lactoferrin exerts its biologic activity by deprivation of iron from the ocular surface and thereby limiting the availability of iron for microorganisms, thus exerting firm control of the bacterial flora at these sites.

Lysozyme is found in tear fluid and account for about 40% of the protein in tears. Lysozyme disrupts the cell wall of susceptible bacteria by hydrolysis (McClellan, 1997).

Together lactoferrin, lysozyme, and a third group of antimicrobial peptides, named defensins, form the nonspecific defense at the ocular surface. Defensins are known to be of significance for the elimination of pathogens such as bacteria, virus, and fungi (McClellan, 1997).

Defensins belong to the group of cationic antimicrobial peptides of less than 100 amino acids long. They are found widely distributed in mammalian epithelial cells and phagocytes, and they are present in high (up to millimolar) concentrations. There are two main subfamilies of defensins, α- and β-defensin, based on structural differences (Ganz, 2003). Defensin-α 1, 2, and 3 are produced by the lacrimal gland, they are present in tears and are regarded as important for the initial defense together with lysozyme and lactoferrin. Corneal epithelium has been shown to express β defensin 1 and 2 (Haynes et al., 1999). Intraocular expression of β defensin 1 and inducible expression of β defensin 2 in freshly isolated ciliary body epithelium (CBE) and RPE has been found by RT-PCR, but no expression of β defensin 5 and 6 has been demonstrated (Haynes et al., 2000). However, β defensin 1 is found in fluid from vitreous and AqH only in very low subbacteriocidal concentrations (Haynes et al., 2000; Lehmann et al., 2000). The low-intraocular expression of defensin may be advantageous since defensin has been demonstrated to cause cell proliferation and fibrin formation (Higazi et al., 1995, 1996), which potentially are harmful to vision.

V. COMPLEMENT

The complement system is known to be of great importance eliminating pathogens such as bacteria, parasites, and viruses. Complement components are known to form a terminal pore forming complex C5b-C9, named membrane attack complex, destroying the target efficiently by lysis. This process can be initiated by C1 complement (classical pathway), mannose-binding lectin (lectin pathway), or by C3 (alternative pathway). Activation of complement leads to the generation of bioactive fragments (C3a, C4a, and C5a) and the formation of the membrane attack complex (C5b-C9). These two components will in turn lead to inflammation. To avoid excessive complement activation leading to an inflammatory response damaging the intraocular structures, a strict control of complement activation in the intraocular environment is needed. This control is achieved by inhibition of complement activation at a variety of levels. A number of different molecules have been identified in the AqH, among these, a low-molecular inhibitor of C1q, which is capable of inhibiting complement mediated lysis (Goslings et al., 1998).

In the eye there is widespread expression of complement regulatory proteins: membrane cofactor protein (MCP, CD46), decay-acceleration factor (DAF, CD55), and membrane inhibitor of reactive lysis (MIRL, CD59) are present both soluble in AqH and as cell bound forms (Chandler et al., 1974; Bora et al., 1993; Sohn et al., 2000a,b). CD46 and CD55 inhibit all three pathways (classical, lectin, and alternative) by inhibiting the formation of C3 convertase. CD59 binds to C5b678 and blocks the binding of C9, and thereby inhibits formation of the terminal complex. The cell surface regulator of complement, Crry (5I2 antigen) found to inhibit activation of C3 in the mouse (Molina, 2002) has also been demonstrated in the eye (Bardenstein et al., 1999; Sohn et al., 2000a). All three proteins are found to be differentially expressed in the human eye. Antibodies to MCP (CD46) show strong expression in the corneal epithelium and weak expression in the corneal keratocytes and photoreceptor cells of CD46. Anti-DAF staining (CD55) is very strong in the corneal epithelium and the ciliary body but moderate in the corneal stroma (keratocytes) and iris. In contrast, anti-CD59 antibodies show strong expression in the corneal epithelium, corneal stroma (keratocytes), iris, choroid, and all layers of the retina but only moderate expression in the ciliary body of CD59 (Bora et al., 1993; Sohn et al., 2000a,b).

Table 3. Factors of Importance for the Immune Privileged of the Eye

Passive	Active
Blood: ocular barrier	Cell surface expression of: CD95 Ligand CD46 (MCP), CD55(DAF), CD59 (MIRL), Crry
Deficient lymphatic drainage	Soluble factors in AqH: TGF-β, α-MSH, VIP, CGRP, MIF, IL-1R α, soluble CD95 Ligand, thrombospondin
Intravenous drainage of AqH Reduced intraocular expression of MHC class I and MHC class II Few APC	

Source: Adapted from Streilein (1999, 2003b).

Antibodies are known to be of great importance for the activation of complement, but not all isotypes of IgG antibodies are capable of initiating complement activation. When a humoral response against intraocular antigens is mounted, the antibody response is biased toward noncomplement fixing antibodies, which will not activate the complement system by the classical pathway (Wilbanks et al., 1990). This results in lower inflammatory responses (Table 3).

VI. NEUTROPHILS, MACROPHAGES, AND NK CELLS

Degranulation of neutrophils releases a number of bioactive substances, including lactoferrin, lysozyme, peroxidase, proteases, and collagenase, which can cause significant damage to ocular tissues. Protection against the actions of neutrophil granulocytes includes inhibition of their activation by α-melanocyte stimulating hormone (α-MSH) (Catania et al., 1996) and interleukin-1 receptor antagonist (IL-1Rα), present in AqH (Kennedy et al., 1995). The activation of neutrophils by FasLigand (CD95L) expressed on ocular epithelium are counteracted by soluble FasLigand (CD95L) and transforming growth factor-β (TGF-β) present in AqH (Gregory et al., 2002). Calcitonin gene-related peptide (CGRP) present in AqH inhibits the production of nitric oxide (NO) by activated macrophages (Taylor et al., 1998).

NK cells have the capability to detect absence of "self" by lysis of cells expressing low amounts of MHC class I on the cell surface. Corneal endothelial cells have been shown to express low amounts of MHC class I, and this renders these cells prone to lysis by NK cells. Counteracting the activity of NK cells entering the anterior chamber is the presence of macrophage migration inhibitory factor (MIF) in AqH. Migration inhibitory factor has been shown to inhibit NK cells from lysing their targets (Apte et al., 1998).

VII. THE ADAPTIVE IMMUNE SYSTEM

The adaptive immune system consists of two arms: humoral (specific antibodies) and cellular (specific T cells). Of fundamental importance in the afferent part of the immune response is the APC that includes B cells, macrophages, and dendritic cells. These cells capture antigens and present them bound to MHC class I or MHC class II antigens on the cell surface. The dendritic cells have the ability to migrate selectively through different tissues and present the captured and processed antigen as peptide fragments for $CD8^+$ and $CD4^+$ T cells bound to MHC class I and MHC class II, respectively. The immune regulatory properties of dendritic cells rely both on their expression of costimulatory molecules (CD40, CD80, and CD86) and their production of chemokines and cytokines (TNF-α, IL-12, IL-10, IFN-γ, and TGF-β) (Banchereau et al., 2000). Immature dendritic cells are characterized by a high capacity to capture and process antigen, but due to their low MHC class II expression and their lack of costimulatory molecules, they are poor stimulators of T cells (Banchereau and Steinman, 1998). Mature dendritic cells expressing CD40, CD80, and CD86 do not capture and process antigens efficiently but are potent stimulators of T cells (O'Keeffe et al., 2003).

Under normal conditions, granulocytes are absent from the eye, and only very few T and B cells can be detected. However, a network of macrophages and dendritic cells capable of antigen capture and delivery to secondary lymphoid organs has been found present in stroma and epithelium surrounding the anterior segment of the eye, including the cornea, iris, and the ciliary body. In contrast, the retina contains few dendritic cells but has a network of microglial cells ($CD45^+$, MHC class II^+) and macrophages ($CD68^+$), of which the majority is MHC class II positive (Yang et al., 2000, 2002). These are the cells responsible for antigen capture and presentation in the retina.

Despite the long-standing observation that allogeneic tissue placed in the anterior chamber can survive for a prolonged period of time or indefinitely, the mechanism remains elusive. However, studies through the last decades using mouse models have contributed significantly to our understanding of how the protection of the eye from a full blown immune response to pathogens is achieved. This includes a deviant systemic immune response, which leaves out the generation of complement fixing antibodies and $CD4^+$ T cells but includes primed $CD8^+$ T_{reg} cells and the formation of noncomplement fixing antibodies (Wilbanks et al., 1990; Streilein, 2003b). This is named anterior chamber associated immune deviation (ACAID). A comparable immune response can be observed when antigens are placed in the subretinal space or vitreous cavity (Jiang et al., 1991; Wenkel et al., 1999).

VIII.   THE SPLEEN IS REQUIRED
FOR THE INDUCTION OF ACAID

The mechanism responsible for ACAID has, to a large extent, been described. After sequestering of antigens within the eye, the APCs will migrate through the trabecular meshwork directly to the venous system, and the APCs will end up in the spleen, which is considered the "regional lymph node" of eye. It has thus been demonstrated that ACAID cannot be induced in an animal when the spleen has been removed. Furthermore, a period of 3–4 days is needed in order to induce ACAID (Kaplan and Streilein, 1974; Streilein and Niederkorn, 1981). This is demonstrated by the fact that removal of either the spleen or the eye up to 4–5 days after injection of antigen into the eye will abort the ACAID induction. This presumably reflects the time needed for APCs to capture, process, and present antigen derived from the eye in the spleen (Wilbanks and Streilein, 1991). Generation of ACAID has most extensively been described within rodents but as well been demonstrated to be present in rabbits and monkeys, and indication of the existence of ACAID in humans has been reported (Kezuka et al., 2001).

IX.   THE CELLULAR BASIS FOR GENERATION OF ACAID

In the spleen, APCs will home to the marginal zone where they secrete chemokines capable of recruiting $\gamma\delta$-T cells, NKT cells, marginal zone B cells together with CD4$^+$ and CD8$^+$ $\alpha\beta$-T cells specific for the processed antigens (Faunce et al., 2001; Sonoda and Stein-Streilein, 2002). The antigen is presented to the CD4$^+$ and CD8$^+$ $\alpha\beta$-T cells as peptides bound to MHC class II and MHC class I antigens on the cell surface of APCs. The APCs that induce ACAID are characterized as F4/80$^+$ (Streilein, 2003a), and they are known to express normal levels of MHC class I and MHC class II molecules, CD80 and CD86, ICAM-1 and enhanced levels of CD1 on the cell surface. Notably, the expression of CD40 is low, and IL-12 production is suppressed (Takeuchi et al., 1998; Streilein, 2003a). These APCs are tolerogenic instead of immunogenic, and the subsequent response will generate regulatory T cells (T$_{reg}$), induce suppression of both Th1- and Th2-mediated immunity and promote generation of noninflammatory adaptive immune effectors (Streilein, 2003b). The CD4$^+$ T$_{reg}$ cells generated will suppress the initial activation and differentiation of naive T cells in the spleen into T$_H$1 effector cells. The CD8$^+$ T$_{reg}$ generated will inhibit the expression of T$_H$1-mediated immunity in the periphery, including the eye. Apart from $\alpha\beta$-T cells, both $\gamma\delta$-T cells and NKT cells are needed for the induction of ACAID. This is supported by the finding that depletion of $\gamma\delta$-T cells or NKT cells in animals by monoclonal antibodies, or the use of genetic disruption abrogates the induction of ACAID (Sonoda et al., 1999; Skelsey et al., 2001).

X. THE MOLECULAR BASIS FOR INDUCTION OF ACAID

The molecular basis for the response has been partly dissected. It is known that the environment, in which the APCs collect and process antigens, is of importance for the response generated. Investigation has shown that a low level of complement activation through the alternative pathway is present within the anterior chamber and that generation of tolerance is dependent on complement C3 fragment iC3b binding to APCs (Sohn et al., 2003). This C3 fragment can prime the APCs to produce TGF-β and inhibits production of interleukin 12 (IL-12). Several studies have shown that the presence of TGF-β in AqH is central for ACAID induction and homing of the APCs to the spleen (Streilein et al., 2002; Streilein, 2003a). This is supported by the finding that APCs exposed to antigen *in vitro* in the presence of AqH, home to the spleen and induce ACAID after intravenous injection. This homing response can also be achieved by treating APCs with active TGF-β alone.

XI. THROMBOSPONDIN (TSP) AND THE GENERATION OF ACAID

Treating APCs with TGF-β2 causes differential regulation of several genes, among these coding for TSP. TSP constitutes a family of four (TSP-1,-2,-3, and -4) multimeric, multidomain glycoproteins that are present in platelets (Streilein et al., 2002). TSP can, in addition, be produced by an array of different cell types depending on the environmental conditions. TSP-1 is constitutively present in AqH, and the gene is also expressed in pigment epithelial (PE) cells and corneal endothelium (Sheibani et al., 2000). TSP-1 is considered to be of fundamental importance for inhibition of angiogenesis in the normal eye, since it displays strong antiangiogenic properties. Studies on the effect of TSP-1 on APCs have revealed that TSP resembles TGF-β2, including down-regulation of the IL-12 and CD40 genes. A role for TSP in generation of ACAID is supported by the finding that antigen-pulsed APCs treated with TSP instead of TGF-β induce ACAID when injected intravenously into naive mice. Furthermore, antigen-pulsed APCs from TSP "knockout" mice fail to induce ACAID, when treated *in vitro* in the presence of TGF-β and subsequently injected intravenously into normal, wild-type mice (Masli et al., 2002).

The effect of TSP is proposed to be mediated through its binding to the cell surface receptors of T cells and APCs, where TSP can bind to CD47 present on APCs and T cells, and to CD36 present on APCs. Binding of TSP to CD47 can mediate inhibition of IL-12 gene activation and IL-12 production. In addition, TSP can be of importance by mediating binding and activation of latent TGF-β to APCs expressing CD36 on their surface (Streilein et al., 2002).

XII.   PEPTIDE HORMONES IN AQH EXERTING IMMUNE REGULATORY ACTIVITIES

A number of other molecules found in AqH are of importance for a restricted T cell response within the eye, these include α-MSH that both suppresses the activation of inflammatory activity by primed T cells and mediates the induction of CD4$^+$CD25$^+$ in regulatory T cells (T$_{reg}$ cells) (Taylor et al., 1992; Namba et al., 2002). Aqueous humor made devoid of α-MSH is no longer able to suppress the production of IFN-γ by Th1 cells. The neuropeptides vasoactive intestinal peptide (VIP) and somatostatin are both able to suppress antigen- and mitogen-induced T cell proliferation (Taylor et al., 1994; Taylor and Yee, 2003).

XIII.   IMMUNE REGULATORY ACTIVITIES OF EPITHELIAL CELLS LINING THE INTERIOR OF THE EYE

Pigment epithelium cells lining the interior of the eye contribute to the immune privilege. Pigment epithelium cells include iris pigment epithelium (IPE), ciliary pigment epithelium (CPE), and RPE (Yoshida et al., 2000a,b; Ishida et al., 2003). Together they form a physical barrier by the existence of tight junctions between the individual cells. Recent research has documented that the epithelial cells in the eye not only form physical barriers but also actively contribute to maintain the immune privilege of the eye. Pigment epithelial cells are capable of modulating the adaptive immune response by soluble or contact mediated mechanisms. Several reports have documented that PE cells derived from the eye can inhibit activation of resting T cells and inhibit proliferation or induce apoptosis of activated T cells (Rezai et al., 1999; Kaestel et al., 2002).

The expression of FasLigand (CD95L) by multiple cells in the eye is of significance for the induction of apoptosis in activated CD95 positive T cells entering the eye (Griffith et al., 1995). The expression of FasLigand (CD95L) on corneal grafts has also been shown to be of importance for successful transplantation (Stuart et al., 1997). Recent findings have shown that IPE constitutively expresses CD86 (B7.2) that normally is expressed on APCs, and it acts here as a strong costimulatory molecule for evolvement of the immune response. Expressed on the surface of IPE, CD86 interacts with cytotoxic T lymphocyte antigen-4 (CTLA-4) on the effector T cells and prevents their activation, proliferation, and production of IFN-γ, converting these T cells into regulators (Sugita and Streilein, 2003).

Different mechanisms have been proposed for mediating RPE-cell induction of programmed cell death and prevention of T cell proliferation. Liversidge et al. (1993) reported that RPE cells produced significant amounts of prostaglandin E_2 (PGE_2) in the presence of activated lymphocytes and that the addition of indometacin could reverse the inhibition of lymphocyte activation. Recently, another molecule, galectin-1 has been reported to be of significance for the inhibition of T cell activation mediated by RPE cells. Galectins are a family of 14 different carbohydrate binding proteins widely expressed throughout the body. All of the galectins are characterized by a carbohydrate binding domain with a canonical amino acid sequence and an affinity for β-galactosides (Hernandez and Baum, 2002). Galectins are found to be expressed both intra- and extracellularly. Expression of galectin has been demonstrated in ocular tissues, including RPE cells, where galectin-1 is found. Retinal pigment epithelium cells that do not express galectin-1 have a reduced capacity to inhibit CD3-mediated T cell activation (Ishida et al., 2003). However, inhibition of T cell activation must be conferred by molecules other than galectin-1, since RPE cells devoid of galectin-1 still inhibit CD3-mediated T cell activation to some extent. The immune regulating properties of RPE cells are probably mediated by several mechanisms yet to be revealed.

XIV. CONCLUSIONS

Avoiding intraocular infection and inflammation are of paramount importance to preserve the vision. Different strategies are employed to achieve this goal. Intraocular elimination of pathogens requires a fine balance of the immune system in strictly controlled manner in order to avoid permanent damage to the delicate structures of the eye. Experimental work during the last decades has disclosed a surprising complexity of regulation. These investigations have primarily focused on the anterior segment of the eye so far, but recent experiments have elucidated some of the strategies that have been used for the posterior segment of the eye.

Understanding the mechanisms, by which ocular immune privilege is maintained can provide helpful information of how to deal with transplantation of foreign tissue and autoimmune diseases.

REFERENCES

Apte, R.S., Sinha, D., Mayhew, E., Wistow, G.J. and Niederkorn, J.Y. (1998). Cutting edge: Role of macrophage migration inhibitory factor in inhibiting NK cell activity and preserving immune privilege. J. Immunol. 160, 5693–5696.

Banchereau, J., Briere, F., Caux, C., Davoust, J., Lebecque, S., Liu, Y.J., Pulendran, B. and Palucka, K. (2000). Immunobiology of dendritic cells. Annu. Rev. Immunol. 18, 767–811.

Banchereau, J. and Steinman, R.M. (1998). Dendritic cells and the control of immunity. Nature 392, 245–252.

Bardenstein, D.S., Cheyer, C., Okada, N., Morgan, B.P. and Medof, M.E. (1999). Cell surface regulators of complement, 5I2 antigen, and CD59, in the rat eye and adnexal tissues. Invest. Ophthalmol. Vis. Sci. 40, 519–524.

Bora, N.S., Gobleman, C.L., Atkinson, J.P., Pepose, J.S. and Kaplan, H.J. (1993). Differential expression of the complement regulatory proteins in the human eye. Invest. Ophthalmol. Vis. Sci. 34, 3579–3584.

Catania, A., Rajora, N., Capsoni, F., Minonzio, F., Star, R.A. and Lipton, J.M. (1996). The neuropeptide alpha-MSH has specific receptors on neutrophils and reduces chemotaxis in vitro. Peptides 17, 675–679.

Chandler, J.W., Leder, R., Kaufman, H.E. and Caldwell, J.R. (1974). Quantitative determinations of complement components and immunoglobulins in tears and aqueous humor. Invest. Ophthalmol. 13, 151–153.

Faunce, D.E., Sonoda, K.H. and Stein-Streilein, J. (2001). MIP-2 recruits NKT cells to the spleen during tolerance induction. J. Immunol. 166, 313–321.

Ganz, T. (2003). Defensins: Antimicrobial peptides of innate immunity. Nat. Rev. Immunol. 3, 710–720.

Goslings, W.R., Prodeus, A.P., Streilein, J.W., Carroll, M.C., Jager, M.J. and Taylor, A.W. (1998). A small molecular weight factor in aqueous humor acts on C1q to prevent antibody-dependent complement activation. Invest. Ophthalmol. Vis. Sci. 39, 989–995.

Gregory, M.S., Repp, A.C., Holhbaum, A.M., Saff, R.R., Marshak-Rothstein, A. and Ksander, B.R. (2002). Membrane Fas ligand activates innate immunity and terminates ocular immune privilege. J. Immunol. 169, 2727–2735.

Griffith, T.S., Brunner, T., Fletcher, S.M., Green, D.R. and Ferguson, T.A. (1995). Fas ligand-induced apoptosis as a mechanism of immune privilege. Science 270, 1189–1192.

Haynes, R.J., McElveen, J.E., Dua, H.S., Tighe, P.J. and Liversidge, J. (2000). Expression of human beta-defensins in intraocular tissues. Invest. Ophthalmol. Vis. Sci. 41, 3026–3031.

Haynes, R.J., Tighe, P.J. and Dua, H.S. (1999). Antimicrobial defensin peptides of the human ocular surface. Br. J. Ophthalmol. 83, 737–741.

Hernandez, J.D. and Baum, L.G. (2002). Ah, sweet mystery of death! Galectins and control of cell fate. Glycobiology 12, 127R–136R.

Higazi, A.A., Barghouti, I.I. and Abu-Much, R. (1995). Identification of an inhibitor of tissue-type plasminogen activator-mediated fibrinolysis in human neutrophils. A role for defensin. J. Biol. Chem. 270, 9472–9477.

Higazi, A.A., Ganz, T., Kariko, K. and Cines, D.B. (1996). Defensin modulates tissue-type plasminogen activator and plasminogen binding to fibrin and endothelial cells. J. Biol. Chem. 271, 17650–17655.

Ishida, K., Panjwani, N., Cao, Z. and Streilein, J.W. (2003). Participation of pigment epithelium in ocular immune privilege. 3. Epithelia cultured from iris, ciliary body, and retina suppress T cell activation by partially non-overlapping mechanisms. Ocul. Immunol. Inflamm. 11, 91–105.

Jiang, L.Q. and Streilein, J.W. (1991). Immune privilege extended to allogeneic tumor cells in the vitreous cavity. Invest. Ophthalmol. Vis. Sci. 32, 224–228.

Kaestel, C.G., Jorgensen, A., Nielsen, M., Eriksen, K.W., Odum, N., Holst, N.M. and Ropke, C. (2002). Human retinal pigment epithelial cells inhibit proliferation and IL2R expression of activated T cells. Exp. Eye Res. 74, 627–637.

Kaplan, H.J. and Streilein, J.W. (1974). Do immunologically privileged sites require a functioning spleen? Nature 251, 553–554.

Kaplan, H.J. and Streilein, J.W. (1977). Immune response to immunization via the anterior chamber of the eye. I. F. lymphocyte-induced immune deviation. J. Immunol. 118, 809–814.

Kaplan, H.J. and Streilein, J.W. (1978). Immune response to immunization via the anterior chamber of the eye. II. An analysis of F1 lymphocyte-induced immune deviation. J. Immunol. 120, 689–693.

Kennedy, M.C., Rosenbaum, J.T., Brown, J., Planck, S.R., Huang, X., Armstrong, C.A. and Ansel, J.C. (1995). Novel production of interleukin-1 receptor antagonist peptides in normal human cornea. J. Clin. Invest. 95, 82–88.

Kezuka, T., Sakai, J., Usui, N., Streilein, J.W. and Usui, M. (2001). Evidence for antigen-specific immune deviation in patients with acute retinal necrosis. Arch. Ophthalmol. 119, 1044–1049.

Lehmann, O.J., Hussain, I.R. and Watt, P.J. (2000). Investigation of beta defensin gene expression in the ocular anterior segment by semiquantitative RT-PCR. Br. J. Ophthalmol. 84, 523–526.

Liversidge, J., McKay, D., Mullen, G. and Forrester, J.V. (1993). Retinal pigment epithelial cells modulate lymphocyte function at the blood-retina barrier by autocrine PGE2 and membrane-bound mechanisms. Cell Immunol. 149, 315–330.

Masli, S., Turpie, B., Hecker, K.H. and Streilein, J.W. (2002). Expression of thrombospondin in TGFbeta-treated APCs and its relevance to their immune deviation-promoting properties. J. Immunol. 168, 2264–2273.

McClellan, K.A. (1997). Mucosal defense of the outer eye. Surv. Ophthalmol. 42, 233–246.

Molina, H. (2002). The murine complement regulator Crry: New insights into the immunobiology of complement regulation. Cell Mol. Life Sci. 59, 220–229.

Namba, K., Kitaichi, N., Nishida, T. and Taylor, A.W. (2002). Induction of regulatory T cells by the immunomodulating cytokines alpha-melanocyte-stimulating hormone and transforming growth factor-beta2. J. Leukoc. Biol. 72, 946–952.

O'Keeffe, M., Hochrein, H., Vremec, D., Scott, B., Hertzog, P., Tatarczuch, L. and Shortman, K. (2003). Dendritic cell precursor populations of mouse blood: Identification of the murine homologues of human blood plasmacytoid pre-DC2 and CD11c$^+$ DC1 precursors. Blood 101, 1453–1459.

Rezai, K.A., Semnani, R.T., Farrokh-Siar, L., Hamann, K.J., Patel, S.C., Ernest, J.T. and van Seventer, G.A. (1999). Human fetal retinal pigment epithelial cells induce apoptosis in allogenic T cells in a Fas ligand and PGE2 independent pathway. Curr. Eye Res. 18, 430–439.

Sheibani, N., Sorenson, C.M., Cornelius, L.A. and Frazier, W.A. (2000). Thrombospondin-1, a natural inhibitor of angiogenesis, is present in vitreous and aqueous humor, and is modulated by hyperglycemia. Biochem. Biophys. Res. Commun. 267, 257–261.

Skelsey, M.E., Mellon, J. and Niederkorn, J.Y. (2001). Gamma delta T cells are needed for ocular immune privilege and corneal graft survival. J. Immunol. 166, 4327–4333.

Sohn, J.H., Bora, P.S., Suk, H.J., Molina, H., Kaplan, H.J. and Bora, N.S. (2003). Tolerance is dependent on complement C3 fragment iC3b binding to antigen-presenting cells. Nat. Med. 9, 206–212.

Sohn, J.H., Kaplan, H.J., Suk, H.J., Bora, P.S. and Bora, N.S. (2000a). Chronic low level complement activation within the eye is controlled by intraocular complement regulatory proteins. Invest. Ophthalmol. Vis. Sci. 41, 3492–3502.

Sohn, J.H., Kaplan, H.J., Suk, H.J., Bora, P.S. and Bora, N.S. (2000b). Complement regulatory activity of normal human intraocular fluid is mediated by MCP, DAF, and CD59. Invest. Ophthalmol. Vis. Sci. 41, 4195–4202.

Sonoda, K.H., Exley, M., Snapper, S., Balk, S.P. and Stein-Streilein, J. (1999). CD1-reactive natural killer T cells are required for development of systemic tolerance through an immune-privileged site. J. Exp. Med. 190, 1215–1226.

Sonoda, K.H. and Stein-Streilein, J. (2002). CD1d on antigen-transporting APC and splenic marginal zone B cells promotes NKT cell-dependent tolerance. Eur. J. Immunol. 32, 848–857.

Streilein, J.W. (1999). Regional immunity and ocular immune privilege. Chem. Immunol. 73, 11–38.

Streilein, J.W. (2003a). Ocular immune privilege: The eye takes a dim but practical view of immunity and inflammation. J. Leukoc. Biol. 74, 179–185.

Streilein, J.W. (2003b). Ocular immune privilege: Therapeutic opportunities from an experiment of nature. Nat. Rev. Immunol. 3, 879–889.

Streilein, J.W., Masli, S., Takeuchi, M. and Kezuka, T. (2002). The eye's view of antigen presentation. Hum. Immunol. 63, 435–443.

Streilein, J.W. and Niederkorn, J.Y. (1981). Induction of anterior chamber-associated immune deviation requires an intact, functional spleen. J. Exp. Med. 153, 1058–1067.

Stuart, P.M., Griffith, T.S., Usui, N., Pepose, J., Yu, X. and Ferguson, T.A. (1997). CD95 ligand (FasL)-induced apoptosis is necessary for corneal allograft survival. J. Clin. Invest. 99, 396–402.

Sugita, S. and Streilein, J.W. (2003). Iris pigment epithelium expressing CD86 (B7-2) directly suppresses T cell activation in vitro via binding to cytotoxic T lymphocyte-associated antigen 4. J. Exp. Med. 198, 161–171.

Takeuchi, M., Alard, P. and Streilein, J.W. (1998). TGF-beta promotes immune deviation by altering accessory signals of antigen-presenting cells. J. Immunol. 160, 1589–1597.

Taylor, A.W., Streilein, J.W. and Cousins, S.W. (1992). Identification of alpha-melanocyte stimulating hormone as a potential immunosuppressive factor in aqueous humor. Curr. Eye Res. 11, 1199–1206.

Taylor, A.W., Streilein, J.W. and Cousins, S.W. (1994). Immunoreactive vasoactive intestinal peptide contributes to the immunosuppressive activity of normal aqueous humor. J. Immunol. 153, 1080–1086.

Taylor, A.W. and Yee, D.G. (2003). Somatostatin is an immunosuppressive factor in aqueous humor. Invest. Ophthalmol. Vis. Sci. 44, 2644–2649.

Taylor, A.W., Yee, D.G. and Streilein, J.W. (1998). Suppression of nitric oxide generated by inflammatory macrophages by calcitonin gene-related peptide in aqueous humor. Invest. Ophthalmol. Vis. Sci. 39, 1372–1378.

Wenkel, H., Chen, P.W., Ksander, B.R. and Streilein, J.W. (1999). Immune privilege is extended, then withdrawn, from allogeneic tumor cell grafts placed in the subretinal space. Invest. Ophthalmol. Vis. Sci. 40, 3202–3208.

Wilbanks, G.A. and Streilein, J.W. (1991). Studies on the induction of anterior chamber-associated immune deviation (ACAID). 1. Evidence that an antigen-specific, ACAID-inducing, cell-associated signal exists in the peripheral blood. J. Immunol. 146, 2610–2617.

Yang, P., Chen, L., Zwart, R. and Kijlstra, A. (2002). Immune cells in the porcine retina: Distribution, characterization and morphological features. Invest. Ophthalmol. Vis. Sci. 43, 1488–1492.

Yang, P., Das, P.K. and Kijlstra, A. (2000). Localization and characterization of immunocompetent cells in the human retina. Ocul. Immunol. Inflamm. 8, 149–157.

Yoshida, M., Kezuka, T. and Streilein, J.W. (2000a). Participation of pigment epithelium of iris and ciliary body in ocular immune privilege. 2. Generation of TGF-beta-producing regulatory T cells. Invest. Ophthalmol. Vis. Sci. 41, 3862–3870.
Yoshida, M., Takeuchi, M. and Streilein, J.W. (2000b). Participation of pigment epithelium of iris and ciliary body in ocular immune privilege. 1. Inhibition of T cell activation *in vitro* by direct cell-to-cell contact. Invest. Ophthalmol. Vis. Sci. 41, 811–821.

FURTHER READING

Wilbanks, G.A. and Streilein, J.W. (1990). Distinctive humoral immune responses following anterior chamber and intravenous administration of soluble antigen. Evidence for active suppression of IgG2-secreting B lymphocytes. Immunology 71, 566–572.

DRUG DELIVERY TO THE EYE

Ashim K. Mitra, Banmeet S. Anand and
Sridhar Duvvuri

Advances in Organ Biology
Volume 10, pages 307–351.
© 2006 Elsevier Inc. All rights reserved.
ISBN: 0-444-50925-9
DOI: 10.1016/S1569-2590(05)10013-5

I. INTRODUCTION

Ophthalmic drug delivery is one of the most challenging tasks facing scientists in the area of ophthalmology and pharmacology. The unique anatomy, physiology, and biochemistry of the eye protect the organ from various exogenous and endogenous insults. The major challenge in ocular drug delivery is the circumvention of the blood–ocular barriers with minimal damage to the eye. It is now well known that ocular disposition and elimination of a therapeutic agent is dependent on its physicochemical properties and its ability to access ocular fluids and tissues.

The aim of this chapter is to provide a general insight into the past, present, and future trends in the area of ocular drug delivery. Discussions on ocular barriers to drug delivery, modes of drug administration to the eye, effect of ocular fluid dynamics, transporter targeted drug delivery, and strategies to exploit transporters in enhancing ocular drug bioavailability have been included in this chapter, which will provide the reader with an overall understanding of ocular drug delivery.

II. STRUCTURE AND FUNCTION OF THE EYE

The eye is a spherical structure with a wall consisting of three layers: (1) external layer formed by cornea and sclera; (2) intermediate layer, divided into two parts: anterior (iris–ciliary body) and posterior (choroid); and (3) internal layer, retina (Figure 1). Eye primarily contains two following fluid chambers: (1) aqueous humor—between cornea and iris; and (2) vitreous humor—between lens and retina.

The choroid layer, residing inside the sclera, possesses numerous blood vessels and is modified in the front of the conjunctiva as the pigmented iris. The biconvex lens is situated just behind the iris. The chamber behind the lens is filled with vitreous humor, a gelatinous substance occupying 80% of the eyeball. The anterior and posterior chambers are positioned between the cornea and iris, and iris and lens, respectively and are filled with aqueous humor. At the back of the eye resides the light-detecting retina.

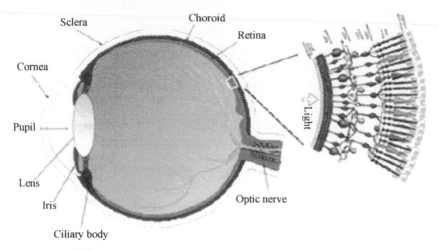

Figure 1. Cross-sectional view of the eye (adapted from http://webvision.med.utah. edu/sretina.html).

The globe is held in position in the orbital cavity by various ligaments and muscles. Conjunctiva is a protective layer that covers the eyeball, which is in contact with the environment. It consists of a thin mucous membrane layer in the inside of the eyelids and anterior sclera. In common parlance the anterior segment refers to the cornea, aqueous humor, iris–ciliary body whereas the vitreous humor, retina, choroid, and sclera together comprise the posterior segment of the eye.

So far a bulk of research activity has been devoted for improving drug penetration into the tissues and fluids of the anterior segment following systemic and topical administrations. Few such investigations have been targeted to the posterior segment. With the advent of diseases like CMV retinitis, bacterial and fungal endophthalmitis, and diabetic and proliferative vitreal and retinal pathologies, requirement of achieving and maintaining effective drug levels in the retina/choroid and vitreous has gained high clinical significance. A close examination of the local physiology and anatomy of the eye along with a discussion on ocular fluid dynamics, which play an important role in ocular drug disposition will aid in the understanding of current and new approaches to deliver drugs to the internal eye structures.

A. Structures of Cornea and Retina

Cornea

The cornea is an optically transparent tissue that conveys images to the back of the eye and covers about one-sixth of the total surface area of the eyeball.

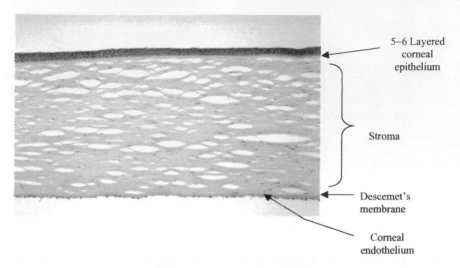

Figure 2.   Cross section of cornea.

This avascular tissue receives nutrients and oxygen via bathing in lachrymal fluid and aqueous humor as well as from blood vessels that line the junction between the cornea and sclera. The cornea is considered to be the main pathway for the permeation of drugs into the anterior chamber (Robinson, 1993). It is approximately 0.5 mm thick in the central region, increasing to approximately 0.7 mm at the periphery and is composed of five layers (Figure 2).

- Epithelium consists of 5–6 layers of squamous stratified cells (increasing to 8–10 layers at the periphery), has a total thickness of around 50–100 μm and a turnover rate of about one cell layer per day. Tight junctions and hydrophobic domains in this layer render this barrier most significant in drug absorption.
- Bowman's membrane is an acellular homogenous sheet, about 8–14 μm thick. It is positioned between the basement membrane of the epithelium and the stroma.
- Stroma, or substantia propria, accounts for about 90% of the corneal thickness. It contains approximately 85% water, and about 200–250 collagenous lamellae that are superimposed onto one another and run parallel to the surface. The lamellae provide physical strength while permitting optical transparency. The stroma has a relatively open structure and will normally allow the diffusion of hydrophilic solutes.
- Descemet's membrane that is secreted by the endothelium lies between the stroma and the endothelium.

- Endothelium responsible for maintaining normal corneal hydration consists of a single layer of flattened hexagonal cells 5 μm high and 20 μm wide. The endothelium is in direct contact with the anterior chamber and is subject to a passive influx of water from the aqueous humor toward the stroma. For a drug to cross the cornea effectively, it has to possess both hydrophilic and lipophilic properties, and be of low molecular size to pass through tight junctions.

Retina

Retina is highly perfused tissue with a complex cellular organization. It is a multilayered membrane of neuroectodermal origin, occupying the internal space of the outer layer of the eye. Retina can be divided into two sections (i.e., neural retina and retinal pigmented epithelium (RPE)). The early optic vesicle invaginates to form a cup, the inner layer of which becomes the neurosensory retina and the outer layer becomes the RPE. Subretinal space separates neural retina from RPE.

The neural retina is a seven-layered structure (Figure 3) that is involved in signal transduction. Light enters the retina through the ganglion cell layer (GCL) and must penetrate all the cell types before reaching the rods and cones. The transduced signal is then conducted out of the retina by various neuronal cells to the optic nerve, which subsequently conducts the signal to the brain where it is registered and an image is formed.

Retinal pigmented epithelium, on the other hand, is a single cell layer structure that separates the outer surface of neural retina from choroid (Rizzolo, 1997). It appears as a uniform and continuous layer extending through the entire retina playing a vital role in supporting and maintaining the viability of the neural retina (Marmor, 1998). Retinal pigmented epithelium cells are differentiated into an apical layer, which faces the neural retina, and a basolateral layer, facing the choroid. The apical side contains numerous microvilli while the basaolateral membrane possesses small-convoluted infoldings that increase the surface area for absorption of nutrients from the choroid side. Retinal pigmented epithelium cells are polarized, expressing specific nutrient transport systems on the apical and basolateral sides that maintain the environment of the neural retina (Marmor, 1998).

The neural elements of the retina are separated from blood supply at two levels. An outer level RPE separates the entire neural retina from the vascular layers of choroid, and at an inner level endothelial cells of the retinal blood vessels, separate the neural retina from retinal blood supply (Cunha-Vaz, 1976). These two components together constitute the BRB, which regulates the diffusion of substances from blood into the vitreous and retina.

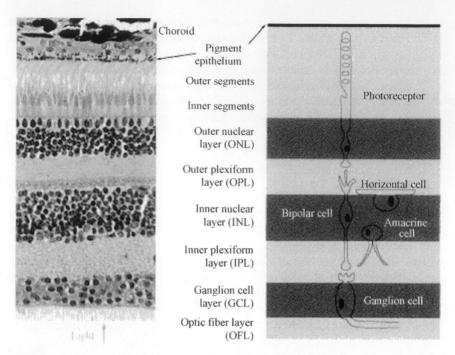

Figure 3. Cross-sectional view of retina (adapted from http://thalamus.wustl.edu/course/eyeret.html).

Blood vessels in choroid possess large fenestrations that allow free diffusion of substances into and out of choroidal stroma. Bruch's membrane (basement membrane separating choroid from the RPE) acts as a diffusional barrier only to macromolecules like proteins, oligonucleotides, and genes. However, RPE exhibits tight junctions creating a barrier to the entry of even small molecules into the retina from choroidal stroma (Cunha-Vaz, 1976). Recently, efflux pumps, such as P-glycoprotein and multidrug resistance proteins (MRP) have been identified on the RPE (Aukunuru et al., 2001; Kennedy et al., 2002) which may limit the permeation of various xenobiotics and endogenous compounds from the choroidal side into the vitreous. Thus, both transcellular and paracellular passages of molecules across the RPE are restricted and only selected nutrients are exchanged between the neural retina and the choroid. These characteristics result in the formation of outer BRB (oBRB).

Retinal capillaries may be considered as endothelial cell barrier. In humans major blood vessels lie in the innermost retinal layers and can usually be found just beneath the inner limiting membrane (a layer formed by neural retinal cells forming a diffusional barrier between vitreous and retina) although in a few animals capillaries can be observed as overlying on the retina, in contact with the vitreous. The ultrastructure of retinal blood capillaries is uniform throughout the retina and consists of a continuous endothelium surrounded by a thick basement membrane. Unlike capillary endothelium in other tissues, retinal endothelial vessels do not exhibit signs of fenestrations and express tight junctions in a conserved manner, surrounding every endothelial cell (Cunha-Vaz, 1976). Astrocytes are reported to be present in close apposition with the retinal blood vessels, similar to brain capillaries that constitute the blood–brain barrier (Holash et al., 1993). A combination of these cellular structures and functions may cause retinal blood vessel endothelial cells to form the inner blood–retinal barrier (iBRB). Recent discovery of a multidrug efflux pump, P-gp, on the retinal endothelium adds to the barrier properties of iBRB (Greenwood, 1992).

Thus, both RPE and retinal endothelial blood vessels collectively form BRB and act in conjunction to prevent unrestricted exchange of molecules between the systemic circulation and the posterior segment of the eye.

III. OCULAR FLUID DYNAMICS

A. Aqueous Humor Dynamics

Aqueous humor is derived from plasma within the capillary network of the ciliary processes by the following three mechanisms.

- *Diffusion*: lipid soluble substances are transported through the lipid portions of the cell membrane according to a concentration gradient across the membrane.
- *Ultrafiltration*: water and water-soluble substances, limited by size and charge, flow through micropores in the cell membrane in response to an osmotic gradient or hydrostatic pressure-influenced by IOP, blood pressure in the ciliary capillaries and plasma oncotic pressure.
- *Active transport*: this process accounts for the majority of aqueous humor production; water-soluble substances of larger size or higher charge are actively transported across the cell membrane requiring expenditure of energy (Na^+, K^+-ATPase and glycolytic enzymes).

Aqueous humor, relative to plasma is slightly hypertonic and acidic. Compared to plasma, it has a significant excess of ascorbate (15 times greater than arterial plasma), marked deficit of protein (0.02% in aqueous vs. 7% in plasma), slight excess of chloride and lactic acid, slight deficit of sodium, bicarbonate, carbon dioxide, and glucose. Albumin/globulin ratio is similar to plasma, although there is less γ-globulin.

The rate of aqueous humor production is about 2–3 μl/min and it completely turns over in about 2–3 h. The volume of the anterior chamber is approximately 250 μl. Aqueous humor formation decreases with sleep, advancing age, uveitis, retinal detachment, and ciliochoroidal detachment. Decreased aqueous humor production with increased IOP has been disputed by recent studies, which indicate that the rate of aqueous humor production is pressure insensitive. Functions of the aqueous humor range from maintaining IOP to provide substrates (glucose, oxygen, and electrolytes) for metabolic requirements of avascular cornea and lens. It also aids in removing metabolic products, such as lactate, pyruvate, and carbon dioxide. The primary path for aqueous humor flow is from the ciliary body along the posterior surface of the iris, through the pupil, across the anterior surface of the iris, and finally out of the globe via the trabecular meshwork (Figure 4). Since the aqueous humor is flowing from posterior to anterior direction the pressure must be highest on the posterior side, causing the iris to be pushed anteriorly by the humor. The effect is relatively small in normal eyes but can be quite pronounced in certain pathological cases, and sometimes the iris even exhibits a confusing posterior deflection.

Normal IOP has been defined as the pressure, which does not result in glaucomatous damage. Average IOP is around 15.8 mmHg (with Schiotz tonometry) and 16 mmHg (with applanation tonometry). The factors that influence IOP include genetics, age, sex, refractive error, race, diurnal variation, postural influence, Valsalva maneuver, blinking, and lid squeezing.

Tonometry is the measurement of the IOP by relating a deformation of the globe to the force responsible for the deformation and is of the following three types:

- Indentation tonometry;
- Applanation tonometry; and
- Noncontact tonometry.

Tonography is a dynamic test of the ability of the eye to recover from the elevation of IOP induced by Schiotz tonometry (continuous IOP measurement by electronic Schiotz tonometer) and measure of the facility of aqueous humor outflow (C). Tonography derives its mathematical basis from Grant's equation, which is described as follows:

$$P_o = \frac{F}{C} + P_v \tag{1}$$

F is the rate of aqueous outflow, C is the coefficient of outflow facility, P_o is the baseline IOP, P_v is the episcleral venous pressure (EVP).

Sources of error with tonography include increased resistance to aqueous outflow with increased IOP, increased episcleral venous pressure with

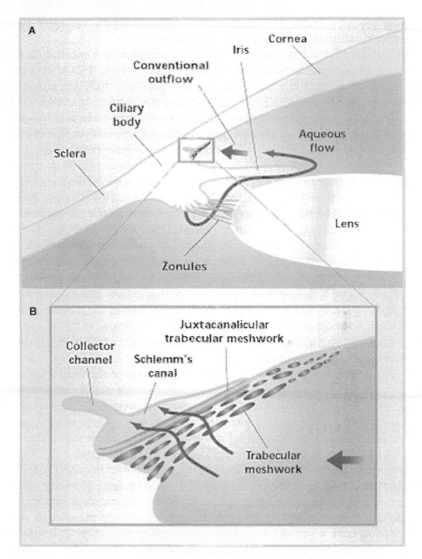

Figure 4. Eye structures involved in aqueous humor dynamics. (A) Aqueous humor is produced by the ciliary body and passes from the posterior chamber through the pupil into the anterior chamber. The conventional outflow pathway for the aqueous humor includes trabecular meshwork, Schlemm's canal and episcleral venous system. (B) The enlarged area of trabecular meshwork and Schlemm's canal. The suggested region of the maximum resistance to aqueous humor outflow is marked by red color. The activation of the ELAM-1 expression through the IL-1/NF-B pathway was observed exactly in this area in the glaucomatous but not normal eyes. Arrows show the main aqueous humor pathway. (Reprinted with permission from Nature Medicine, www.nature.com.)

increased IOP, effects of ocular rigidity and expulsion of uveal blood, instrument calibration, and patient cooperation.

Researchers have long recognized the importance of understanding the detailed physiological processes inside and around the eye. However, aspects of fluid dynamics within the eye have not yet been fully examined and quantitatively explained. Flow in the anterior chamber is a well-known phenomenon that has been observed clinically. Fluorescein when injected into the body appeared in the anterior chamber of animal eyes in a centrally placed vertical line. Such molecular movement may have resulted from fluid flow in the anterior chamber by a temperature gradient. Later it was concluded that convection in the anterior chamber was responsible for asymmetries in pupil shape and placement during pharmacologic pupil dilation (Wyatt, 1996). The flow appears to take place in a single convection cell, rising near to the back of the chamber and falling toward the front. Little or no lateral movement of the fluid was noticed. The flow transit times have been best estimated around 1 min and that time period appears to be normal. Fluid present in the anterior segment of the eye is produced continuously by the ciliary body. It flows throughout the pupil aperture in the center of the back wall of the chamber. Aqueous humor appears to be a linear viscous fluid with a viscosity, density, and expansivity identical to that of water. Fluid drains from the chamber through channels in the angle between the iris and the cornea (Canning et al., 2002).

Canning et al. recently developed a model to analyze the fluid flow in the anterior chamber of the human eye. It was shown that under normal conditions, such flow inevitably occurs and is viscosity dominated. The observed flow is driven by the buoyant convection arising from the temperature gradient across the anterior chamber of the eye. This difference exists due to a temperature difference between the back of the anterior chamber (37°C) and the outside of the cornea (room temperature, i.e., approximately 25°C). Usually the flow in the anterior chamber remains fairly constant. However, it may change in a significant manner in pathologies where particulate matter is present in the anterior chamber. This may happen due to the presence of red or white blood cells, which may be present as a result of a variety of diseases or conditions and also in other circumstances where pigment particles may detach from the iris. Movement and distribution of such particles can be predicted by well-established models in various ocular conditions, such as hyphemas, keratic precipitates, hypopyons, and Krukenberg spindles (Canning et al., 2002).

B. Fluid Flow in Vitreous Body

Vitreous is a clear, avascular connective tissue that fills the space between lens and retina. In adults, volume of vitreous is about 4 ml and it weighs about 4 g (Lee et al., 1992). Vitreous possesses a gel-like consistency and is primarily composed of water (99.9%). The remaining 0.1% consists of collagen, hyaluronic acid (HA) and other ions, such as sodium, calcium, potassium, magnesium, and chloride. With age, concentrations of collagen and HA become elevated; however, even at elevated state concentrations are still relatively low (i.e., 0.13 and 0.4 mg/ml, respectively) (Stuart et al., 2003). Individual components contribute significantly to the rheological properties of the vitreous (Lee et al., 1994).

Collagen is the major structural protein of the vitreous. Collagen fibers consisting of 20–200 nm fibrils assembled from 3 to 5 nm microfibrils, are aggregates of tropocollagen molecules, composed of a triple helix of three identical polypeptide chains (Styrer, 1981). Collagen molecules in vitreous do not branch out and are of uniform thickness. The fibres are organized in a network where they can slip relative to each other (Balazs et al., 1959). Small diameter collagen fibers may minimize light scattering (Sebag, 1989). In humans and pigs, collagen is in the central vitreous at 6–10-fold less amounts or concentrations than in the posterior region (Lee et al., 1994). Hyaluronic acid is the other major macromolecular component of vitreous. Hyaluronic acid, a glycosaminoglycan, consists of repeating disaccharide units of hexosamine glycosydically linked to either uronic acid or galactose. Glycosaminoglycans are known to interlink with collagen molecules (Mathews, 1965; Asakura, 1985). Hyaluronic acid can stabilize the helical structure of collagen by increasing the thermal stability of cartilaginous collagen (Gelman et al., 1974). Moreover, large HA molecules may keep the collagen fibrils widely dispersed, which may also serve as a barrier to the influx of cells and macromolecules from the surrounding tissues (Swann, 1980). The spatial distribution of collagen and HA is not homogenous in the vitreous. Concentration of HA in the vitreous exhibits a monotonous increase from the anterior region to the posterior region of the vitreous in humans and other experimental animals like pigs and cows (Lee et al., 1994). The ratio of HA to collagen is approximately two in the anterior portion near the lens and grows to 10 in the center of the eye, and is about 20-fold in the posterior segment near the retina (Buchsbaum et al., 1984).

Electrolyte concentrations, which might play an important role in the vitreous rheology, have been studied in detail. Electrolyte gradients appear to develop in the vitreous and also between vitreous, aqueous and plasma suggesting possible electrolyte exchange among vitreous and neighboring tissues (Reddy et al., 1961). Potassium, sodium and bicarbonate ion concentrations appear to be elevated in the anterior portion of vitreous, adjacent to the ciliary body, when compared to the posterior vitreous (Naumann et al., 1959). Calcium levels in both vitreous and aqueous are similar to plasma. Chloride ion levels in vitreous appear to be greater than in the posterior chamber and anterior chamber or plasma due to the exchanges across the retina (Lee et al., 1994). Magnesium concentrations are normally higher in vitreous than aqueous humor and plasma (Frank et al., 1986).

Liquefaction and shrinkage of the vitreous usually lead to the loosening and separation of the vitreal cortex from the inner limiting membrane resulting in vitreal detachment (Schepens et al., 1987). Retinal tears and eventually retinal detachment result from traction caused by the shrinkage of a strand located inside the vitreous, which is strongly attached to retina. Mechanism of vitreal liquefaction is not fully understood, but is thought to be dependent on age and degree of myopia (Schepens, 1954; Sebag, 1989).

Various models have been devised to explain the behavior of vitreous many of which are complicated. A mathematical model was applied to explain the movement of particles inside the vitreous, assuming a spherical geometry for the vitreous (Fatt, 1975). Analysis revealed that fluid flow in the vitreous is too slow to seriously influence diffusive flux of small solute tracers. In case of larger particles the diffusion constant is very low and the particle movement can no longer be described by simplified flux equations because such particles move as a front along with the convective flow of the vitreous (Fatt, 1975). A simple model illustrating the behavior of the vitreous assumes that the large hydrodynamic resistance of HA allows the gel to produce effective yield stress. Such stress developments prevent bulk fluid flow within the gel under normal circumstances, although movement causes the gel to be spread and elastic waves may propagate (Canning et al., 2002). Drug movement inside the vitreous is primarily due to their diffusion inside the vitreous as the convective flow is too slow to affect diffusional properties.

One of the major routes of drug administration to the posterior segment is the intravitreal route. Vitreous diffusivity plays an important role in the retinal permeation of drugs. Diffusivity within the vitreous to a large extent depends on its pathophysiological state and molecular weight of the drugs being introduced. It has long been accepted that the diffusivity of solutes in the vitreous is unrestricted. A relationship was developed to determine diffusivity of drug molecules in the vitreous (Stuart et al., 2003). Diffusivity of a substance in a hydrogel can be estimated relative to its free aqueous value using (Eq. 2),

$$\frac{D_p}{D_o} = \exp^{[-(5+10^{-4}(Mw)]C_p} \tag{2}$$

D_p and D_o represent the diffusivity in vitreous and in a polymer free aqueous solution, respectively, Mw denotes the molecular weight of the diffusing species, and C_p refers to the concentration of collagen and HA in grams per gram of vitreous. Dependence of vitreous diffusivity with various molecular weight markers indicates that even for molecular weight of 100,000 Da, the ratio of D_p to D_o is still 0.992, indicating that diffusivity in aqueous solution is an accurate representation of vitreous diffusivity. This theory holds true for compounds that do not bind or interact with collagen or HA molecules in the vitreous. Molecular weight of administered drugs primarily falls within the range of approximately 100–10,000 Da (Stuart et al., 2003).

IV.   FACTORS AFFECTING DRUG DELIVERY TO EYE

Ocular bioavailability of drugs applied topically as eyedrops is extremely low (<1%). Absorption of drugs in the eye is severely limited by several protective mechanisms that ensure the proper functioning of the eye, and by other concomitant factors, such as:

- drainage of the instilled solution;
- lacrimation and tear turnover;
- metabolism;
- tear evaporation;
- nonproductive absorption/adsorption;
- limited corneal area and poor corneal permeability; and
- binding by the lacrimal proteins.

The drainage of the administered dose via the nasolacrimal system into the nasopharynx and the gastrointestinal tract takes place when the volume of fluid in the eye exceeds the normal lacrimal volume of 7–10 µl. Thus, a portion of the instilled dose (one to two drops, corresponding to 50–100 µl.) that is not eliminated by spillage from the palpebral fissure is drained quickly and the contact time of the dose with the absorbing surfaces (cornea and sclera) is reduced to a maximum of 2 min. The lacrimation and the physiological tear turnover (16% per minute in humans in normal conditions) can be stimulated and increased by the instillation even of mildly irritating solutions. The net result is a dilution of the applied medication and an acceleration of drug loss. The rate at which instilled solutions are removed from the eye varies linearly with instilled volume. In other words, the larger the instilled volume, the more rapidly the instilled solution is drained from the precorneal area. Ideally, a high concentration of drug in a minimum drop volume would be desirable. However, there is a practical limit to the concept of minimum dosage volume. Droppers delivering small volumes are difficult to design and produce. In addition, their practical usefulness could be reduced by the fact that most patients cannot detect the administration of small volumes. The conjunctival absorption, which occurs via the vessels of the palpebral and scleral conjunctiva, further lowers the drug absorption into the eye. Any instilled drug that has not been swept away from the precorneal area by the drainage apparatus is subject to protein binding and metabolic degradation in the tear film. All of these factors may result in transcorneal absorption of 1% or less of the drug applied topically as a solution. In summary, the rate of loss of drug from the eye can be 500–700 times greater than the rate of absorption into the anterior chamber following topical administration. Drugs applied topically are potentially available for absorption by the scleral and palpebral conjunctiva (the so-called "nonproductive" absorption). Although direct transscleral access to intraocular tissues cannot be excluded, drugs that penetrate the conjunctiva are rapidly removed from the eye by local circulation and may undergo systemic absorption, which may range, for example, from 65% for dipivalylepinephrine to 74% for flurbiprofen and 80% for timolol (Lee, 1993). These effects are frequently not anticipated, recognized, or treated appropriately. In conclusion, the fluid dynamics in the precorneal area of the eye can profoundly influence ocular drug absorption and disposition. As the normal fluid dynamics are altered by (e.g., tonicity, pH, or irritant drugs or vehicles), the situation becomes more complex. The formulation of ophthalmic drug products must take into account not only the stability and compatibility of a drug in a given formulation, but also the influence of that formulation on precorneal fluid dynamics (Himmelstein et al., 1978).

Drug delivery to the posterior segment of the eye, composed of the retina, choroid, vitreous humor, and sclera, is one of the most interesting and challenging tasks currently facing the pharmaceutical scientists. The choroid is richly perfused with blood vessels lined by endothelial cells that allow easy exchange of molecules between blood and the tissue. Such rapid drug uptake allows effective systemic therapy of infections involving the choroid, such as choroiditis. However, diseases commonly affecting the retina (i.e., bacterial endopthalmitis and cytomegaloviral (CMV) retinitis) may lead to severe sight-threatening pathologies particularly due to poor drug penetration into the retina. Two main components of the barrier that prevent entry of molecules from the systemic circulation into the ocular compartments are the blood–retinal barrier (BRB), in the posterior segment, and the blood–aqueous barrier (BAB) in the anterior segment of the eye. Blood–retinal barrier and BAB together form the blood–ocular barrier. Blood–retinal barrier prevents free passage of xenobiotics from the choroid into the retina and vitreous. Circumventing this barrier in an efficient manner without causing no or minimal damage to the tissue, is a significant challenge facing the ocular drug delivery scientists. Passage of drugs from the anterior segment to the posterior segment may not be efficient because of the continuous drainage of the aqueous humor. Thus, overcoming the BAB may not be a useful strategy to enhance drug concentrations in the retina and vitreous. Conventional ophthalmic dosage forms (i.e., solutions, ointments, and creams are unable to provide an answer to the problem of drug delivery to the retina and vitreous). Development of newer dosage forms like intravitreal injections, implants, subconjunctival implants etc. have had only limited success. To understand the barriers for drug delivery to the retina, a closer look at the structure and anatomy of eye and retina is necessary.

V. CURRENT MODES OF ADMINISTRATION

A. Topical

Eye diseases can cause discomfort and anxiety in patients, with the ultimate fear of loss of vision or even facial disfigurement. Many regions of the eye are relatively inaccessible to systemically administered drugs and, as a result, topical drug delivery remains the preferred route in most cases. Drugs may be delivered to treat the precorneal region for infections, such as conjunctivitis and blepharitis, or to provide intraocular treatment via the cornea for diseases, such as glaucoma and uveitis.

Topical drug delivery is complicated by effective removal mechanisms (which operate to keep the ocular surface free from foreign substances) and the barriers of the precorneal area. These include the blinking reflex, tear turnover, and low corneal permeability. Conventional eyedrops require frequent instillation. Only 1–2% of an instilled dose of pilocarpine hydrochloride, a drug used in the treatment of glaucoma, reaches its target tissue in the eye. In addition, eyedrops are often difficult for patients to self-administer, particularly the elderly. Thus, in recent years a significant effort has been directed toward the development of new systems for ophthalmic drug delivery. Various conditions treated by the topical application of ophthalmic drugs include following.

- *Glaucoma*—buildup of pressure in the anterior and posterior chambers of the choroid layer that occurs when the aqueous humor fails to drain properly.
- *Conjunctivitis*—an inflammation of the conjunctiva that may be caused by bacterial and viral infection, pollen and other allergens, smoke, and pollutants.
- *Dry eye syndrome*—inadequate wetting of the ocular surface.
- *Keratitis*—an inflammation of the cornea, caused by bacterial, viral, or fungal infection.
- *Iritis* (anterior uveitis)—commonly has an acute onset with the patient suffering pain and inflammation of the eye.
- Other conditions include ophthalmic complications of Rosacea, blepharitis (inflammation of the lid margins), and chalazia (Meibomian cysts of the eyelid).

Table 1. Examples of Drugs Used in Ocular Drug Delivery

Classification	Examples	Typical Indications
Antibacterials	Chloramphenicol, gentamicin, fusidic acid	Conjunctivitis, keratitis, blepharitis
Antivirals	Aciclovir, Idoxuridine	Dendritic corneal ulcers, keratitis
Corticosteroids	Betamethasone, prednisolone, hydrocortisone	Uveitis, screitis
Antiinflammatory agents	Cromoglycate, nedocromil, antihistamines, such as Antazoline, Azelastine	Inflammation and allergic conjunctivitis
Miotics	Pilocarpine	Glaucoma
Sympathomimetics	Adrenaline, cipefrine	Glaucoma
Beta-blockers	Timolol, betaxolol	Glaucoma
Local anesthetics	Amethocaine, ignocaine	Anesthesia during drug treatment procedures

Examples of drugs that may be administered topically are given in Table 1. Most ophthalmic drugs contain functional groups, such as alcohol, carboxylic acid, and phenol, which lend themselves to simple derivatization. Prodrugs (pharmacologically inactive derivatives of drugs that are chemically or enzymatically converted to their active parent compound after administration) of pilocarpine and β-blocker have been used to enhance bioavailability.

Instillation of drops into the lower cul-de-sac is the most common method of ocular drug delivery. Concentration of drug in the precorneal area provides the driving force for its transport across the cornea via passive diffusion. Thus, efficient ocular drug absorption requires good corneal penetration as well as prolonged contact time with the corneal tissue. Iontophoresis, prodrugs, ion pair formation, and cyclodextrins have all been used as means of enhancing ocular drug absorption. An ideal topical ophthalmic formulation would enhance bioavailability by sustaining drug release, while remaining in the precorneal pocket for prolonged periods of time (Le Bourlais et al., 1995). A wide variety of ophthalmic drug delivery systems are currently available on the market (Lang, 1995). Nevertheless, about 70% of prescriptions for eye medication are for conventional eyedrops. Such high use may be related to factors, such as low cost, ease in bulk manufacture, excellent patient compliance, efficacy, and stability. Various types of vehicles currently being employed in ophthalmic dosage forms are liquids—eye drops/lotions, eye ointments, aqueous gels, and solid matrices/devices. In all cases, one key requirement is that the formulation must be capable of being sterilized or manufactured in a sterile environment.

A number of inserts are currently available on the market or in the latter stages of development. These inserts have been classified as degradable or nondegradable (i.e., those that need to be removed on completion of therapy). Various materials have been utilized in the development of degradable inserts, including polyvinyl alcohol, hydroxypropylcellulose, polyvinylpyrrolidone, and hyaluronic acid. Nondegradable inserts have been shown to provide more predictable release rates than soluble inserts (Canning et al., 2002) and are mainly prepared from insoluble materials, such as ethylene vinyl acetate copolymers and styrene-isoprene-styrene block copolymers.

B. Systemic Administration

Systemic administration of drugs for treatment of ocular disorders has found limited success primarily due to the exclusion of the organ from systemic circulation. Major pathologies affecting the retina are fungal and bacterial endophthalmitis, CMV retinitis, and diabetic, glaucomatic, and proliferative vitreal retinopathies. Local drug delivery to the posterior segment of the eye, particularly to the retina is a challenging task due to the presence of BRB. Various strategies have been employed to deliver drugs to the retina. A temporary increase in the BRB permeability has been attempted by transient modification of the barrier properties of the RPE and endothelial cells of the retinal blood vessels by intracarotid infusion of a hyperosmotic solution, such as mannitol or arabinose. However, osmotic gradient method results in nonspecific opening of BRB and is associated with retinal and nervous system toxicity (Frank et al., 1986; Millay et al., 1986). Chemical modification of drug molecules is more commonly employed to overcome the BRB. This strategy involves the use of lipophilic prodrugs of therapeutic entities so as to increase their transcellular diffusion into the neural retina (Macha et al., 2002).

Drug levels attained in the posterior chamber of the eye also depend on the mode of administration of the drug. Commonly employed techniques of delivering drugs to the retinal tissue include systemic and topical administrations, intravitreal and local injections, and ocular implants.

Systemic administration of drugs to treat ocular conditions has been attempted by various researchers. However, only 1–2% of the plasma drug levels are achieved in the vitreous humor. Such low concentrations warrant frequent administrations of high doses in order to maintain the drug levels in the ocular tissues two to three times above the minimum therapeutic concentrations. In the treatment of acute human cytomegalovirus (HCMV) retinitis, intravenous therapy includes IV foscarnet sodium or ganciclovir (Costabile, 1998). An advantage of systemic administration is that it controls the spread of infection to other tissues. Disadvantages that outweigh the advantages of intravenous therapy include nonspecific absorption by other tissues, which could result in serious toxicities. Taking both advantages and disadvantages into consideration, systemic administration of drugs for treatment of ocular pathologies is not an ideal strategy unless the drug can be targeted primarily to the eye.

C. Intravitreal Drug Delivery

Intravitreal delivery is primarily employed to treat posterior segment diseases that involve light sensitive structures like neural retina, retinal pigmented epithelium, and retinal blood vessels.

Over the past 20 years, intravitreal injections have become a mainstay treatment of posterior segment infections/diseases. Ocular antivirals like ganciclovir (GCV), acyclovir (ACV), cidofovir, foscarnet, and antibiotics like cephalexin, cefazolin, ceftazidime, gentamicin etc, have been administered intravitreally to assess their pharmacokinetic parameters so that proper dose and dosing frequency can be designed to effectively treat CMV retinitis and endophthalmitis (Ben-Nun et al., 1989; Cundy et al., 1996; Hughes et al., 1996; Waga et al., 1999; Lopez-Cortes et al., 2001; Macha et al., 2001b).

Penetration studies with various drugs, particularly antibiotics, in the direction of blood to ocular fluids have confirmed that passage into vitreous is much more restrictive than penetration into the aqueous humor, probably by one order of magnitude (Cunha-Vaz, 1976). Recently our laboratory reported on the penetration of a paracellular marker fluorescein into the aqueous and vitreous humors of rabbits. It was delineated that a tighter barrier surrounds the vitreous in comparison to aqueous humor. Only about 1–2 % of the plasma levels of fluorescein were detected in vitreous humor (Macha et al., 2001a). Thus, in order to achieve sufficiently high drug levels in the retina and vitreous, higher plasma levels are required. In an attempt to increase the plasma levels, clinicians either increased the oral dose administered or administered the drugs by rapid intravenous infusions (Lopez-Cortes et al., 2000; Lalezari et al., 2002). Even though these procedures resulted in increased ocular levels, long-term toxicity concerns remain (Henderly et al., 1987; Jabs et al., 1987).

Intravitreal injection offers several advantages over systemic and topical administration for drug delivery to the retina and vitreous. It may result in increased drug levels in the posterior segment without causing any systemic side effects. Therapeutic levels may be maintained with lower doses, as this route of administration circumvents the blood–retinal barrier. Intravitreal injections are a mainstay therapy in severe cases of endophthalmitis and CMV retinitis (Diamond, 1981; Baum et al., 1982; Macha et al., 2001b). As discussed earlier, vitreous diffusivity of a drug molecule following intravitreal injection is similar to its free aqueous diffusivity. Thus, for most drugs, injected intravitreally, diffusion across the vitreous is not a major rate-limiting step.

Intravitreal injections suffer from several deficiencies, such as patient noncompliance, need for repeated injections (once a week), injection-associated infections like endophthalmitis and retinal detachment (Cantrill et al., 1989; Baudouin et al., 1996; Stone et al., 2000; Martin et al., 2002). To overcome some of these shortcomings researchers have attempted to increase the drug residence time in the vitreous following a single injection. Intravitreal injection of liposomes, microparticles, lipophilic ester prodrugs have met with mixed results (Peyman et al., 1987; Diaz-Llopis et al., 1992; Veloso et al., 1997; Stone et al., 2000; Herrero-Vanrell et al., 2001; Macha et al., 2002). Administration of particulate substances carries the risk of infections and blurred vision (Maurice, 2001). Our laboratory has reported on the vitreous disposition of ganciclovir (GCV) and its various monoester prodrugs. Ganciclovir monoester prodrugs generated sustained GCV levels in the vitreous for time periods longer than those from direct injection of the parent drug (Macha et al., 2002). Despite all the reported advantages, intravitreal injections are not well tolerated due to the associated pain and discomfort.

D. Intravitreal Implants

Direct intraocular injections overcome the problem of high intravenous dose-related toxicity. However, multiple injections are associated with an increased risk of cataract, astigmatism, endophthalmitis, retinal detachment, and vitreous hemorrhage (Cantrill et al., 1989). Intravitreal implants have been developed as an alternative to multiple intravitreal injections. Vitrasert® is a nonbiodegradable intraocular implant. It is inserted surgically in the posterior segment of the eye where it delivers GCV locally over a period of 5–8 months. Once the device is depleted of drug it is replaced with a new device. Data from Chiron Vision phase III clinical trial demonstrated that the disease progression was significantly delayed in patients implanted with Vitrasert® compared to patients receiving intravenous doses of GCV [201] (http://www.centerwatch.com/patient/drugs/dru67.html). However, risks associated with intravitreal injections are not completely eliminated. Adverse reactions reported for Vitrasert® include loss of visual acuity, vitreous hemorrhage, retinal detachment, cataract formation, lens opacities, macular abnormalities, IOP spikes, optic disk/nerve changes, hyphemas, uveitis, and acute or delayed onset endophthalmitis (Sakurai et al., 2001). Also, repeated surgeries are required for implantation and removal of the implant. Like intravitreal injections, intravitreal implants release drug locally into the vitreous humor and may not be able to stem the spread of infections to the contralateral eye. Thus, systemic administration is recommended in conjunction with intravitreal injections and implants.

E. Scleral Drug Delivery

A major risk associated with intravitreal injections and implants is retinal detachment, which generally occurs as a result of retinal piercing due to surgery or multiple injections. To overcome this injury to the retina, researchers have investigated sustained release devices implanted into the sclera. Previous work has revealed that sclera does not constitute a significant barrier toward diffusion of drug molecules into the vitreous chamber. It appears to be five times more permeable than cornea for hydrocortisone (Unlu et al., 1998). Moreover, primary mechanism of transscleral drug transport involves passive diffusion through the intercellular aqueous pore pathway. As a result molecular size appears to be the controlling factor in drug transport rather than lipophilicity of the molecule (Unlu et al., 1998). Scleral drug delivery to the posterior segment of the eye has been attempted in two ways: (1) scleral plugs and implants; and (2) subconjunctival injections.

Scleral plugs and implants

To lower the surgery related risk, biodegradable devices have been developed (Sakurai et al., 2001). The inserts were prepared from poly (DL-lactide-co-glycolide) (PLGA) and poly (DL-lactide) (PLA) polymers. Inserts were active for a period of 3 weeks in HCMV inoculated eyes and the vitreo-retinal lesions were reduced in comparison to single intravitreal injection of GCV (Sakurai et al., 2001). In another study, various blends of PLA and PLGA were examined for their suitability as vitreal implants (Yasukawa et al., 2001). Poly (DL-lactide)-70,000 and 5000 (80/20) resulted in sustained GCV release for a period of 24-weeks with drug levels maintained well above the MIC during the entire study period.

In conclusion, sustained drug release could be achieved with scleral implants. The main disadvantages of these implants are repeated need for surgical intervention and patient discomfort associated with the presence of a foreign body in the eye.

F. Subconjunctival

Subconjunctival injections may be employed to target drugs to the posterior segment of the eye. The primary absorption surfaces available are the sclera and conjunctiva. Absorption of Tilisolol, a β-blocker into ocular structures and plasma was studied following corneal, conjunctival, and scleral applications. Scleral application resulted in higher intraocular levels relative to conjunctival administration (Sasaki et al., 1996). Reverse was true for plasma concentrations (i.e., higher plasma levels of the drug were observed upon conjunctival application). It may be concluded from this report that conjunctival absorption results in drainage of drug into systemic circulation. As a consequence scleral application resulted in increased drug levels in intraocular tissues relative to conjunctival administration. Ocular pharmacokinetics of an anticancer drug, dacarbazine was studied following systemic and subconjunctival administrations. Elevated levels of dacarbazine were observed in aqueous and vitreous humors upon subconjunctival injections relative to intravenous injections (Kalsi et al., 1991).

An attempt was made to achieve sustained release of drug upon subconjunctival injections of polymer-based ophthalmic formulations. Commonly employed polymers for such purposes must be biocompatible and biodegradable. Such a material should not induce or cause any inflammatory reactions in the sensitive ocular tissues. The polymer should erode or degrade in a controlled manner so that a zero-order release profile can be attained over a prolonged period. Polymers employed in ocular drug delivery include alginates (Hatefi et al., 2002), pluronics and poloxamers (Geroski et al., 2001), polyorthoesters (Zignani et al., 2000; Einmahl et al., 2002), and polyvinyl alcohol (Smith et al., 1996). Efficacy of these polymers in providing sustained ocular drug levels following subconjunctival administration is still under investigation.

In spite of recent progress, retinal drug delivery is still challenging due to various factors including drug lipohilicity, presence of multidrug resistance (MDR) gene products and drug ionization. In conditions like CMV retinitis, drug entry into retinal cells could be restricted due to their polar and hydrophilic properties (GCV, cidofovir, and foscarnet). Membrane transporters have been targeted to gain cellular entry by an active transport process. This approach has met with considerable success in enhancing permeation of several molecules across biological membranes. Such a strategy can be employed to increase the intracellular drug concentrations in the retina following systemic, intravitreal, and transscleral administration. A brief background discussion on transporters and transporter-mediated drug delivery is presented in the following section prior to the discussion of its application to retinal drug delivery.

VI. CARRIER-MEDIATED DRUG DELIVERY

Bioavailability of drugs is restricted due to undesirable physicochemical properties that may present as pharmacological, pharmacokinetic, or pharmaceutical barriers in drug delivery. Cell membranes impose a barrier to the free movement of molecules. Simple diffusion, facilitated diffusion, primary active transport, secondary active transport, transcytosis, and group translocation constitute the primary mechanisms of solute transport across a membrane (Saier, 2000). A solute can be transported either through the cells or between the cells across any continuous epithelium. Cellular pathway across lipid bilayer is usually referred to as the transcellular route and the intercellular pathway is known as the paracellular route. Transcellular movement requires the interaction of solute with the components of the cell membrane. Paracellular movement on the other hand is limited only by size and charge of the permeating species. Prodrugs are designed to overcome the undesirable properties of drugs and are themselves biologically inactive. Ideally the prodrug should be converted to the parent drug as soon as the barrier is circumvented, followed by rapid elimination of the released inactive derivatizing group (Sinkula et al., 1975; Stella et al., 1985) (Figure 5).

A traditional prodrug design consists of the chemical modification of the active ingredient to alter the physicochemical properties, such as solubility, stability, and lipophilicity (Burton et al., 1991). Lipophilic modification can enhance drug transport to an extent by improving the transcellular permeability of the otherwise hydrophilic molecules, but is often limited by the solubility of the modified drug (Hughes et al., 1993).

Recent progress on molecular cloning of transporter genes and functional analysis by expressing those genes in cultured cells has greatly contributed to our mechanistic understanding of structure and function of membrane transporters. Membrane transporters were initially thought to be responsible for transferring endogenous compounds across the cell membranes, thereby strictly regulating the exchange of these agents between intracellular and extracellular spaces (Lee, 2000). However it is now clear that some of the transporters are involved in drug transport across various tissues, and may play a key role in intestinal absorption, tissue distribution, and elimination. Significant evidence regarding transporter-mediated drug permeation in tissues, such as liver, kidney, and small intestine has been accumulated (Beauchamp et al., 1992; Ganapathy et al., 1995; Balimane et al., 1998; Guo et al., 1999; Zhu et al., 2000). The structure–activity relationships of various drug molecules have been developed in order to define the structural specificity of substrates of the transporters (Inui et al., 2000; Terada et al., 2000).

The utility of carrier-mediated absorption via membrane transporters/ receptors is particularly important when the parent drug or the prodrug is polar or ionized, where passive transcellular absorption is negligible. Therefore, the use of prodrugs has been actively pursued to achieve very precise and direct effects at the "site of action".

The principal membrane barriers of the anterior segment of the eye are located in the cornea, iris–ciliary body, lens epithelium, and retina (Figure 1).

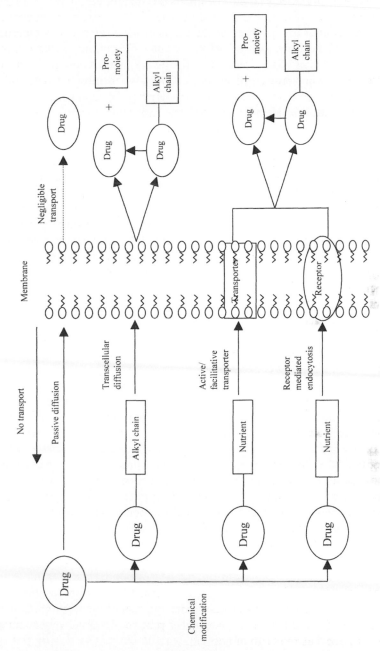

Figure 5. Mechanism of drug transport across cellular membranes.

A. Drug Efflux Pumps in Ocular Tissues

Nearly 200 proteins are involved in the efflux of substrates across biological membranes. These efflux pumps belong to an ATP-binding cassette (ABC) superfamily of proteins, also termed as traffic ATPases. These efflux pumps consist of two major efflux transporters: (1) P-glycoprotein (P-gp); and (2) multidrug resistance associated proteins (MRPs).

P-glycoprotein, considered as a versatile xenobiotic pump, was initially discovered in cancer cells. This efflux pump is a member of a highly conserved group of the energy-dependent ABC transporters found in cells from various tissues. Mammalian P-gps display approximately 60–65% homology with most P-gps from other species, suggesting that their role in drug trafficking is highly conserved throughout their evolution (Paulsen et al., 1996; Goffeau et al., 1997). Sequence analyses have revealed P-gp to be a member of ABC superfamily and hence ATP hydrolytic energy powers the efflux action of P-gp (Doige et al., 1993; Paulsen et al., 1996; Boyum et al., 1997). It is an integral membrane protein composed of two homologous halves, each consisting of an N-terminal hydrophobic domain, with six transmembrane segments, that is separated from a hydrophilic domain, containing a nucleoside binding fold, by a flexible linker polypeptide. P-glycoprotein encoding genes from hamster, mice, and human have also been identified in other species (Ruetz et al., 1994; Loo et al., 1995; Kast et al., 1996). This transporter is encoded by a small multigene family (*mdr* Class I, II, and III). P-glycoprotein belonging to all three classes is present in rodents, while human cells express P-gp belonging to classes I and III. Transfection studies have shown that the class I and II isoforms confer MDR, while class III isoform represents phosphatidylcholine translocase responsible for secretion of phospholipids into bile (Smit et al., 1993; Ruetz et al., 1994; Smith et al., 1994). Functional studies with P-gp substrates like cyclosporine A, verapamil, propranolol and dexamethasone have confirmed the presence of P-gp activity in conjunctival epithelium (Saha et al., 1998; Yang et al., 2000). In the eye, P-gp has been found in cornea (Kawazu et al., 1999), retina (Holash et al., 1993), and iris (Schlingemann et al., 1998). Cyclosporin A (CsA) uptake and transport in cultured rabbit corneal epithelial cells (RCECs) was studied and it was concluded that p-glycoprotein activity was present in rabbit corneal epithelial cells (RCECs) (Kawazu et al., 1999). A recent repot from our laboratory confirmed the molecular presence and functional expression of P-gp (MDR1) in human and rabbit cornea (Dey, 2002). Demonstration of the functional activity of the efflux pump may be instrumental in the design of efficient drug delivery strategies to the anterior segment. Blood–tissue barrier characteristics of iris and ciliary body have also been studied and it was found out that P-gp is expressed in ciliary muscles and iris capillaries (Schlingemann et al., 1998).

Volume-activated chloride current has been found to be associated with the endogenous expression of P-gp in pigmented ciliary epithelial cells (Chen et al., 2002). Very recently P-gp has been identified in rat lens and functional expression of P-gp (*mdr1a* and *mdr2*) was also reported (Merriman-Smith et al., 2002). The initial cellular damage phenotype of extracellular space dilations caused by P-glycoprotein inhibitors was identical to that caused by chloride channel inhibitors, indicating that P-glycoprotein may play a role in regulating cell volume in the lens.

Human multidrug resistance-associated proteins (MRPs) may play an important role in protecting the cells against carcinogenic drugs. Currently, the MRP family consists of seven members, MRP1–7 (Borst et al., 2000). The ability of these membrane proteins to efflux a wide range of anticancer drugs out of cell and their presence in many forms of tumors render them prime suspects in unexplained cases of drug resistance. Multidrug resistance-associated proteins are organic anion transporters (i.e., transport anionic drugs and neutral drugs conjugated to acidic ligands). Multidrug resistance-associated proteins 1–3 also cause resistance to neutral organic drugs that are not known to be conjugated to acidic ligands (Gottesman et al., 1996; Sharom, 1997; Liscovitch et al., 2000). Multidrug resistance-associated proteins also play a vital role in resistance development against nucleoside analogs in cancer chemotherapy. MRP1–3 and MRP6 have an extra N-terminal domain that lacks P-gp (Borst et al., 2000). Also, MRP4 and MRP5 are much more homologs to the other MRPs than to P-gp or other classes of ABC transporters. MRP1 has been in rabbit conjunctival epithelial cells as indicated by a ∼190 kDa protein corresponding to MRP1 in Western blots (Saha et al., 1998; Yang et al., 2000). Existence of MRPs in other ocular tissues has not been extensively explored.

Recently there have been reports of P-gp expression of RPE and retinal blood vessels oriented toward the blood side, preventing substances from entering the ocular structures from systemic circulation.

B. Strategy for Site-Specific Drug Delivery

Carrier-mediated drug transport is relatively unexplored in comparison to passive transcellular and paracellular drug transport. Prodrugs have been used to achieve site directed delivery for enhanced absorption and reduced toxicity. It has been suggested that at least three factors should be optimized for the site-specific drug delivery by the prodrug approach (Stella et al., 1980).

- The prodrug must be readily transported to the site of action and its uptake by the target cells site must be rapid and perfusion rate limited.

- Once at the site, the prodrug must be selectively cleaved to the active drug.
- The active drug once regenerated must be sufficiently retained by the target tissue to cause complete regeneration of the active drug.

As mentioned earlier the prodrugs can either be targeted to the transporters or enzymes for site-specific drug delivery. Enzymes can be recognized as presystemic metabolic sites or prodrug–drug *in vivo* reconversion sites (Han et al., 2000). The enzyme-targeted prodrug approach can be used to improve the oral drug absorption as well as site-specific drug delivery. An improvement in oral drug absorption can be achieved by targeting the gastrointestinal enzymes and using a nutrient moiety as a derivatizing group in order to permit a more specific targeting to these enzymes (Bannerjee et al., 1985). Targeting of enzymes for improved oral absorption and site-specific drug delivery has been made easy by the extensive literature present on the gastrointestinal enzymes, which provide necessary information, such as enzyme distribution, activity, and specificity for prodrug design (Amidon et al., 1980; Bannerjee et al., 1985; Fleisher et al., 1985; Bai et al., 1992; Han et al., 2000). Recently new approaches, such as antibody directed enzyme prodrug therapy (ADEPT) (Bagshawe, 1987, 1989, 1993; Bagshawe et al., 1988; Connor et al., 1995) and gene directed enzyme prodrug therapy (GDEPT) (Harris et al., 1994; Huber et al., 1994) have been proposed, which attempt the localization of prodrug activation enzymes into specific cancer cells prior to prodrug administration.

On the other hand prodrugs can be designed to resemble the intestinal nutrients structurally and be absorbed by specific carrier proteins. Prodrug targeting toward transporter/receptor requires considerable knowledge of the carrier proteins, including their distribution and substrate specificity.

C. Prodrugs Targeted toward Transporters

A host of transporter proteins can be targeted for improving drug absorption. Generally, membrane transport mechanisms exist for amino acids (Smith et al., 1988; Christensen, 1990), dipeptides (Dantzig et al., 1990; Hashimoto et al., 1994; Han et al., 2000; Kiss et al., 2000), monosaccharides (Hediger et al., 1987; Tamai et al., 2000), monocarboxylic acids (Tiruppathi et al., 1988), organic anions (Sekine et al., 1997), organic cations (Okuda et al., 1996), phosphates (Feild et al., 1999), nucleosides/nucleobases (Ritzel et al., 1997; Patel et al., 2000; Hamilton et al., 2001), bile acids (Wilson, 1981), ascorbic acid (Tsukaguchi et al., 1999; Wang et al., 1999), and urea (Kim et al., 2002).

Among the various transporters listed above, peptide transporters have received the most attention, as they are known to be versatile and robust in their choice of substrates. A summary of all transporters/receptors present in

Table 2. Transporters and Receptors in Various Ocular Tissues

Transporter	Cornea	Conjunctiva	Iris–ciliary	Lens
Peptide	+$^?$	–	–	–
Amino acids	+	–	–	–
Nucleoside	+	–	+	–
Glucose	+	+	+	+
Vitamin C	–	–	–	+
Acid/base	+	+	–	–
Glutathione	–	–	–	+
Efflux pump				
P-glycoprotein, MDR1	+	+	+	+
MRP	NA	+	NA	NA
Receptor				
Insulin/insulin-like growth factor	+	+	+	+
Growth Factors	+	+	+	+
Prostanoid		+	+	
Bradykinin/tachykinin	+	+	+	–
Muscarinic	+	+	+	+
Adrenergic	+	+	+	+
Histamine	+	+	+	+
Estrogen	+	–	+	+
Progesterone	+	–	+	+
Prostanoid	+	+	+	+
Serotonin	+	–	+	–
Glucocorticoid	+	NA	+	+
Mineralocorticoid	+	NA	+	+
Tumor necrosis factor	+	–	+	–
VEGF	+	+	+	–
Hyaluronan	+	+	+$^?$	+

(+) denotes the presence of the transporter/receptor.
(–) denotes the absence of the transporter/receptor.
(?) denotes the presence where the exact subtype of the transporter/receptor is unknown.
NA denotes where no data is available or unknown.

various tissues on the anterior segment has been presented in Table 2. Various nutrient transporters expressed on the retina include peptide (Berger et al., 1999), glucose (Ban et al., 2000), amino acids (i.e., glycine, GABA, glutamine, taurine, arginine, proline, and tryptophan) (Pow, 2001), and nucleoside like adenosine (Williams et al., 1994). Among the transporters identified, only the very few that could be targeted to enhance corneal and retinal drug delivery are discussed in the following sections.

D. Transporters/Receptors in Ocular Tissues

Peptide transporters have been identified on various epithelia (Kramer et al., 1990a; Daniel et al., 1993; Smith, 1993; Dieck et al., 1999; Chen et al., 2001; Groneberg et al., 2002; Putnam et al., 2002). Considerable information is available on their mechanism and substrate specificity (Kramer et al., 1990b;

Terada et al., 1996; Fei et al., 1997; Han et al., 1998; Meredith et al., 2000; Yang et al., 2001b; Doring et al., 2002). These transporters (i.e., PepT1 and PepT2) have broad substrate specificity particularly for smaller peptides like di- and tri-peptides. Through prodrug modification, a drug molecule may be converted to a substrate for this transporter to be easily ferried across lipophilic membranes (Yang et al., 2001b). At the same time the solubility of the compound may be enhanced depending on the amino acids in the peptide promoiety. Following its transport, the enzymes, specifically the cholinesterases, dipeptidases, and aminopeptidases, present in the vitreal and retinal tissue may cleave the prodrug to regenerate the parent drug. Peptide prodrugs of nucleosides like GCV, zidovudine, and ACV are substrates of peptide transporter, which result in a significant increase in the bioavailability of the parent drugs upon oral administration (Han et al., 1998; Pescovitz et al., 2000). The presence of a peptide transporter on the retina facing the vitreous (Berger et al., 1999) on the RPE side facing blood (Atluri et al., 2002) has been reported. Transporter on the retina facing the vitreous could be targeted to increase the intracellular drug concentrations following intravitreal administration. On the other hand, targeting to the peptide transporter present on the RPE can enhance retinal and vitreal drug concentrations following systemic administration. A recent report from our laboratory provided functional evidence for the presence of an oligopeptide transport system on the intact rabbit cornea (Anand et al., 2002). The study describes a carrier-mediated transport for a nonpeptidic drug L-Val-ACV, a prodrug of ACV, across freshly excised rabbit cornea. Transport of L-Val-ACV was found to be higher than that of the parent drug ACV, which is saturable at higher concentrations, pH-dependent, and inhibited competitively by other known hPEPT1 substrates. L-Val-ACV has also been shown to be a substrate for PEPT1/PEPT2 resulting in a fivefold higher bioavailability in comparison to ACV following oral delivery (Beauchamp et al., 1992).

Amino acid transporters on the blood–brain barrier and intestinal mucosa have been studied extensively (Hargreaves et al., 1988; Kilberg et al., 1993; Fukasawa et al., 2000; Kanai et al., 2000; Munck et al., 2000; Yang et al., 2001a). These transporters are very substrate specific and lead to passage of only specific amino acids. Various amino acid transporters including glutamate, glycine, GABA, tryptophan, and proline have been identified on the retina (Pow, 2001). Expression of such transport systems on the retina or RPE once again opens up the opportunity of designing prodrugs targeted toward these carrier-proteins. Functional evidence of a high affinity, Na^+ independent pheylalanine (Vakkalagadda et al., 2002) and tyrosine (Balakrishnan et al., 2002) carrier systems with characteristics similar to the LAT1 transporter have also been identified on SIRC cell line as well as rabbit cornea. RT-PCR on the RNA extracted from human cornea showed a 520-bp band and a 754-bp band, which were confirmed to be hLAT1 and hATB$^{0,+}$, respectively by subcloning and sequencing.

Mammalian cells take up and excrete lactate, pyruvate, and other monocarboxylic acids by means of proton coupled monocarboxylic acid transporters (Nord et al., 1983; Alm et al., 1985; Kang et al., 1990; Terasaki et al., 1991; Bergersen et al., 1999; Gerhart et al., 1999; Tamai et al., 1999). Various MCT isoforms (MCT 1–9) have been identified (Halestrap et al., 1999). Monocarboxylic drugs like salicylic acid and pravastatin have been reported to be transported via MCTs in the small intestine (Takanaga et al., 1994, 1996; Tamai et al., 1995). In the eye, MCTs are expressed on the retina and conjunctiva (Horibe et al., 1998; Hosoya et al., 2001). MCT 1 has been reported to be present on the apical membrane and MCT3 on the basolateral membrane of rat RPE (Yoon et al., 1997). MCT can thus translocate substrates (carboxylate drugs and prodrugs) from the blood/choroid into the retina following intravenous and transscleral administrations.

A recent study has identified GLUT1, one of the glucose transporters, in the bovine corneal epithelium and bovine corneal tissues (Bildin et al., 2001). The expression pattern of GLUT1 in the cornea varies between control and diabetic rats (Takahashi et al., 2000), implying that GLUT1 has minimal influence on delayed healing of corneal wounds in diabetes. In rats, GLUT1 mRNA expression is enhanced during corneal epithelial wound repair. Corneal damage facilitates GLUT1 mRNA expression and protein synthesis. Higher expression of GLUT1 may be responsible for increased transport of glucose, providing the metabolic energy necessary for cell migration and proliferation (Takahashi et al., 1996). Although the glucose transporters exhibit significantly higher capacity than other nutrient transporters they are not amenable to drug delivery (Pardridge, 1983; Silverman, 1991; Thorens, 1993; Wright, 1993; Shepherd et al., 1999), because of their extremely rigid structural specificity requirements.

Two types of transport processes have been identified for folate entry into mammalian cells. One process involves the folate receptors that bind and internalize the bound folate via receptor-mediated endocytosis. A second process utilizes the reduced folate transporter, which is a typical transporter protein with multiple membrane spanning domains. The protein interacts with the reduced folate more efficiently than folate itself (Antony, 1992; Sirotnak et al., 1999; Smith et al., 1999; Chancy et al., 2000; Reddy et al., 2000). Folate receptors have been targeted for facilitating anticancer drug entry into cancer cells and also for liposome based gene delivery (Sudimack et al., 2000). Thus, targeting the folate transporter could result in enhanced drug levels in the vitreous. Since the folate receptor is generally absent in most normal tissues with the exception of choroid plexus and placenta, or is expressed at very low levels (lung, thyroid, and kidney) (Sudimack et al.,

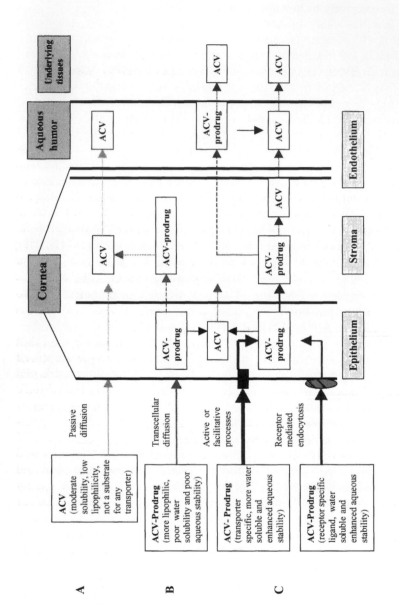

2000), it may be possible to achieve targeted delivery of the folate ester prodrugs to the retina, following IV administration.

VII. STRATEGIES TO IMPROVE OCULAR BIOAVAILABILITY BY TRANSPORTER-MEDIATED DRUG DELIVERY

A. Anterior Segment

Topical ocular drug delivery is clinically significant in the treatment of diseases of the eye that affect the anterior chamber like corneal epithelial and stromal keratitis, glaucoma, conjunctivitis, and possibly in the posterior chamber including bacterial endophthalmitis, retinitis, and macular degeneration. One of the major challenges in ocular drug delivery has been to deliver drugs to the anterior and posterior chambers of the eye using topical dosage forms. The main disadvantages are related to the facts that the drug is diluted and washed off almost immediately due to a high tear turnover rate. Precorneal loss results in poor drug absorption and subtherapeutic drug concentrations in the eye. Several approaches have been employed to increase bioavailability of drugs in the eye. Targeted prodrug design has proven to be an efficient strategy for site-directed delivery.

Various lipophilic prodrugs have shown increased ocular absorption thus delivering more of the parent drug to the interior of the eye. Acyclovir (ACV) is used in the treatment of herpes simplex virus (HSV) keratitis and is the drug of choice for viral infection in the eye. Various modes of ACV transport through the cornea have been outlined in Figure 6. Owing to its limited bioavailability the drug has shown moderate antiviral efficacy following topical administration (Sanitato et al., 1984). Various lipophilic ACV prodrugs have been examined for improved ocular absorption and enhanced efficacy in HSV keratitis. These prodrugs although exhibiting an elevated permeability across the cornea lacked adequate aqueous solubility necessary for their formulation into 1–3% eyedrops (Hughes et al., 1993). Since the cornea consists of multiple tissue layers with lipophilic and hydrophilic regions (Figure 2), a balance between the lipophilicity and the hydrophilicity has to be reached for a compound to penetrate into the deeper tissues and to

---

Figure 6. A schematic of transport of acyclovir across cornea: (A) conventional approach involves topical delivery of ACV. (B) Improved delivery of ACV by enhancing its lipophilicity and increasing ACV permeation across the corneal epithelium. (C) Improved delivery of ACV and permeation across corneal epithelium and stroma by utlizing transporters and receptors present on the corneal epithelium and by enhancing its aqueous solubility.

be effective especially in cases where the stroma and underlying tissues have been affected. Lipophilicity to a certain degree lowers epithelial resistance, the primary barrier to corneal permeation, but if one continues to increase the lipophilicity, permeation across cornea decreases (Narurkar et al., 1989). For a compound to be effective topically and to be formulated into eyes drops it must possess sufficient hydrophilicity and at the same time exhibit sufficient permeation through the cornea to reach three to five times the MIC levels. In order to overcome problems of insufficient solubility and transport of the parent drug ACV dipeptide prodrugs (US patent pending) were synthesized in order to target the oligopeptide transporter on the cornea (Nashed et al., 2002). These prodrugs exhibited excellent stability, antiviral activity, permeability across cornea (Anand et al., 2002), and affinity toward hPEPT1 (Anand et al., 2003). The dipeptide prodrugs of ACV were found to be substrates of hPEPT1 and were transported across intestinal cell line and isolated rabbit cornea owing to their recognition by the oligopeptide transporters. Therefore, the dipeptide prodrugs of ACV seem to be promising candidates for the treatment of epithelial and stromal HSV keratitis, a clinical indication not adequately treated by current therapy with trifluridine because of cytotoxicity.

B. Posterior Segment

Expression of transporters on the retina, particularly on RPE and endothelial cells of the retinal blood vessels, provides us with an opportunity to increase the retinal and vitreal levels of various drugs thereby increasing their efficacy and decreasing the required dose. The strategies that can be employed to target these transporters on retina are illustrated in Figure 7. Compounds can be targeted to the transporter by delivering through any of the three main routes: (1) systemic; (2) intravitreal, and (3) subconjunctival.

Systemic administration of prodrugs

As discussed earlier, vitreal drug levels achieved upon systemic administration is around 1–2% of the plasma levels. If a drug is targeted to a transporter expressed on the BRB, it could result in markedly higher concentrations in the retina. Prodrug cleavage in the retinal cells also may generate drug levels in the vitreous humor. The strategy is shown in Figure 7A.

Main disadvantage with this strategy is that drug uptake will be nonspecific due to possible distribution of the transporter at multiple tissues.

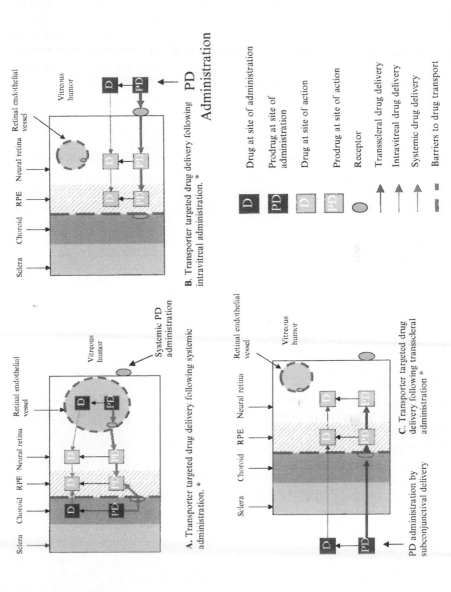

Figure 7. Strategies involved in enhancing drug delivery to posterior chamber by transporter-mediated drug delivery.

Intravitreal administration

In order to avoid the nonspecific absorption of drug into nontarget tissues and to avoid systemic toxicity, intravitreal administration of the prodrugs may be justified. Figure 7B describes this strategy where the prodrug is injected directly into the vitreous humor thereby targeting to the transporter expressed on the retinal cell membranes. Prodrug uptake by neural retina and RPE cells may be more efficient than the parent drug itself.

Subconjunctival administration

Subconjunctival administration usually results in higher vitreal concentrations than systemic administration. Moreover, subconjunctival injections of particulate dosage forms may result in sustained drug release over a prolonged period of time thereby decreasing the frequency of administration. This route may be used to deliver prodrugs targeted to specific transporters expressed on the basolateral side of the RPE. Following subconjunctival administration the prodrug first diffuses into the sclera and then into the choroidal circulation where it interacts with transporters expressed on the RPE. These transporters will then carry the prodrug into the retinal tissue where it is cleaved into the parent drug. This mechanism is schematically depicted in Figure 7C. If the drug is incorporated into a polymeric vehicle, which controls the release of the prodrug, a sustained delivery of drug to retina and the vitreous layers may be possible.

Different ocular routes that have been used to deliver drugs and prodrugs include cornea, conjunctiva, iris–ciliary body, and lens epithelium. Transporters and receptors have been identified in these tissues and have been utilized to improve ocular drug delivery. On the other hand discovery of efflux transporters (P-glycoprotein) identified on the cornea and conjunctiva will lead to a better understanding of restricted transport of drugs (antiviral, antineoplastic etc.) to the inner compartment of the eye.

VIII.   CONCLUSIONS

Ocular pathologies can cause discomfort and anxiety in patients, with the ultimate fear of loss of vision or even facial disfigurement. In spite of the continued effort directed toward the improvement and optimization of ocular drug delivery systems a progress in this area did not appear to take place at a fast pace that is typical of other delivery routes (oral, transdermal, transmucosal, etc.). A cautious advancement is evidently imposed by the delicate nature of the eye and many restraints imposed by its anatomy and physiology.

Fluid dynamics in the precorneal area of the eye has a profound effect on ocular drug absorption and disposition. When the normal fluid dynamics are altered (i.e., tonicity, pH or irritant drugs or vehicles), the situation becomes more complex. Formulations of ophthalmic drug products must take into account not only the stability and compatibility of a drug in a given formulation, but also the influence of the formulation on precorneal fluid dynamics.

Drug delivery to retina has always been a challenge to ophthalmologists. The complex physiology of the retina coupled with low blood flow makes it even more difficult. Several routes of administration including systemic, topical, intravitreal, and scleral have been attempted and have met with varying degrees of success.

The prodrug approach has been successfully employed to overcome the undesirable pharmacokinetic properties. Recent developments in the cloning and expression of the membrane transporters from human and animal species have added a new dimension to the understanding of the transporter structure/function and their precise structural interactions with the substrates. Design of prodrugs to target specific transporters is a sound strategy for the development of efficient and selective drug delivery systems targeted to the ocular anterior and posterior segments.

ACKNOWLEDGMENTS

This work was supported by NIH grants RO1 EY09171, RO1 EY10659, RO1 GM64320, and RO1 AI 36624.

REFERENCES

Alm, A. and Tornquist, P. (1985). Lactate transport through the blood-retinal and the blood-brain barrier in rats. Ophthalmic. Res. 17, 181–184.

Amidon, G.L., Leesman, G.D. and Elliott, R.L. (1980). Improving intestinal absorption of water-insoluble compounds: A membrane metabolism strategy. J. Pharm. Sci. 69, 1363–1368.

Anand, B.S. and Mitra, A.K. (2002). Mechanism of corneal permeation of L-valyl ester of acyclovir: Targeting the oligopeptide transporter on the rabbit cornea. Pharma. Res. 19, 1194–1202.

Anand, B.S., Patel, J. and Mitra, A.K. (2003). Interactions of the dipeptide ester prodrugs of acyclovir with the intestinal oligopeptide transporter: Competitive inhibition of glycylsarcosine transport in human intestinal cell line-Caco-2. J. Pharmacol. Exp. Ther. 304, 781–791.

Antony, A.C. (1992). The biological chemistry of folate receptors. Blood 79, 2807–2820.

Asakura, A. (1985). Histochemistry of hyaluronic acid of the bovine vitreous body by electronmicroscopy. Nippon Ganka Gakkai Zasshi 89, 179–191.

Aukunuru, J.V., Sunkara, G., Bandi, N., Thoreson, W.B. and Kompella, U.B. (2001). Expression of multidrug resistance-associated protein (MRP) in human retinal pigment epithelial cells and its interaction with BAPSG, a novel aldose reductase inhibitor. Pharm. Res. 18, 565–572.

Bagshawe, K.D. (1987). Antibody directed enzymes revive anti-cancer prodrugs concept. Br. J. Cancer 56, 531–532.

Bagshawe, K.D. (1989). The first Bagshawe lecture. Towards generating cytotoxic agents at cancer sites. Br. J. Cancer 60, 275–281.

Bagshawe, K.D. (1993). Antibody-directed enzyme prodrug therapy (ADEPT). Adv. Pharmacol. 24, 99–121.

Bagshawe, K.D., Springer, C.J., Searle, F., Antoniw, P., Sharma, S.K., Melton, R.G. and Sherwood, R.F. (1988). A cytotoxic agent can be generated selectively at cancer sites. Br. J. Cancer 58, 700–703.

Bai, J.P. and Amidon, G.L. (1992). Structural specificity of mucosal-cell transport and metabolism of peptide drugs: Implication for oral peptide drug delivery. Pharm. Res. 9, 969–978.

Balakrishnan, A., Jain, B., Yang, C., Pal, D. and Mitra, A.K. (2002). Carrier mediated uptake of L-tyrosine and its competitive inhibition by model tyrosine linked compounds in a rabbit corneal cell line (SIRC)—strategy for the design of transporter/receptor targeted prodrugs. Int. J. Pharm. 247(1–2), 115–125.

Balazs, E.A., Laurent, T.C., Laurent, U.B.G., Deroche, M.H. and Bunney, D.M. (1959). Studies on the structure of the vitreous body. VII. Comparative biochemistry. Arch. Biochem. Biophys. 81, 464–479.

Balimane, P.V., Tamai, I., Guo, A., Nakanishi, T., Kitada, H., Leibach, F.H., Tsuji, A. and Sinko, P.J. (1998). Direct evidence for peptide transporter (PepT1)-mediated uptake of a nonpeptide prodrug, valacyclovir. Biochem. Biophys. Res. Commun. 250, 246–251.

Ban, Y. and Rizzolo, L.J. (2000). Regulation of glucose transporters during development of the retinal pigment epithelium. Brain Res. Dev. Brain Res. 121, 89–95.

Bannerjee, P.K. and Amidon, G.L. (Eds.) (1985). Design of Prodrugs Based on Enzymes-Substrate Specificity. Elsevier, New York.

Baudouin, C., Chassain, C., Caujolle, C. and Gastaud, P. (1996). Treatment of cytomegalovirus retinitis in AIDS patients using intravitreal injections of highly concentrated ganciclovir. Ophthalmologica 210, 329–335.

Baum, J., Peyman, G.A. and Barza, M. (1982). Intravitreal administration of antibiotic in the treatment of bacterial endophthalmitis. III. Consensus. Surv. Ophthalmol. 26, 204–206.

Beauchamp, L.M., Orr, G.F., Miranda, P.D. and Krenitsky, T.A. (1992). Amino acid ester prodrugs of acyclovir. Antiviral Chem. Chemotherap. 3, 157–164.

Ben-Nun, J., Joyce, D.A., Cooper, R.L., Cringle, S.J. and Constable, I.J. (1989). Pharmacokinetics of intravitreal injection. Assessment of a gentamicin model by ocular dialysis. Invest. Ophthalmol. Vis. Sci. 30, 1055–1061.

Berger, U.V. and Hediger, M.A. (1999). Distribution of peptide transporter PEPT2 mRNA in the rat nervous system. Anat. Embryol. (Berl.) 199, 439–449.

Bergersen, L., Johannsson, E., Veruki, M.L., Nagelhus, E.A., Halestrap, A., Sejersted, O.M. and Ottersen, O.P. (1999). Cellular and subcellular expression of monocarboxylate transporters in the pigment epithelium and retina of the rat. Neuroscience 90, 319–331.

Bildin, V.N., Iserovich, P., Fischbarg, J. and Reinach, P.S. (2001). Differential expression of Na: K:2Cl cotransporter, glucose transporter 1, and aquaporin 1 in freshly isolated and cultured bovine corneal tissues. Exp. Biol. Med. (Maywood) 226, 919–926.

Borst, P., Evers, R., Kool, M. and Wijnholds, J. (2000). A family of drug transporters: The multidrug resistance-associated proteins. J. Natl. Cancer Inst. 92, 1295–1302.

Boyum, R. and Guidotti, G. (1997). Effect of ATP binding cassette/multidrug resistance proteins on ATP efflux of *Saccharomyces cerevisiae*. Biochem. Biophys. Res. Commun. 230, 22–26.

Buchsbaum, G., Sternklar, M., Litt, M., Grunwald, J.E. and Riva, C.E. (1984). Dynamics of an oscillating viscoelastic sphere: A model of the vitreous humor of the eye. Biorheology 21, 285–296.

Burton, P.S., Conradi, R.A. and Hilgers, R.A. (1991). Mechanism of peptide and protein absorption. 2. Transcellular mechanism of peptide and protein absorption: Passive aspects. Adv. Drug Deliv. Rev. 7, 365–386.

Canning, C.R., Greaney, M.J., Dewynne, J.N. and Fitt, A.D. (2002). Fluid flow in the anterior chamber of a human eye. IMA J. Math. Appl. Med. Biol. 19, 31–60.

Cantrill, H.L., Henry, K., Melroe, N.H., Knobloch, W.H., Ramsay, R.C. and Balfour, H.H., Jr. (1989). Treatment of cytomegalovirus retinitis with intravitreal ganciclovir. Long-term results. Ophthalmology 96, 367–374.

Chancy, C.D., Kekuda, R., Huang, W., Prasad, P.D., Kuhnel, J.M., Sirotnak, F.M., Roon, P., Ganapathy, V. and Smith, S.B. (2000). Expression and differential polarization of the reduced-folate transporter-1 and the folate receptor alpha in mammalian retinal pigment epithelium. J. Biol. Chem. 275, 20676–20684.

Chen, L., Wright, L.R., Chen, C.H., Oliver, S.F., Wender, P.A. and Mochly-Rosen, D. (2001). Molecular transporters for peptides: Delivery of a cardioprotective epsilonPKC agonist peptide into cells and intact ischemic heart using a transport system, R(7). Chem. Biol. 8, 1123–1129.

Chen, L.X., Wang, L.W. and Jacob, T. (2002). The role of MDR1 gene in volume-activated chloride currents in pigmented ciliary epithelial cells. Sheng Li Xue Bao 54, 1–6.

Christensen, H.N. (1990). Role of amino acid transport and countertransport in nutrition and metabolism. Physiol. Rev. 70, 43–77.

Connor, T.A. and Knox, R.J. (1995). Prodrugs in cancer chemotherapy. Stem Cells 13, 501–511.

Costabile, B.S. (1998). Treatment of cytomegalovirus retinitis with intraocular implants. AORN J. 67, 356–360, 362–355, 368; quiz 369–372.

Cundy, K.C., Lynch, G., Shaw, J.P., Hitchcock, M.J. and Lee, W.A. (1996). Distribution and metabolism of intravitreal cidofovir and cyclic HPMPC in rabbits. Curr. Eye Res. 15, 569–576.

Cunha-Vaz, J.G. (1976). The blood-retinal barriers. Doc. Ophthalmol. 41, 287–327.

Daniel, H. and Adibi, S.A. (1993). Transport of beta-lactam antibiotics in kidney brush border membrane. Determinants of their affinity for the oligopeptide/H+ symporter. J. Clin. Invest. 92, 2215–2223.

Dantzig, A.H. and Bergin, L. (1990). Uptake of the cephalosporin, cephalexin, by a dipeptide transport carrier in the human intestinal cell line, Caco-2. Biochim. Biophys. Acta 1027, 211–217.

Dey, S., Patel, J., Anand, B.S., Jain-Vakkalagadda, B., Kaliki, P., Pal, D., Ganapathy, V. and Mitra, A.K. (2003). Molecular evidence and functional expression of P-Glycoprotein (MDR1) in cornea and corneal epithelial cell line. Invest. Ophthalmol. Vis. Sci. 44(7), 2909–2918.

Diamond, J.G. (1981). Intraocular management of endophthalmitis. A systematic approach. Arch. Ophthalmol. 99, 96–99.

Diaz-Llopis, M., Martos, M.J., Espana, E., Cervera, M., Vila, A.O., Navea, A., Molina, F.J. and Romero, F.J. (1992). Liposomally-entrapped ganciclovir for the treatment of cytomegalovirus retinitis in AIDS patients. Eperimental toxicity and pharmacokinetics, and clinical trial. Doc. Ophthalmol. 82, 297–305.

Dieck, S.T., Heuer, H., Ehrchen, J., Otto, C. and Bauer, K. (1999). The peptide transporter PepT2 is expressed in rat brain and mediates the accumulation of the fluorescent dipeptide derivative beta-Ala-Lys-Nepsilon-AMCA in astrocytes. Glia 25, 10–20.

Doige, C.A. and Ames, G.F. (1993). ATP-dependent transport systems in bacteria and humans: Relevance to cystic fibrosis and multidrug resistance. Annu. Rev. Microbiol. 47, 291–319.

Doring, F., Martini, C., Walter, J. and Daniel, H. (2002). Importance of a small N-terminal region in mammalian peptide transporters for substrate affinity and function. J. Membr. Biol. 186, 55–62.

Einmahl, S., Savoldelli, M., D' Hermies, F., Tabatabay, C., Gurny, R. and Behar-Cohen, F. (2002). Evaluation of a novel biomaterial in the suprachoroidal space of the rabbit eye. Invest. Ophthalmol. Vis. Sci. 43, 1533–1539.

Fatt, I. (1975). Flow and diffusion in the vitreous body of the eye. Bull. Math. Biol. 37, 85–90.

Fei, Y.J., Liu, W., Prasad, P.D., Kekuda, R., Oblak, T.G., Ganapathy, V. and Leibach, F.H. (1997). Identification of the histidyl residue obligatory for the catalytic activity of the human H+/peptide cotransporters PEPT1 and PEPT2. Biochemistry 36, 452–460.

Fleisher, D., Stewart, B.H. and Amidon, G.L. (1985). Design of prodrugs for improved gastrointestinal absorption by intestinal enzyme targeting. Methods Enzymol. 112, 360–381.

Frank, J.A., Dwyer, A.J., Girton, M., Knop, R.H., Sank, V.J., Gansow, O.A., Magerstadt, M., Brechbiel, M. and Doppman, J.L. (1986). Opening of blood-ocular barrier demonstrated by contrast-enhanced MR imaging. J. Comput. Assist. Tomogr. 10, 912–916.

Fukasawa, Y., Segawa, H., Kim, J.Y., Chairoungdua, A., Kim, D.K., Matsuo, H., Cha, S.H., Endou, H. and Kanai, Y. (2000). Identification and characterization of a Na(+)-independent neutral amino acid transporter that associates with the 4F2 heavy chain and exhibits substrate selectivity for small neutral D- and L-amino acids. J. Biol. Chem. 275, 9690–9698.

Ganapathy, M.E., Brandsch, M., Prasad, P.D., Ganapathy, V. and Leibach, F.H. (1995). Differential recognition of beta-lactam antibiotics by intestinal and renal peptide transporters, PEPT 1 and PEPT 2. J. Biol. Chem. 270, 25672–25677.

Gelman, R.A. and Blackwell, J. (1974). Collagen-mucopolysaccharide interactions at acid pH. Biochim. Biophys. Acta 342, 254–261.

Gerhart, D.Z., Leino, R.L. and Drewes, L.R. (1999). Distribution of monocarboxylate transporters MCT1 and MCT2 in rat retina. Neuroscience 92, 367–375.

Geroski, D.H. and Edelhauser, H.F. (2001). Transscleral drug delivery for posterior segment disease. Adv. Drug Deliv. Rev. 52, 37–48.

Goffeau, A., Park, J., Paulsen, I.T., Jonniaux, J.L., Dinh, T., Mordant, P. and Saier, M.H., Jr. (1997). Multidrug-resistant transport proteins in yeast: Complete inventory and phylogenetic characterization of yeast open reading frames with the major facilitator superfamily. Yeast 13, 43–54.

Gottesman, M.M., Pastan, I. and Ambudkar, S.V. (1996). P-glycoprotein and multidrug resistance. Curr. Opin. Genet. Dev. 6, 610–617.

Greenwood, J. (1992). Characterization of a rat retinal endothelial cell culture and the expression of P-glycoprotein in brain and retinal endothelium in vitro. J. Neuroimmunol. 39, 123–132.

Groneberg, D.A., Doring, F., Theis, S., Nickolaus, M., Fischer, A. and Daniel, H. (2002). Peptide transport in the mammary gland: Expression and distribution of PEPT2 mRNA and protein. Am. J. Physiol. Endocrinol. Metab. 282, E1172–E1179.

Guo, A., Hu, P., Balimane, P.V., Leibach, F.H. and Sinko, P.J. (1999). Interactions of a nonpeptidic drug, valacyclovir, with the human intestinal peptide transporter (hPEPT1) expressed in a mammalian cell line. J. Pharmacol. Exp. Ther. 289, 448–454.

Halestrap, A.P. and Price, N.T. (1999). The proton-linked monocarboxylate transporter (MCT) family: Structure, function and regulation. Biochem. J. 343 (Pt 2), 281–299.

Hamilton, S.R., Yao, S.Y., Ingram, J.C., Hadden, D.A., Ritzel, M.W., Gallagher, M.P., Henderson, P.J., Cass, C.E., Young, J.D. and Baldwin, S.A. (2001). Subcellular distribution membrane topology of the mammalian concentrative Na$^+$-nucleoside cotransporter rCNT1. J. Biol. Chem. 276, 27981–27988.

Han, H., de Vrueh, R.L., Rhie, J.K., Covitz, K.M., Smith, P.L., Lee, C.P., Oh, D.M., Sadee, W. and Amidon, G.L. (1998). 5′-Amino acid esters of antiviral nucleosides, acyclovir, and AZT are absorbed by the intestinal PEPT1 peptide transporter. Pharm. Res. 15, 1154–1159.

Han, H.K. and Amidon, G.L. (2000). Targeted prodrug design to optimize drug delivery. AAPS PharmSci. 2, E6.

Hargreaves, K.M. and Pardridge, W.M. (1988). Neutral amino acid transport at the human blood-brain barrier. J. Biol. Chem. 263, 19392–19397.

Harris, J.D., Gutierrez, A.A., Hurst, H.C., Sikora, K. and Lemoine, N.R. (1994). Gene therapy for cancer using tumour-specific prodrug activation. Gene Ther. 1, 170–175.

Hashimoto, N., Fujioka, T., Toyoda, T., Muranushi, N. and Hirano, K. (1994). Renin inhibitor: Transport mechanism in rat small intestinal brush-border membrane vesicles. Pharm. Res. 11, 1448–1451.

Hatefi, A. and Amsden, B. (2002). Biodegradable injectable *in situ* forming drug delivery systems. J. Control Rel. 80, 9–28.

Hediger, M.A., Coady, M.J., Ikeda, T.S. and Wright, E.M. (1987). Expression cloning and cDNA sequencing of the Na$^+$/glucose co-transporter. Nature 330, 379–381.

Henderly, D.E., Freeman, W.R., Causey, D.M. and Rao, N.A. (1987). Cytomegalovirus retinitis and response to therapy with ganciclovir. Ophthalmology 94, 425–434.

Herrero-Vanrell, R. and Refojo, M.F. (2001). Biodegradable microspheres for vitreoretinal drug delivery. Adv. Drug Deliv. Rev. 52, 5–16.

Himmelstein, K.J., Guvenir, I. and Patton, T.F. (1978). Preliminary pharmacokinetic model of pilocarpine uptake and distribution in the eye. J. Pharm. Sci. 67, 603–606.

Holash, J.A. and Stewart, P.A. (1993). The relationship of astrocyte-like cells to the vessels that contribute to the blood-ocular barriers. Brain Res. 629, 218–224.

Horibe, Y., Hosoya, K., Kim, K.J. and Lee, V.H. (1998). Carrier-mediated transport of monocarboxylate drugs in the pigmented rabbit conjunctiva. Invest. Ophthalmol. Vis. Sci. 39, 1436–1443.

Hosoya, K., Kondo, T., Tomi, M., Takanaga, H., Ohtsuki, S. and Terasaki, T. (2001). MCT1-mediated transport of L-lactic acid at the inner blood-retinal barrier: A possible route for delivery of monocarboxylic acid drugs to the retina. Pharm. Res. 18, 1669–1676.

Huber, B.E., Richards, C.A. and Austin, E.A. (1994). Virus-directed enzyme/prodrug therapy (VDEPT). Selectively engineering drug sensitivity into tumors. Ann. N.Y. Acad. Sci. 716, 104–114; discussion 140–103.

Hughes, P.M., Krishnamoorthy, R. and Mitra, A.K. (1996). Vitreous disposition of two acycloguanosine antivirals in the albino and pigmented rabbit models: A novel ocular microdialysis technique. J. Ocul. Pharmacol. Ther. 12, 209–224.

Hughes, P.M. and Mitra, A.K. (1993). Effect of acylation on the ocular disposition of acyclovir. II: Corneal permeability and anti-HSV 1 activity of 2′-esters in rabbit epithelial keratitis. J. Ocul. Pharmacol. 9, 299–309.

Inui, K., Terada, T., Masuda, S. and Saito, H. (2000). Physiological and pharmacological implications of peptide transporters, PEPT1 and PEPT2. Nephrol. Dial. Transplant. 15, 11–13.

Jabs, D.A., Newman, C., De Bustros, S. and Polk, B.F. (1987). Treatment of cytomegalovirus retinitis with ganciclovir. Ophthalmology 94, 824–830.

Kalsi, G.S., Silver, H.K. and Rootman, J. (1991). Ocular pharmacokinetics of dacarbazine following subconjunctival versus intravenous administration in the rabbit. Can. J. Ophthalmol. 26, 247–251.

Kanai, Y., Segawa, H., Chairoungdua, A., Kim, J.Y., Kim, D.K., Matsuo, H., Cha, S.H. and Endou, H. (2000). Amino acid transporters: Molecular structure and physiological roles. Nephrol. Dial. Transplant. 15, 9–10.

Kang, Y.S., Terasaki, T. and Tsuji, A. (1990). Acidic drug transport in vivo through the blood-brain barrier. A role of the transport carrier for monocarboxylic acids. J. Pharmacobiodyn. 13, 158–163.

Kast, C., Canfield, V., Levenson, R. and Gros, P. (1996). Transmembrane organization of mouse P-glycoprotein determined by epitope insertion and immunofluorescence. J. Biol. Chem. 271, 9240–9248.

Kawazu, K., Yamada, K., Nakamura, M. and Ota, A. (1999). Characterization of cyclosporin A transport in cultured rabbit corneal epithelial cells: P-glycoprotein transport activity and binding to cyclophilin. Invest. Ophthalmol. Vis. Sci. 40, 1738–1744.

Kilberg, M.S., Stevens, B.R. and Novak, D.A. (1993). Recent advances in mammalian amino acid transport. Annu. Rev. Nutr. 13, 137–165.

Kim, Y.H., Kim, D.U., Han, K.H., Jung, J.Y., Sands, J.M., Knepper, M.A., Madsen, K.M. and Kim, J. (2002). Expression of urea transporters in the developing rat kidney. Am. J. Physiol. Renal. Physiol. 282, F530–F540.

Kiss, A., Farah, K., Kim, J., Garriock, R.J., Drysdale, T.A. and Hammond, J.R. (2000). Molecular cloning and functional characterization of inhibitor-sensitive (mENT1) and inhibitor-resistant (mENT2) equilibrative nucleoside transporters from mouse brain. Biochem. J. 352 (Pt 2), 363–372.

Kramer, W., Dechent, C., Girbig, F., Gutjahr, U. and Neubauer, H. (1990a). Intestinal uptake of dipeptides and beta-lactam antibiotics. I. The intestinal uptake system for dipeptides and beta-lactam antibiotics is not part of a brush border membrane peptidase. Biochim. Biophys. Acta 1030, 41–49.

Kramer, W., Durckheimer, W., Girbig, F., Gutjahr, U., Leipe, I. and Oekonomopulos, R. (1990b). Influence of amino acid side-chain modification on the uptake system for beta-lactam antibiotics and dipeptides from rabbit small intestine. Biochim. Biophys. Acta 1028, 174–182.

Lalezari, J.P., Friedberg, D.N., Bissett, J., Giordano, M.F., Hardy, W.D., Drew, W.L., Hubbard, L.D., Buhles, W.C., Stempien, M.J., Georgiou, P., Jung, D.T. and Robinson, C.A. (2002). High dose oral ganciclovir treatment for cytomegalovirus retinitis. J. Clin. Virol. 24, 67–77.

Lang, J.C. (1995). Ocular drug delivery conventional ocular formulations. Adv. Drug Del. Rev. 16, 39–43.

Le Bourlais, C.A., Treupel-Acar, L., Rhodes, C.T., Sado, P.A. and Leverge, R. (1995). New ophthalmic drug delivery systems. Drug Develop. Industrial Pharmacy 21, 19–59.

Lee, B., Litt, M. and Buchsbaum, G. (1992). Rheology of the vitreous body. Part I: Viscoelasticity of human vitreous. Biorheology 29, 521–533.

Lee, B., Litt, M. and Buchsbaum, G. (1994). Rheology of the vitreous body: part 3. Concentration of electrolytes, collagen and hyaluronic acid. Biorheology 31, 339–351.

Lee, V.H. (2000). Membrane transporters. Eur. J. Pharm. Sci. 11 (suppl. 2), S41–S50.

Lee V.H.L. (Ed.) (1993). In Precorneal corneal and postcorneal factors. In: Ophthalmic Drug Delivery Systems. (Mitra, A.K., Ed.). Mercel Dekker, Inc., New York.

Liscovitch, M. and Lavie, Y. (2000). Multidrug resistance: A role for cholesterol efflux pathways? Trends Biochem. Sci. 25, 530–534.

Loo, T.W. and Clarke, D.M. (1995). P-glycoprotein. Associations between domains and between domains and molecular chaperones. J. Biol. Chem. 270, 21839–21844.

Lopez-Cortes, L.F., Pastor-Ramos, M.T., Ruiz-Valderas, R., Cordero, E., Uceda-Montanes, A., Claro-Cala, C.M. and Lucero-Munoz, M.J. (2001). Intravitreal pharmacokinetics and retinal concentrations of ganciclovir and foscarnet after intravitreal administration in rabbits. Invest. Ophthalmol. Vis. Sci. 42, 1024–1028.

Lopez-Cortes, L.F., Ruiz-Valderas, R., Lucero-Munoz, M.J., Cordero, E., Pastor-Ramos, M.T. and Marquez, J. (2000). Intravitreal, retinal, and central nervous system foscarnet concentrations after rapid intravenous administration to rabbits. Antimicrob. Agents Chemother. 44, 756–759.

Macha, S. and Mitra, A.K. (2001a). Ocular pharmacokinetics in rabbits using a novel dual probe microdialysis technique. Exp. Eye Res. 72, 289–299.

Macha, S. and Mitra, A.K. (2001b). Ocular pharmacokinetics of cephalosporins using microdialysis. J. Ocul. Pharmacol. Ther. 17, 485–498.

Macha, S. and Mitra, A.K. (2002). Ocular disposition of ganciclovir and its monoester prodrugs following intravitreal administration using microdialysis. Drug Metab. Dispos. 30, 670–675.

Marmor, M.F. (1998). Structure, Function and Disease of the Retinal Pigment Epithelium. Oxford University Press, New York.

Martin, D.F., Sierra-Madero, J., Walmsley, S., Wolitz, R.A., Macey, K., Georgiou, P., Robinson, C.A. and Stempien, M.J. (2002). A controlled trial of valganciclovir as induction therapy for cytomegalovirus retinitis. N. Engl. J. Med. 346, 1119–1126.

Mathews, M.B. (1965). The interaction of collagen and acid mucopolysaccharides. A model for connective tissue. Biochem. J. 96, 710–716.

Maurice, D. (2001). Review: Practical issues in intravitreal drug delivery. J. Ocul. Pharmacol. Ther. 17, 393–401.

Meredith, D., Temple, C.S., Guha, N., Sword, C.J., Boyd, C.A., Collier, I.D., Morgan, K.M. and Bailey, P.D. (2000). Modified amino acids and peptides as substrates for the intestinal peptide transporter PepT1. Eur. J. Biochem. 267, 3723–3728.

Merriman-Smith, B.R., Young, M.A., Jacobs, M.D., Kistler, J. and Donaldson, P.J. (2002). Molecular identification of p-glycoprotein: A role in lens circulation? Invest. Ophthalmol. Vis. Sci. 43, 3008–3015.

Millay, R.H., Klein, M.L., Shults, W.T., Dahlborg, S.A. and Neuwelt, E.A. (1986). Maculopathy associated with combination chemotherapy and osmotic opening of the blood–brain barrier. Am. J. Ophthalmol. 102, 626–632.

Munck, L.K., Grondahl, M.L., Thorboll, J.E., Skadhauge, E. and Munck, B.G. (2000). Transport of neutral, cationic and anionic amino acids by systems B, b(o,+), X(AG), and ASC in swine small intestine. Comp. Biochem. Physiol. A Mol. Integr. Physiol. 126, 527–537.

Narurkar, M.M. and Mitra, A.K. (1989). Prodrugs of 5-iodo-2′-deoxyuridine for enhanced ocular transport. Pharm. Res. 6, 887–891.

Nashed, Y.E., Anand, B.S. and Mitra, A.K. (2003). Synthesis and characterization of novel dipeptide ester prodrugs of acyclovir. Spectrochim. Acta A. Mol. Biomol. Spectrosc. 59(9), 2033–2039.

Naumann, H.N. and Memphis, M.D. (1959). Postmortem chemistry of the vitreous in man. A M A Arch. Ophthol. 62, 356–363.

Nord, E.P., Wright, S.H., Kippen, I. and Wright, E.M. (1983). Specificity of the Na^+-dependent monocarboxylic acid transport pathway in rabbit renal brush border membranes. J. Membr. Biol. 72, 213–221.

Okuda, M., Saito, H., Urakami, Y., Takano, M. and Inui, K. (1996). cDNA cloning and functional expression of a novel rat kidney organic cation transporter, OCT2. Biochem. Biophys. Res. Commun. 224, 500–507.

Pardridge, W.M. (1983). Brain metabolism: A perspective from the blood-brain barrier. Physiol. Rev. 63, 1481–1535.

Patel, D.H., Crawford, C.R., Naeve, C.W. and Belt, J.A. (2000). Cloning, genomic organization and chromosomal localization of the gene encoding the murine sodium-dependent, purine-selective, concentrative nucleoside transporter (CNT2). Gene 242, 51–58.

Paulsen, I.T., Brown, M.H. and Skurray, R.A. (1996). Proton-dependent multidrug efflux systems. Microbiol. Rev. 60, 575–608.

Pescovitz, M.D., Rabkin, J., Merion, R.M., Paya, C.V., Pirsch, J., Freeman, R.B., O' Grady, J., Robinson, C., To, Z., Wren, K., Banken, L., Buhles, W. and Brown, F. (2000). Valganciclovir results in improved oral absorption of ganciclovir in liver transplant recipients. Antimicrob. Agents Chemother. 44, 2811–2815.

Peyman, G.A., Khoobehi, B., Tawakol, M., Schulman, J.A., Mortada, H.A., Alkan, H. and Fiscella, R. (1987). Intravitreal injection of liposome-encapsulated ganciclovir in a rabbit model. Retina 7, 227–229.

Pow, D.V. (2001). Amino acids and their transporters in the retina. Neurochem. Int. 38, 463–484.

Putnam, W.S., Pan, L., Tsutsui, K., Takahashi, L. and Benet, L.Z. (2002). Comparison of bidirectional cephalexin transport across MDCK and caco-2 cell monolayers: Interactions with peptide transporters. Pharm. Res. 19, 27–33.

Reddy, D.V.N., Rosenberg, C. and Kinsey, V.E. (1961). Steady state distribution of free amino acids with in aqueous humor, vitreous body, and plasma of the rabbit. Exp. Eye Res. 1, 175–181.

Reddy, J.A. and Low, P.S. (2000). Enhanced folate receptor mediated gene therapy using a novel pH-sensitive lipid formulation. J. Control Rel. 64, 27–37.

Ritzel, M.W., Yao, S.Y., Huang, M.Y., Elliott, J.F., Cass, C.E. and Young, J.D. (1997). Molecular cloning and functional expression of cDNAs encoding a human Na^+-nucleoside cotransporter (hCNT1). Am. J. Physiol. 272, C707–C714.

Rizzolo, L.J. (1997). Polarity and the development of the outer blood-retinal barrier. Histol. Histopathol. 12, 1057–1067.

Ruetz, S. and Gros, P. (1994). Phosphatidylcholine translocase: A physiological role for the mdr2 gene. Cell 77, 1071–1081.

Saha, P., Yang, J.J. and Lee, V.H. (1998). Existence of a p-glycoprotein drug efflux pump in cultured rabbit conjunctival epithelial cells. Invest. Ophthalmol. Vis. Sci. 39, 1221–1226.

Saier, M.H., Jr. (2000). A functional-phylogenetic classification system for transmembrane solute transporters. Microbiol. Mol. Biol. Rev. 64, 354–411.

Sakurai, E., Matsuda, Y., Ozeki, H., Kunou, N., Nakajima, K. and Ogura, Y. (2001). Scleral plug of biodegradable polymers containing ganciclovir for experimental cytomegalovirus retinitis. Invest. Ophthalmol. Vis. Sci. 42, 2043–2048.

Sanitato, J.J., Asbell, P.A., Varnell, E.D., Kissling, G.E. and Kaufman, H.E. (1984). Acyclovir in the treatment of herpetic stromal disease. Am. J. Ophthalmol. 98, 537–547.

Sasaki, H., Ichikawa, M., Kawakami, S., Yamamura, K., Nishida, K. and Nakamura, J. (1996). In situ ocular absorption of tilisolol through ocular membranes in albino rabbits. J. Pharm. Sci. 85, 940–943.

Schepens, C.L. (1954). Clinical aspects of pathological changes in the vitreous body. Am. J. Ophthalmol. 38, 8–21.

Schepens, C.L. and Neetens, A.N. (1987). The Vitreous and Vitreoretinal Interface. Springer-Verlag, New York.

Schlingemann, R.O., Hofman, P., Klooster, J., Blaauwgeers, H.G., Van der Gaag, R. and Vrensen, G.F. (1998). Ciliary muscle capillaries have blood-tissue barrier characteristics. Exp. Eye Res. 66, 747–754.

Sebag, J. (1989). The Vitreous: Structure, Function and Pathobiology. Springer-Verlag, New York.

Sekine, T., Watanabe, N., Hosoyamada, M., Kanai, Y. and Endou, H. (1997). Expression cloning and characterization of a novel multispecific organic anion transporter. J. Biol. Chem. 272, 18526–18529.

Sharom, F.J. (1997). The P-glycoprotein efflux pump: How does it transport drugs? J. Membr. Biol. 160, 161–175.

Shepherd, P.R. and Kahn, B.B. (1999). Glucose transporters and insulin action—implications for insulin resistance and diabetes mellitus. N. Engl. J. Med. 341, 248–257.

Silverman, M. (1991). Structure and function of hexose transporters. Annu. Rev. Biochem. 60, 757–794.

Sinkula, A.A. and Yalkowsky, S.H. (1975). Rationale for design of biologically reversible drug derivatives: Prodrugs. J. Pharm. Sci. 64, 181–210.

Sirotnak, F.M. and Tolner, B. (1999). Carrier-mediated membrane transport of folates in mammalian cells. Annu. Rev. Nutr. 19, 91–122.

Smit, J.J., Schinkel, A.H., Oude Elferink, R.P., Groen, A.K., Wagenaar, E., van Deemter, L., Mol, C.A., Ottenhoff, R., van der Lugt, N.M. and van Roon, M.A., et al. (1993). Homozygous disruption of the murine mdr2 P-glycoprotein gene leads to a complete absence of phospholipid from bile and to liver disease. Cell 75, 451–462.

Smith, A.J., Timmermans-Hereijgers, J.L., Roelofsen, B., Wirtz, K.W., van Blitterswijk, W.J., Smit, J.J., Schinkel, A.H. and Borst, P. (1994). The human MDR3 P-glycoprotein promotes translocation of phosphatidylcholine through the plasma membrane of fibroblasts from transgenic mice. FEBS Lett. 354, 263–266.

Smith, Q.R. (1993). Drug delivery to brain and the role of carrier-mediated transport. Adv. Exp. Med. Biol. 331, 83–93.

Smith, Q.R. and Stoll, J. (Eds.) (1988). Blood–Brain Barrier Amino Acid Transport. Cambridge University Press, Cambridge.

Smith, S.B., Kekuda, R., Gu, X., Chancy, C., Conway, S.J. and Ganapathy, V. (1999). Expression of folate receptor alpha in the mammalian retinol pigmented epithelium and retina. Invest. Ophthalmol. Vis. Sci. 40, 840–848.

Smith, T.J. and Ashton, P. (1996). Sustained-release subconjunctival 5-fluorouracil. Ophthalmic. Surg. Lasers 27, 763–767.

Stella, V.J., Charman, W.N. and Naringrekar, V.H. (1985). Prodrugs. Do they have advantages in clinical practice? Drugs 29, 455–473.

Stella, V.J. and Himmelstein, K.J. (1980). Prodrugs and site-specific drug delivery. J. Med. Chem. 23, 1275–1282.

Stone, T.W. and Jaffe, G.J. (2000). Reversible bull's-eye maculopathy associated with intravitreal fomivirsen therapy for cytomegalovirus retinitis. Am. J. Ophthalmol. 130, 242–243.

Stuart, F., Bradely, S. and Cheng, Y.L. (2003). Mathematical modelling of drug distribution in the vitreous humor. In: (Mitra, A.K., Ed.), Ophthalmic Drug Delivery Systems Second Edition, Revised and Expanded (Drugs and the Pharmaceutical Sciences). Marcel Dekker, New York, NY, Vol. 130, March 2003.

Styrer, L. (1981). Biochemistry. W.H. Freeman Co, San Francisco.

Sudimack, J. and Lee, R.J. (2000). Targeted drug delivery via the folate receptor. Adv. Drug. Deliv. Rev. 41, 147–162.

Swann, D.A. (1980). Chemistry and biology of the vitreous body. Int. Rev. Exp. Pathol. 22, 1–64.

Takahashi, H., Kaminski, A.E. and Zieske, J.D. (1996). Glucose transporter 1 expression is enhanced during corneal epithelial wound repair. Exp. Eye Res. 63, 649–659.

Takahashi, H., Ohara, K., Ohmura, T., Takahashi, R. and Zieske, J.D. (2000). Glucose transporter 1 expression in corneal wound repair under high serum glucose level. Jpn. J. Ophthalmol. 44, 470–474.

Takanaga, H., Maeda, H., Yabuuchi, H., Tamai, I., Higashida, H. and Tsuji, A. (1996). Nicotinic acid transport mediated by pH-dependent anion antiporter and proton cotransporter in rabbit intestinal brush-border membrane. J. Pharm. Pharmacol. 48, 1073–1077.

Takanaga, H., Tamai, I. and Tsuji, A. (1994). pH-dependent and carrier-mediated transport of salicylic acid across Caco-2 cells. J. Pharm. Pharmacol. 46, 567–570.

Tamai, I., Sai, Y., Ono, A., Kido, Y., Yabuuchi, H., Takanaga, H., Satoh, E., Ogihara, T., Amano, O., Izeki, S. and Tsuji, A. (1999). Immunohistochemical and functional characterization of pH-dependent intestinal absorption of weak organic acids by the monocarboxylic acid transporter MCT1. J. Pharm. Pharmacol. 51, 1113–1121.

Tamai, I., Takanaga, H., Maeda, H., Ogihara, T., Yoneda, M. and Tsuji, A. (1995). Proton-cotransport of pravastatin across intestinal brush-border membrane. Pharm. Res. 12, 1727–1732.

Tamai, I. and Tsuji, A. (2000). Transporter-mediated permeation of drugs across the blood-brain barrier. J. Pharm. Sci. 89, 1371–1388.

Terada, T., Saito, H., Mukai, M. and Inui, K.I. (1996). Identification of the histidine residues involved in substrate recognition by a rat H+/peptide cotransporter, PEPT1. FEBS Lett. 394, 196–200.

Terada, T., Sawada, K., Irie, M., Saito, H., Hashimoto, Y. and Inui, K. (2000). Structural requirements for determining the substrate affinity of peptide transporters PEPT1 and PEPT2. Pflugers Arch. 440, 679–684.

Terasaki, T., Takakuwa, S., Moritani, S. and Tsuji, A. (1991). Transport of monocarboxylic acids at the blood–brain barrier: Studies with monolayers of primary cultured bovine brain capillary endothelial cells. J. Pharmacol. Exp. Ther. 258, 932–937.

Thorens, B. (1993). Facilitated glucose transporters in epithelial cells. Annu. Rev. Physiol. 55, 591–608.

Tiruppathi, C., Balkovetz, D.F., Ganapathy, V., Miyamoto, Y. and Leibach, F.H. (1988). A proton gradient, not a sodium gradient, is the driving force for active transport of lactate in rabbit intestinal brush-border membrane vesicles. Biochem. J. 256, 219–223.

Tsukaguchi, H., Tokui, T., Mackenzie, B., Berger, U.V., Chen, X.Z., Wang, Y., Brubaker, R.F. and Hediger, M.A. (1999). A family of mammalian Na$^+$-dependent L-ascorbic acid transporters. Nature 399, 70–75.

Unlu, N. and Robinson, J.R. (1998). Scleral permeability to hydrocortisone and mannitol in the albino rabbit eye. J. Ocul. Pharmacol. Ther. 14, 273–281.

Vakkalagadda, B.J., Dey, S., Pal, D. and Mitra, A.K. (2003). Identification and Functional Characterization of a Na$^+$- Independent Large Neutral Amino Acid Transporter LAT1 on Rabbit Cornea. Invest. Ophthalmol. Vis. Sci., Vol. 44(7), 2919–2927.

Veloso, A.A., Jr., Zhu, Q., Herrero-Vanrell, R. and Refojo, M.F. (1997). Ganciclovir-loaded polymer microspheres in rabbit eyes inoculated with human cytomegalovirus. Invest. Ophthalmol. Vis. Sci. 38, 665–675.

Waga, J., Nilsson-Ehle, I., Ljungberg, B., Skarin, A., Stahle, L. and Ehinger, B. (1999). Microdialysis for pharmacokinetic studies of ceftazidime in rabbit vitreous. J. Ocul. Pharmacol. Ther. 15, 455–463.

Wang, H., Dutta, B., Huang, W., Devoe, L.D., Leibach, F.H., Ganapathy, V. and Prasad, P.D. (1999). Human Na$^{(+)}$-dependent vitamin C transporter 1 (hSVCT1): Primary structure, functional characteristics and evidence for a non-functional splice variant. Biochim. Biophys. Acta 1461, 1–9.

Williams, E.F., Ezeonu, I. and Dutt, K. (1994). Nucleoside transport sites in a cultured human retinal cell line established by SV-40 T antigen gene. Curr. Eye Res. 13, 109–118.

Wilson, F.A. (1981). Intestinal transport of bile acids. Am. J. Physiol. 241, G83–G92.

Wright, E.M. (1993). The intestinal Na$^+$/glucose cotransporter. Annu. Rev. Physiol. 55, 575–589.

Wyatt, H.J. (1996). Ocular pharmacokinetics and convectional flow: Evidence from spatio-temporal analysis of mydriasis. J. Ocul. Pharmacol. Ther. 12, 441–459.

Yang, C. and Mitra, A.K. (2001a). Nasal absorption of tyrosine-linked model compounds. J. Pharm. Sci. 90, 340–347.

Yang, C., Tirucherai, G.S. and Mitra, A.K. (2001b). Prodrug based optimal drug delivery via membrane transporter/receptor. Expert Opin. Biol. Ther. 1, 159–175.

Yang, J.J., Kim, K.J. and Lee, V.H. (2000). Role of P-glycoprotein in restricting propranolol transport in cultured rabbit conjunctival epithelial cell layers. Pharm. Res. 17, 533–538.

Yasukawa, T., Kimura, H., Tabata, Y. and Ogura, Y. (2001). Biodegradable scleral plugs for vitreoretinal drug delivery. Adv. Drug Deliv. Rev. 52, 25–36.

Yoon, H., Fanelli, A., Grollman, E.F. and Philp, N.J. (1997). Identification of a unique monocarboxylate transporter (MCT3) in retinal pigment epithelium. Biochem. Biophys. Res. Commun. 234, 90–94.

Zhu, T., Chen, X.Z., Steel, A., Hediger, M.A. and Smith, D.E. (2000). Differential recognition of ACE inhibitors in *Xenopus laevis* oocytes expressing rat PEPT1 and PEPT2. Pharm. Res. 17, 526–532.

Zignani, M., Einmahl, S., Baeyens, V., Varesio, E., Veuthey, J.L., Anderson, J., Heller, J., Tabatabay, C. and Gurny, R. (2000). A poly(ortho ester) designed for combined ocular delivery of dexamethasone sodium phosphate and 5-fluorouracil: Subconjunctival tolerance and *in vitro* release. Eur. J. Pharm. Biopharm. 50, 251–255.

FURTHER READING

Field, J.A., Zhang, L., Brun, K.A., Brooks, D.P. and Edwards, R.M. (1999). Cloning and functional characterization of a sodium-dependent phosphate transporter expressed in human lung and small intestine. Biochem. Biophys. Res. Commun. 258, 578–582.

THE SCLERA

Klaus Trier

ABSTRACT

The sclera is the skeleton of the eye. It defines the size of the eye, provides a stable support for its optical elements, and is essential to the achievement of a focused retinal image. The sclera provides attachment for the extraocular

Advances in Organ Biology
Volume 10, pages 353–373.
© 2006 Elsevier Inc. All rights reserved.
ISBN: 0-444-50925-9
DOI: 10.1016/S1569-2590(05)10013-5

muscles and allows passage of vital structures such as the optic nerve, the arterial blood supply, and the venous drainage system. The overall elastic properties of the sclera neutralize short-term fluctuations of the intra-ocular pressure. More specialized functions of the sclera are the drainage of aqueous humor and the mechanical support provided for the fibers of the optic nerve during their passage through the eye wall. Drainage of aqueous humor from the anterior chamber is controlled partly by a specialized part of the sclera, the trabecular meshwork, and partly by the uveoscleral route of which the final segment involves passive transscleral fluid transport. The lamina cribrosa is the specialized part of the sclera that provides mechanical support for the optic nerve as it leaves the eye.

Disturbances in the biochemistry and biomechanical properties of the sclera can have severe consequences for the visual function by producing an eye that is not spherical, too long, too short, too rigid, or too elastic. Such disturbances can also interfere with the vascular supply of the eye, the control mechanisms of the intraocular pressure, or the resistance of the transscleral volume flow.

I. ANATOMY

There are important species differences among the vertebrates in the anato-my of the sclera. The following description relates to the human eye. The eye wall consists of the cornea and the sclera. The anterior part of the sclera is called the limbus that is a narrow, 0.5–1 mm broad band that forms the junction between the white sclera and the clear cornea. The junction between the cornea and the limbal sclera is particularly well defined internally, where it forms Schwalbe's line. The sclera constitutes the posterior four-fifths of the outer wall of the normal eye; it is approximately spherical with an outer diameter of 22 mm. Figure 1 shows the posterior aspect of the eye.

The tendons of the four rectus muscles (medial, lateral, superior, and anterior) insert 5.5–7.7 mm behind the limbus, along a circle that roughly corresponds to the ora serrata at the internal aspect of the eye. The widest diameter of the eye, the equator, is approximately 12 mm posterior to the limbus, and the posterior pole of the eye, corresponding to the center of macula, is ca. 17 mm further posterior. The thickness of the sclera varies from 0.8 mm near the limbus, 0.3 mm behind the insertion of the rectus muscles, 0.4–0.6 mm at the equator, and 1.0 mm near the optic nerve. The optic nerve passes through the sclera about 3 mm nasally to the post-erior pole of the eye. The cone-shaped opening that allows the exit of the optic nerve has an internal diameter of 1.5–2 mm and an external diameter of 3–3.5 mm. The outer two-thirds of the thickness of the sclera follows the optic nerve backwards and forms part of the dural nerve sheath, while the inner third forms the lamina cribrosa through which the axons of the nerve as well as the central retinal artery and vein pass. The lamina cribrosa,

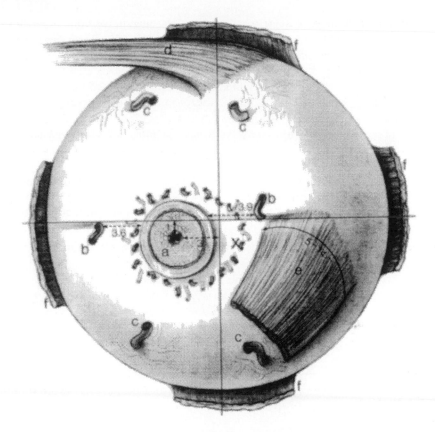

Figure 1. The posterior eye. The optic nerve with its central vessels is seen (a). Surrounding it are the short posterior ciliary arteries. The approximate position of the macula is at x. Along the horizontal meridian are the long posterior ciliary arteries (b). The exit of four vortex veins are shown (c). The cut end of the extraocular muscles are seen (d, e, f). (From Hogan, M.J., Alverado, J.A., and Weddell, J.E. (Eds.) (1971). Histology of the human eye. An atlas and textbook. W.B. Saunders, p. 51, fig. 2-3.)

being thinner and having a higher elasticity than the rest of the sclera, tends to have an outward bulge that is accentuated when the intraocular pressure is raised (Duke-Elder, 1961).

At the corneal junction, the sclera presents a thickening at the internal surface, the projection of which into the interior of the eye is known as the scleral spur. This rigid ring structure stabilizes the anterior part of the eye, which is important for the proper functioning of the corneal optics. It also provides anchoring for the ciliary muscle and via the zonular apparatus for the lens as well.

The trabecular meshwork is a sieve-like specialization of the inner, limbal sclera. It lies as a circumferential band between Schwalbe's line and the scleral spur. Externally the trabecular meshwork is lined by the canal of Schlemm, which is another specialization of the limbal sclera. The trabecular meshwork and the canal of Schlemm form the most important drainage route for the aqueous humor, and is therefore of crucial importance for the maintenance of a normal intraocular pressure.

The blood vessels that supply the inner eye must traverse the sclera. The central retinal artery and central retinal vein that supply the inner retina enter the eye through the lamina cribrosa part of the sclera. The anterior uvea is supplied by the anterior ciliary arteries that travel with the extraocular rectus muscles and traverse the sclera obliquely anterior to the insertion of these muscles. The choroid is supplied by the short and long posterior ciliary arteries, as described in Chapter 10. The vorticose veins are the main venous drainage channels of the uvea, although anteriorly, there are minor anastomitic communications with the episcleral veins. The vorticose veins, four to seven in number, pass the sclera obliquely through approximately 4-mm long tunnels, leaving the sclera about 3 mm posterior to the equator. The sclera contains no, or very few blood vessels of its own.

At its outer aspect, the sclera borders the episclera, a loose vascular connective tissue supplied by the anterior and posterior ciliary arteries. The thickness of the episclera decreases in the anteroposterior direction. At the inner aspect of the sclera, the lamina fusca constitutes a transition zone between the sclera and the ciliary body or the choroid, with collagen bundles intermingling with both structures.

II. STRUCTURE AND ULTRASTRUCTURE

The sclera consists of densely packed collagen bundles, interspersed with fibroblasts and embedded in ground substance.

The bundles, composed of parallel collagen fibrils, branch and intermingle in various planes. The collagen bundles tend to be arranged in a circular fashion around the limbus and the optic nerve, parallel to the direction of the highest tension in the sclera at these locations (Maurice, 1984), while elsewhere, the bundles form a crisscross pattern in the plane parallel to the surface of the eye (Hogan et al., 1971). The collagen bundles form waves in a relaxed state, which are straightened out when the intraocular pressure is raised. This contributes to the elasticity of the tissue (Maurice, 1984).

On the external surface of the sclera, the collagen fibrils are arranged in a reticular fashion, while the collagen fibrils of the internal surface lay in 0.5–6-μm thick bundles forming rhombic patterns (Thale et al., 1996).

The thickness of collagen fibrils in adult sclera varies between 28 and 280 nm, with increasingly thicker fibrils in the outer parts of the sclera. Subfibrils with a diameter of 10 nm and a helicoidal arrangement with a rightward direction and a 5°-inclination angle to the fibril axis have been demonstrated in bovine sclera (Yamamoto et al., 2000).

The opaque appearance of the sclera is probably caused by the variability in thickness of the collagen fibrils. In sclera from a patient with osteogenesis imperfecta, a disease characterized by translucency of the sclera, the collagen fibrils were found to be 50% narrower than in a normal control and much more uniform in size (Mietz et al., 1997).

Beside collagen fibrils, elastin fibers have been identified in human sclera (Marshall, 1995). The scattered scleral fibroblasts are elongated and lay in close apposition to the collagen bundles. The production line of the extracellular matrix components is seen on electron microscopy as rough endoplasmic reticulum and occasional secretory vacuoles (Figure 2).

Myofibroblasts, containing contractile elements like α-actin, have been demonstrated in the scleral spur (Tamm et al., 1995) and in the inner 20% of the thickness of the posterior sclera from individuals older than 17 months. The number of posterior scleral myofibroblasts increases with age (Poukens et al., 1998). Expression of genes for a number of receptor molecules has

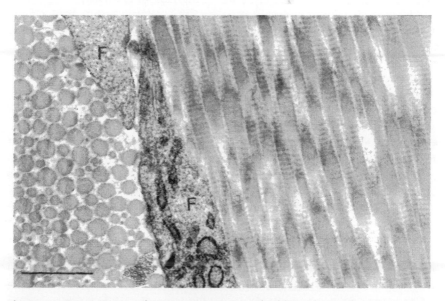

Figure 2. Transmission electron microscopy of rabbit sclera. Fibroblasts (F) are seen with rough endoplasmic reticulum. Collagen fibrils are shown cut longitudinally and transversely. (By courtesy of Takasi Kobayashi, Department of Dermatology, University of Copenhagen, Bispebjerg Hospital, Denmark.)

been detected in scleral fibroblasts, including receptors for prostaglandin and estrogen, nicotinic cholinerg, dopamine, parathyroid hormone, serotonin, and several growth factors. Gene expression is changed in response to mechanical strain (Cui et al., 2004).

III. BIOCHEMISTRY

The water content of sclera is approximately 70%. Collagen, predominantly type I, accounts for approximately 80% of the dry weight of sclera (Bailey, 1987). Collagen type III contributes with less than 5% and is possibly only found in specialized regions of the sclera like the trabecular meshwork and lamina cribrosa (Keeley et al., 1984; Rehnberg et al., 1987). All collagens are formed by polypeptide chains with a left-handed helix structure, of which three chains are organized into a right-handed triple helix. Hydroxyproline, hydroxylysine, and lysine are typical amino acids of collagen (Prockop and Kivirikko, 1995). Genes coding for the different collagen types found in the eye have been identified (Ihanamaki et al., 2004). Mutations in type I or III collagen genes cause syndromes with scleral involvement such as osteogenesis imperfecta (blue sclerae, abnormal scleral thickness and fibril structure) or Ehlers–Danlos syndrome (myopia) (Ihanamaki et al., 2004).

The limited space between the collagen fibrils is occupied by hyaluronan, a nonsulphated glycosaminoglycans or proteoglycans, which are large water-binding glycosaminoglycan molecules attached to core proteins. Glycosaminoglycans are polysaccharides with repeating disaccharide units. The main proteoglycans of human sclera are decorin with one chondroitin-dermatan sulphate side chain and biglycan with two chondroitin-dermatan sulphate side chains. Aggrecan with more than 100 chondroitin sulphate side chains and more than 30 keratan sulphate side chains have also been demonstrated. Decorin, biglycan and aggrecan represent 74%, 20%, and 6%, respectively, of the newly synthesized proteoglycans from sclera organ culture (Rada, 1997). The proteoglycan core proteins constitute around 2% of the dry weight of the sclera (Ayad et al., 1994).

The most abundant scleral proteoglycan, decorin, is a normal finding in fibrous tissue in which it is bound to specific sites on the surface of the mature collagen fibrils (Figure 3), and thought to play a role in the regulation of the diameter of the fibrils (Scott, 1993). *In vitro* experiments have shown that decorin extracted from bovine sclera is capable of inducing increased diameter of collagen fibrils, and in general, tissues rich in dermatan sulphate have thicker collagen fibrils, while tissues with a higher proportion of hyaluronan or chondroitin sulphate are associated with thinner fibrils (Kuc and Scott, 1997). Aggrecan is found in tissues exposed to compression like cartilage or compressed regions of tendons (Vogel et al., 1993).

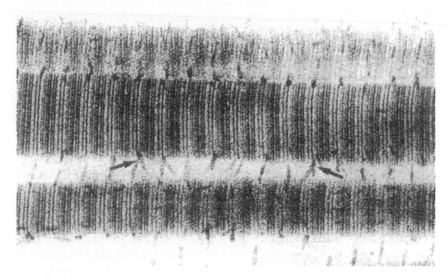

Figure 3. Proteoglycans (decorin) from rabbit sclera visualized by electron microscopy as fine filaments emanating from the d band of the collagen fibril (arrows).
(From Young, R.D. (1985). J. Cell. Sci. 74, 99, fig. 2-3, bottom.)

The glycosaminoglycan content of sclera is 0.5–1% of its dry weight
(Breen et al., 1972). Glycosaminoglycan composition of the human sclera
varies in different locations with 29–49% dermatan sulphate, 28–48% chondroitin sulphate, 2–12% heparan sulphate and 19–33% hyaluronic acid.
Keratan sulphate contributes with around 1.5% (Trier et al., 1990; Rada,
1997). The thin sclera from equator is rich in hyaluronic acid but has a low
content of dermatan sulphate and collagen, while this ratio is reversed in
sclera from around the optic disc. A high proportion of chondroitin sulphate
is seen in the thick sclera from the posterior pole, where the "cartilage
proteglycan" aggrecan is also most abundant (Trier et al., 1990, 1991; Rada
et al., 2000a).

The hydration of the sclera depends largely on its content of the negatively charged, water-binding glycosaminoglycans, but the architecture of the
collagen fibrils and the size of the interfibrillar spaces are probably equally
important. The posterior sclera has a higher content of glycosaminoglycans
(Trier et al., 1991), and a higher degree of hydration (Boubriak et al., 2003).

The synthesis of glycosaminoglycans, which is essential for the organization of the connective tissue and its biomechanical properties, is influenced
by physical factors such as tension or compression, probably transmitted
through changes in fibroblast cytoskeleton and cell shape (Evangelisti et al.,
1993). Integrin receptors on the surface of the scleral fibroblasts are believed

to mediate fibroblast communication with the extracellular matrix (McBrien and Gentle et al., 2003).

In the mature sclera, turnover of collagen is very slow with a half-life of probably many years, but some remodeling takes place, mediated by collagen degrading matrix metalloproteinases (MMPs), which have been demonstrated in adult human sclera (Gaton et al., 1999). The tissue inhibitors of metalloproteinases (TIMPs) balance the activity of MMPs. Cytokines, such as transforming growth factor-β (TGF-β), are capable of upregulating collagen and proteoglycan synthesis in mammalian scleral fibroblasts (Jobling et al., 2004).

Elastin content in sclera is around 2% and increases to 5% in the scleral spur and trabecular meshwork (Moses et al., 1978).

Glycoproteins that are important for cell adhesion, such as fibronectin and laminin, are present in the sclera; laminin is present as part of the microfibrillar sheaths of elastin fibers (Marshall, 1995; Chapman et al., 1998).

A recent study of genes expressed in human sclera has provided an extended list of extracellular matrix proteins that might be of importance for the function of sclera (Young et al., 2004).

IV.   SCLERAL BIOMECHANICS

The biochemical composition and the ultrastructural organization of the sclera endow it with characteristic biomechanical properties that are important for the functions of the eye. The biomechanical properties of the eye wall are largely determined by those of the sclera since this tissue contributes four-fifth of the entire eye wall. The ocular rigidity is the term used to quantify ocular pressure–volume relations. Friedenwald (1957) provided, now classical, empirical tables of the pressure–volume relations based on the variations in Schiotz indentation tonometer readings with different plunger load (i.e., volume displacement). However, these experiments are difficult to interpret in terms of a global ocular rigidity figure, because the eye is not spherical, the wall thickness is neither uniform nor perfectly elastic, and there are local variations in its mechanical properties (Purslow and Karwatowski, 1996). Investigation of the stress–strain relationship (load vs. extension) of isolated human scleral strips have shown that the posterior sclera, with its looser weave of collagen fibers and higher degree of hydration, has only about 60% of the stiffness of the anterior sclera (Friberg and Lace, 1988). Nevertheless, at the normal intraocular pressure of 20 mmHg, the tensile stress (S) in the sclera can be calculated to be around 10 kPa by using the formula $S = P^*R/2t$ in which P is the intraocular pressure, R the radius of the eye, and t the thickness of the eye wall (Alejandro and Amaya, 1986).

The eye has to be sufficiently rigid to retain its size and shape with intraocular pressure fluctuations above and below this normal value. On the other hand, if too rigid, the eye would not be able to neutralize potentially neuron damaging intraocular pressure spikes resulting from external pressure or variation in the production of aqueous humor. The elasticity of the sclera, is to a great extent, dependent on the wave form of the collagen fibrils that at the normal intraocular pressure are in a semirelaxed state. With hypotony of the eye, for example, following trabeculectomy with excess filtration, the posterior curvature of the eye may be flattened because of shortening of the sclera in this particular region. The result is chorioretinal folds and potentially maculopathy. This condition is more likely to develop in younger persons with myopia and relatively thin posterior sclera. Older persons seem less prone to develop hypotony maculopathy, probably because they have a more rigid sclera (Cohen et al., 1995).

Besides elasticity, the eye also displays creep, which is defined as slow, irreversible deformation under a constant load. This phenomenon is most pronounced in the immature eye, and possibly forms the basis of the emmetropization process in which the size of the eye is adjusted to fit the optics of the cornea and lens so that the retinal image is focused when the growth of the eye cedes around the age of 14. The thicknesses of the sclera and the collagen fibrils gradually increase, and the cross-linking of the collagen becomes more pronounced during postnatal development. Hence, the potential for creep decreases in the first postnatal years, and the capacity of the eye to undergo enlargement in response to raised intraocular pressure ceases at the age of 3 years. Experimentally, the creep of scleral strips can be drastically accelerated by treatment with the collagen degrading enzyme collagenase (Shchukin et al., 1997).

V. PERMEABILITY OF THE SCLERA

The sclera is not supplied by blood vessels of its own, and it must, therefore, be permeable to fluids and metabolites in order to receive nutrition from the neighboring choroid and episclera. Transscleral transport of fluid from the suprachoroidal space to the episclera is dependent on the thickness of the sclera and the content and organization of the collagen fibrils, proteoglycans, and glycosaminoglycans. Nanophthalmic and highly hypermetropic eyes, which have an abnormally thick sclera with disorganized collagen fiber bundles and deposits of proteoglycans in the matrix, are associated with the uveal effusion syndrome in which there is an accumulation of fluid in the suprachoroidal space, and choroidal detachments and exudative retinal detachment can occur (Uyama et al., 2000).

Transscleral fluid movement is probably an important component of the uveoscleral pathway, which accounts for up to 40% of the aqueous outflow in young individuals but less with increasing age. Recent interest in possible transscleral delivery of specific antiangiogenic or neuroprotective agent to the choroid and retina has prompted studies on the diffusion properties of large molecules (Geroski and Edelhauser, 2001; Weinreb, 2001). Sclera is permeable to large molecules, such as IgG or dextrans with molecular weight up to 150 kDa, and it appears that molecular radius plays a greater role in determining scleral permeability than molecular weight or charge (Ambati et al., 2000). The posterior sclera is more permeable than the anterior, possibly because of the higher hydration in this region (Boubriak et al., 2003).

Scleral permeability is greatly reduced when the intraocular pressure is raised, possibly because of compression of the tissue with reduction of the spaces between the collagen fibers and extracellular matrix molecules that define the pathways for diffusion (Rudnick et al., 1999).

VI. THE TRABECULAR MESHWORK AND THE LAMINA CRIBROSA

In the trabecular meshwork and lamina cribrosa, the sclera is modified to carry out specialized functions. The two structures share a trabecular nature and some biochemical features. The lamina cribrosa can be divided into an anterior part rich in neural supportive astrocytes and a posterior part rich in extracellular matrix. As shown in Figure 4, the collagen fibrils form pores around the penetrating axons by their circular arrangements in interlacing figures of eight (Thale et al., 1996). The sizes of these pores have a characteristic distribution with smaller pores in the center and in the nasal/temporal quadrant and larger pores in the superior/inferior quadrants. Thus the superior/inferior quadrants have less mechanical support from extracellular matrix, and this may be the reason why these quadrants are most vulnerable to axonal damage in eyes with intraocular hypertension (Birch et al., 1997). The lamina cribrosa is distinguished from the sclera proper by a ring of collagenous fibril bundles (Thale et al., 1996).

The average diameter of collagen fibrils of the lamina cribrosa is much smaller than those of peripapillary or equatorial sclera (Quigley et al., 1991). Elastin fibers together with collagen, predominantly type III, make up the core of the cribriform plates with a covering of the basement membrane specific collagen type IV and laminin (Hernandez et al., 1987; Rehnberg et al., 1987). The astrocytes separate the axons from the supporting connective tissue.

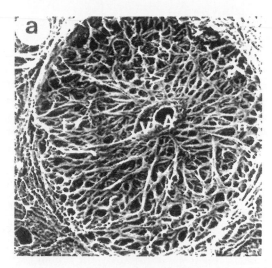

Figure 4. Scanning electron microscopy of collagenous fabirc of the human lamina cribrosa with the retinal artery (A) and vein (V) in the center. Circular arrangements of the collagen fibrils surround the openings for the passage of axons. (From Thale, A., Tillmann, B., and Rochels, R. (1996). Ophthalmologica 210, 144, fig. 1a.)

With aging, there is a decrease in lamina cribrosa collagen type III relative to collagen type I, an increase in total collagen content, and an increase in cross-linking of collagen and elastin. The elastin content increases from 7% to 28%, and the collagen content from 24% to 45% of dry tissue weight in the age interval 0–89 years. Concomitantly, there is a reduction in total glycosaminoglycan content from 3% to 1%. There is a steady decline in the number of astrocytes from birth to senescence (Albon et al., 2000a). The age-related biochemical changes manifest themselves in a much stiffer and less resilient structure after 40–50 years of age. This may increase the stress on the axons from the intraocular pressure (Albon et al., 2000b).

Outflow of aqueous humor from the anterior chamber takes place through some 20,000 pores with a diameter of 0.1–3 μm in the trabecular meshwork. The primary site of outflow resistance is located near the inner wall of the canal of Schlemm (Bill, 1993). The trabecular meshwork consists of beams of extracellular matrix surrounded by a basal lamina and lined by endothelial-like trabecular meshwork cells. Collagen, predominantly of the glycoprotein type IV and VI, have been localized in the trabecular meshwork with contributions from type I, III, and V (Rehnberg et al., 1987; Yue, 1996). Hyaluronan is represented in all layers of the trabecular meshwork but most apparent in association with the trabecular meshwork cells. It is possible that hyaluronan fills the intratrabecular spaces and prevents the fusion of the trabecular lamellae, which gives it a key role in the regulation of the outflow resistance (Lerner et al., 1997). Hyaluronan is present in the aqueous humor (Laurent, 1983a) and may become trapped in the trabecular meshwork.

The changes in the trabecular meshwork with aging consist of reduced density of trabecular cells and accumulation of extracellular matrix with thickening of the trabecular beams, a process that has been coined "hyalinization." The consequent narrowing of the open spaces contributes to the increase in resistance to aqueous outflow and may lead to elevated intraocular pressure. Experimental work points at accumulation of collagen type IV and other glycoproteins as the cause of the thickening of the trabecular beams (Tripathi et al., 1997). Furthermore, the synthesis of hyaluronan by aged trabecular meshwork cells is reduced (Schachtschabel and Binninger, 1990).

VII. EMBRYOLOGY AND DEVELOPMENT

Scleral development seems to be induced by the retinal pigment epithelium and the choroid (Duke-Elder, 1963). The origins of the sclera can be traced to the neural crest with a small contribution from mesoderm (Johnston et al., 1979).

In the human, embryo differentiation of the sclera proceeds backwards from the limbus and inside outwards with increasing condensation of the tissue and cytological maturation. The diameter of the collagen fibrils increases three times from week 6.4 to week 24 (Sellheyer and Spitznas, 1988).

The characteristic transscleral fibril gradient with increasingly thicker collagen fibrils in the outer part of the sclera is not present at birth but develops postnatally, perhaps as a response to differential circumferential stresses across the scleral thickness (McBrien et al., 2001). The sclera in the neonate is somewhat translucent and appears blue because of the underlying uvea. During the first decade of life, it gradually becomes opaque due to increasing thickness and increasingly variable diameter of the collagen fibrils.

Normally, the axial length increases from 17 mm at birth to 24 mm at 14 years after which the dimensions of the eye do not change. From an initial hypermetropia the refraction gradually changes toward emmetropia, the process being modified by a flattening of the corneal curvature and the lens so that there is usually no resultant myopia in spite of the increasing axial length. This process is called emmetropization (Lawrence and Azar, 2002). It has been hypothesized that a growth-control mechanism exists that can either accelerate or slow down axial growth in response to defocusing of the retinal image during the period of normal eye growth. It is possible that this mechanism works by regulating the scleral extensibility through complex changes in the synthesis and degradation of the extracellular matrix. This regulation may involve tissue factors such as metalloproteinases

(Guggenheim and McBrien, 1996) and transforming growth factor-β (Jobling et al., 2004). An increased scleral extensibility will lead to an expansion of the eye through forces exerted by the intraocular pressure (McBrien et al., 2003).

VIII.   AGING OF THE SCLERA

Deposition of calcium salts in a grayish, translucent plaque anterior to the insertion of the medial and lateral rectus muscles is often seen in elderly people. With further calcification, these plaques may become opaque. In addition, there is diffuse deposition of calcium and fat globules among the scleral bundles giving the aged sclera a yellowish hue (Spencer, 1985).

With age there is a steady increase in the ocular rigidity (Friberg and Lace, 1988), probably due to increased cross-linking of the scleral collagen. Figure 5 shows the relationship between age and scleral rigidity.

The collagen in immature sclera is initially stabilized by divalent keto and aldimine cross-links. In the mature sclera, these early cross-links are spontaneously replaced by more stable trivalent cross-links. These cross-linking mechanisms are dependent on the copper-dependent enzyme lysyl oxidase, and inhibition of this enzyme in copper deficiency and lathyrism leads to universal fragility of collagenous tissue. The reduced metabolic turnover of collagen allows another type of cross-links to accumulate through a reaction with glucose and its oxidation products, the so-called glycation, which is

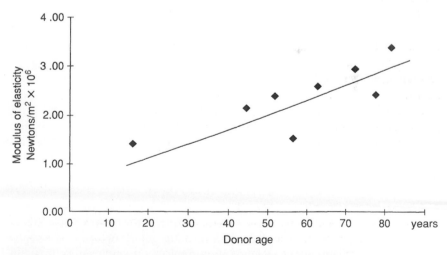

Figure 5. Increase in the stiffness of human sclera as a function of age. (From Friberg, T.R. and Lace, J.W. (1988). Exp. Eye Res. 47, 433, fig. 4.)

nonenzymatic and purely adventitious. Glycation of collagen plays a major role in the pathogenesis of aging with dysfunction of vulnerable tissue, such as the renal basement membrane, the cardiovascular system, and the retinal capillaries. Glycation also accounts for much of the increased morbidity and mortality seen in patients with diabetes. Glycation reduces the flexibility and permeability of the tissues and reduces turnover (Baily et al., 1998).

The hydration of both anterior and posterior sclera decreases by about 1% per decade (Boubriak et al., 2003). Aging is associated with some loss of scleral glycosaminoglycans, in particular, dermatan sulphate. In the anterior sclera, there are also reduced amounts of the small proteglycans decorin and biglycan with age. On the other hand, the large proteoglycan aggrecan is retained in aging posterior sclera (Brown et al., 1994; Rada et al., 2000).

The reduced hydration of posterior sclera with increasing age in spite of little change in the content of water-binding glycosaminoglycans may be explained by the increased cross-linking of collagen making the tissue more compact. The reduced water content is in keeping with the reduced transscleral diffusion of water-soluble compounds found in aging sclera (Boubriak et al., 2003). Experimental cross-linking of collagen in rabbit sclera leads to a reduced scleral permeability for water-soluble compounds. This suggests that cross-linking of collagen is at least in part responsible for the reduced permeability seen in aging sclera (Boubriak et al., 2000).

IX.   THE ROLE OF SCLERA IN DISORDERS OF THE EYE

The sclera itself may be the seat of disease, and alterations in the biomechanical properties of the sclera may cause dysfunction of other ocular tissues.

Scleritis with focal scleral swelling and inflammation is often associated with systemic connective tissue disease, such as rheumatoid arthritis, systemic lupus erythematosus, polyarteritis nodosa, relapsing polychondritis, and Wegeners granulomatosis, the common background being an autoimmune reaction toward components like proteoglycans. Induction of MMPs and other proteolytic enzymes by various cytokines secreted by fibroblasts and infiltrating inflammatory cells may lead to necrotizing scleritis and perforation of the eye globe (Girolamo et al., 1997).

An interesting property of the sclera is its ability to function as a barrier against the spread of malignant melanoma arising in the choroid. Apparently the tumor needs to induce degradation of the sclera before it is able to extent into it. These changes include loosening of the collagen bundles by MMPs, increased hydration and accumulation of glycosaminoglycans

(Alyahya et al., 2003). Decorin that is well-represented in normal sclera binds tumor promoting growth factors, such as TGF-β (Yamaguchi et al., 1990), and inhibits cell proliferation by upregulating p21, an inhibitor of cyclin dependant kinases (De Luca et al., 1996), while increased expression of chondroitin sulphate with a high capacity for water binding is seen in cancer tissue (Olsen et al., 1988).

In myopic children, an excessive growth of the axial length is seen from age 8–12 years until 20 years after which the axial length and the degree of myopia usually remain stable. The axial length at this stage will most often be in the range 24–28 mm. Rarely the myopia presents itself at an earlier age, with congenital, localized weak areas at the posterior pole of the eye (scleral ectasies). In these cases the axial length is prone to increase throughout the life, and the myopia can reach extreme levels, up to 40 diopters, with axial lengths of 30–35 mm.

Both in humans and in animals with experimental myopia, abnormalities in the biochemistry and ultrastructure of the sclera have been found. The diameter of the collagen fibrils in myopic sclera is reduced compared to normal eyes. In addition the contents of collagen and glycosaminoglycans are reduced (Avettisov et al., 1984). It is known that deprivation myopia in mammals is accompanied by a decrease in the scleral content of proteoglycans and collagen (Norton and Rada, 1995; Rada et al., 2000b) with reverse changes taking place during recovery (McBrien et al., 2000). In mammalian models of high myopia, an increased number of small diameter collagen fibrils were found in the longer term, consistent with findings in myopic humans (Funata and Tokoro, 1990; McBrien et al., 2001).

Primary open angle glaucoma is most commonly seen among elderly people and is characterized by increased intraocular pressure with large fluctuations, a reduced outflow facility at the trabecular meshwork, progressive loss of axons at the optic nerve head, and distortion of the lamina cribrosa.

A decrease in collagen density and curling of the elastin fibers is seen in glaucomatous lamina cribrosa and may relate to the decrease in mechanical compliance, which causes the irreversible cupping of the optic disc. The peripapillary sclera also exhibits a loss of collagen in glaucoma (Quigley et al., 1991). The lamina cribrosa from glaucoma patients has lost the orderly circular arrangement of the collagen fibers in exchange of coarse bundles with no preferred alignment (Thale et al., 1996). Experimental work has indicated that the remodeling of the lamina cribrosa with accumulation of nonfunctional elastin (elastosis) is a specific response of the astrocytes to mechanical stress. This effect may be mediated by increased release of cytokines like TGF-β2 (Pena et al., 2001).

The optic nerve head receives much of its blood supply from the short ciliary arteries, piercing the peripapillary sclera. Increased scleral rigidity in this region caused by loss of collagen may reduce the ability of the vessels to dilate and thus interfere with the autoregulation of the arterial blood-flow. Trabecular meshwork from patients with glaucoma also shows increased deposition of elastin (Umihira et al., 1994), and in parallel with findings in the optic nerve head the levels of TGF-β2 in aqueous humor are increased. TGF-β2 increases the synthesis of collagen and glycoprotein in bovine trabecular meshwork but inhibits the synthesis of hyaluronan (Cao et al., 2002).

The total glycosaminoglycan content of trabecular meshwork is unchanged in glaucoma, but the composition is markedly changed with depletion of hyaluronan and accumulation of chondroitin sulphate. This trend is also seen in normal, aged trabecular meshwork and may relate to an increase in the resistance of aqueous outflow. Glycocorticoids cause elevated intraocular pressure in some individuals. In animals with glycocorticoid, induced intraocular hypertension, there is a reduction in trabecular meshwork hyaluronan and an increase in chondroitin sulphate, similar to what is found in primary open angle glaucoma (Knepper et al., 1996). Glycocorticoids reduce the concentration of hyaluronan in aqueous humor in rabbits (Laurent, 1983b).

Trabecular meshwork cells express various MMPs and TIMPs, and disturbances in the balance of these may also play a part in the excessive accumulation of extracellular matrix seen in glaucoma (Pang et al., 2003).

Age-related macular degeneration (AMD) has been related to increased ocular rigidity, and this observation has led to the establishment of the so-called hemodynamic model of the pathogenesis of AMD. According to the model, the stiff sclera causes impairment of the choroidal blood flow, leading to either atrophic retinal lesions or choroidal neovascular membranes (Friedman, 2004). Results from population studies suggesting a mildly increased risk of AMD with increasing degree of hypermetropia may lend some support to this model. Compression of the vorticose veins passing obliquely through the thick and stiff hypermetropic sclera may cause obstruction of the choroidal venous outflow (Ikram et al., 2003). Interestingly, an association between elastosis in both sun-exposed and sun-protected skin, and neovascular AMD has been found, suggesting a generalized susceptibility of elastic fibers to degenerative stimuli (Blumenkrantz et al., 1986).

X. PHARMACEUTICAL MODULATION OF THE SCLERA

Latanoprost, a synthetic prostaglandin analog, reduces the intraocular pressure by enhancing uveoscleral outflow. In monkeys treated for 5 days with prostaglandin F2 isopropyl ester, collagen types I and III in the sclera adjacent to the ciliary body were reduced by 43% and 45%, respectively (Sagara et al., 1999).

In vitro exposure to prostaglandin increases the permeability of human sclera, probably due to increased release of MMPs and remodeling of the extracellular matrix (Kim et al., 2001). Recent data indicate that topical latanoprost penetrates as far as to the posterior pole of the rabbit eye, where it induces biochemical changes in the sclera (Trier and Ribel-Madsen, 2004).

Aspirin-like compounds, and vitamin C and E prevent nonenzymatic cross-linking (glycation) of collagen *in vitro* (Malik and Meek, 1996), and may be useful in an effort to retard the progressive stiffening of the sclera with aging.

Induction of cross-links has been proposed as possible treatment in order to arrest the progression of myopia (Wollensak and Spoerl, 2004).

Systemic methylxanthines increase the concentration of collagen and the diameter of collagens in rabbit sclera and may thus work against axial elongation in childhood myopia. The effect is possibly exerted by way of receptors in the retina or retinal pigment epithelium and transmitted to the sclera (Trier et al., 1999).

Receptors for estrogen are expressed in the scleral fibroblasts. Estradiol and the soy isoflavone genistein enhances the production of hyaluronan in human skin cell culture (Miyazaki et al., 2002) and possibly also in ocular tissues. Loss of hyaluronan from the trabecular meshwork may play a role in the reduction of aqueous humor outflow in aged people, and hyaluronan-stimulating compounds like genistein or vitamin C (Schachtschabel and Binninger, 1993) could therefore be of interest in respect of preventing intraocular hypertension.

REFERENCES

Albon, J., Karwatowski, W.S.S., Easty, D.L., Sims, T.J. and Duance, V.C. (2000a). Age related changes in the non-collagenous components of the extracellular matrix of the human lamina cribrosa. Br. J. Ophthalmol. 84, 311–317.

Albon, J., Purslow, P.P., Karwatowski, W.S.S. and Easty, D.L. (2000b). Age related compliance of the lamina cribrosa in human eyes. Br. J. Ophthalmol. 84, 318–323.

Alyahya, G.A., Ribel-Madsen, S.M., Heegaard, S., Prause, J.U. and Trier, K. (2003). Melanoma-associated spongiform scleropathy: Biochemical changes and possible relation to tumour extension. Acta Ophthalmol. 81, 625–629.

Alejandro, A. and Amaya, L.E. (1986). Mechanical behaviour of the sclera. Ophthalmologica Basel 193, 45–55.

Ambati, J., Canakis, S.C., Miller, J.W., Gragoudas, E.S., Edwards, A., Weisssgold, D.J., Kim, I., Delori, F.C. and Adamis, A.P. (2000). Diffusion of high molecular weight compounds through sclera. Invest. Ophthalmol. Vis. Sci. 41, 1181–1185.

Avettisov, E.S., Savitskaya, N.F. Vinetskaya, M.L., et al. (1984). A study of biochemical and biomechanical properties of normal and myopic eye sclera in humans of different age groups. Metab. Pediatr. Syst. Ophthalmol. 7, 183–188.

Ayad, S., Boot Handford, R., Humphries, M.J., Kadler, K.E. and Shuttleworth, A. (1994). The Extracellular Matrix. Facts Book, 86–87. Academic Press, London, UK.

Bailey, A.J. (1987). Structure, function and ageing of the collagens of the eye. Eye 1, 175–183.

Baily, A.J., Paul, R.G. and Knott, L. (1998). Mechanisms of maturation and ageing of collagen. Mech. Age. Dev. 106, 1–56.

Bill, A. (1993). Some aspects of human humour drainage. Eye 7, 14–19.

Birch, M., Brotchie, D., Roberts, N. and Grierson, I. (1997). The three-dimensional structure of the connective tissue in the lamina cribrosa of the human optic nerve head. Ophthalmologica 211, 183–191.

Blumenkrantz, M.S., Russell, S.R., Robey, M.G., Kott-Blumenkrantz, R. and Penneys, N. (1986). Risk factors in age-related maculopathy complicated by choroidal neovascularization. Ophthalmology 93, 552–557.

Boubriak, O.A., Urban, J.P.G., Akhtar, S., Meek, K.M. and Bron, A.J. (2000). The effect of hydration and matrix composition on solute diffusion in rabbit sclera. Exp. Eye Res. 71, 503–514.

Boubriak, O.A., Urban, J.P.G. and Bron, A.J. (2003). Differential effects of aging on transport properties of anterior and posterior human sclera. Exp. Eye Res. 76, 701–713.

Breen, M., Johnson, R.L., Sittig, R.A., Weinstein, H.G. and Veis, A. (1972). The acidic glycosaminoglycans in human fetal development and adult life: Cornea, sclera and skin. Connective Tissue Res. 1, 291–303.

Brown, C.T., Vural, M., Johnson, M. and Trikaus-Randall, V. (1994). Age-related changes of the scleral hydration and sulphated glycosaminoglycans. Mech. Age Dev. 77, 97–107.

Cao, Y., Wei, H., Zhang, Y., Da, B. and Lu, Y. (2002). Effect of transforming growth factor-beta(2) on the extracellular matrix synthesis in bovine trabecular meshwork cells. Chung Hua Yen Ko Tsa Chih 38, 429–432.

Chapman, S.A., Ayad, S., O' Donoghue, E. and Bonshek, R.E. (1998). Glycoproteins of trabecular meshwork, cornea and sclera. Eye 12, 440–448.

Cohen, S.M., Flynn, H.W., Palmberg, P.F., Gass, J.D.M., Grajewski, A.L. and Parrish, R.K. (1995). Treatment of hypotony maculopathy after trabeculectomy. Ophthalmic Surg. Lasers 26, 435–441.

Cui, W., Bryant, M.R., Sweet, P.M. and McDonnell, P.J. (2004). Changes in gene expression in response to mechanical strain in human scleral fibroblasts. Exp. Eye Res. 78, 275–284.

De Luca, A., Santra, M., Baldi, A., Giodano, A. and Iozzo, R.V. (1996). Decorin-induced growth suppression is associated with up-regulation of p21, an inhibitor of cyclin-dependant kinases. J. Biol. Chem. 271, 18961–18965.

Duke-Elder, S. (1961). System of Ophthalmology, Vol. II, p. 84. Kimpton, London.

Duke-Elder, S. (1963). System of Ophthalmology, Vol. III, p. 162. Kimpton, London.

Evangelisti, R., Becchetti, E., Locci, P., Bodo, M., Arena, N., Lilli, C., De Mattei, M. and Carinci, P. (1993). Coordinate effects of concanavalin A on cytoskeletal organization, cell shape, glycosaminoglycan accumulation and exoglycosidase activity in chick embryonic cultured fibroblasts. Eur. J. Histochem. 37, 161–172.

Friberg, T.R. and Lace, J.W. (1988). A comparison of the elastic properties of human choroids and sclera. Exp. Eye Res. 47, 429–436.

Friedenwald, J.S. (1957). Tonometer calibration. Trans. Am. Acad. Ophthal. Otolaryng. 61, 108–123.

Friedman, E. (2004). Update of the vascular model of AMD. Br. J. Ophthalmol. 88, 161–163.

Funata, M. and Tokoro, T. (1990). Scleral changes in experimentally myopic monkeys. Graefe's Arch. Clin. Exp. Ophthalmol. 228, 174–179.

Gaton, D.D., Sagara, T., Lindsey, J.D. and Weinreb, R.N. (1999). Matrix metalloproteinase-1 localization in the normal human uveoscleral outflow pathway. Invest. Ophthalmol. Vis. Sci. 40, 363–369.

Geroski, D.H. and Edelhauser, H.F. (2001). Transscleral drug delivery for posterior segment disease. Adv. Drug Deliv. Res. 52, 37–48.

Girolamo, N.D., Lloyd, A., McCluskey, P., Filipic, M. and Wakefield, D. (1997). Increased expression of matrix metalloproteinases *in vivo* in scleritis tissue and *in vitro* in cultured human scleral fibroblasts. Am. J. Pathol. 150, 653–666.

Guggenheim, J.A. and McBrien, N.A. (1996). Form-deprivation myopia induces activation of scleral matrix metalloproteinase-2 in tree shrew. Invest. Ophthalmol. Vis. Sci. 37, 1380–1395.

Hernandez, M.R., Luo, X.X., Igoe, F. and Neufeld, A.H. (1987). Extracellular matrix of the human lamina cribrosa. Am. J. Ophthalmol. 104, 567–576.

Hogan, M.J., Alvarado, J.A. and Weddell, J.E. (1971). Histology of the Human Eye. 283–201. WB Saunders, Philadelphia.

Ihanamaki, T., Pelliniemi, L.J. and Vuori, E. (2004). Collagens and collagen-related matrix components in the human and mouse eye. Prog. Retin. Eye Res. 23, 403–434.

Ikram, M.K., van Leeuwen, R., Vingerling, J.R., Hofman, A. and de Jong, T.V.M. (2003). Relationship between refraction and prevalent as well as incident age-related maculopathy: The Rotterdam study. Invest. Ophthalmol. Vis. Sci. 44, 3778–3782.

Jobling, A., Nguyen, M., Gentle, A. and McBrien, N.A. (2004). Isoform-specific changes in scleral TGF-β expression and the regulation of collagen synthesis during myopia progression. J. Biol. Chem. 279, 18121–18126.

Johnston, M.C., Noden, D.M., Hazelton, R.D., Coulombre, J.L. and Coulombre, A.J. (1979). Origins of avian ocular and periocular tissue. Exp. Eye Res. 29, 27–43.

Keeley, F.W., Morin, J.D. and Vesely, S. (1984). Characterization of collagen from normal human sclera. Exp. Eye Res. 39, 533–542.

Kim, J., Lindsey, J.D., Wang, N. and Weinreb, R.N. (2001). Increased Human Scleral Permeability with prostaglandin exposure. Invest. Ophthalmol. Vis. Sci. 42, 1514–1521.

Knepper, P.A., Goossens, W., Hvizd, M. and Palmberg, P.F. (1996). Glycosaminoglycans of the human trabecular meshwork in primary open-angle glaucoma. Invest. Ophthalmol. Vis. Sci. 37, 1360–1367.

Kuc, I.M. and Scott, P.G. (1997). Increased diameter of collagen fibrils precipitated *in vitro* in the presence of decorin from various connective tissues. Connective Tissue Res. 36, 287–296.

Laurent, U.B.G. (1983a). Hyaluronate in human aqueous humor. Arch. Ophthalmol. 101, 129–183.

Laurent, U.B.G. (1983b). Reduction of the hyaluronate concentration in rabbit aqueous humour by topical prednisolone. Acta Ophthalmol. 61, 751–755.

Lawrence, M.S. and Azar, D.T. (2002). Myopia and models and mechanisms of refractive error control. Ophthalmol. Clin. North Am. 15, 127–133.

Lerner, L.E., Polansky, J.R., Howes, E.L. and Stern, R. (1997). Hyaluronan in the trabecular meshwork. Invest. Ophthalmol. Vis. Sci. 38, 1222–1228.

Malik, N.S. and Meek, K.M. (1996). Vitamins and analgesics in the prevention of collagen ageing. Age Ageing 25, 279–284.

Marshall, G.E. (1995). Human scleral elastic system: an immunoelectron microscopic study. Br. J. Ophthalmol. 79, 57–64.

Maurice, D.M. (1984). The cornea and the sclera. In: The Eye. (Davson, H., Ed.), Vol. 1B, pp. 32–33. Academic press, New York.

McBrien, N.A. and Gentle, A. (2003). Role of the sclera in the development and pathological complications of myopia. Prog. Ret. Eye Res. 22, 307–338.

McBrien, N.A., Cornell, L.M. and Gentle, A. (2001). Structural and ultrastructural changes to the sclera in a mammalian model of high myopia. Invest. Ophthalmol. Vis. Sci. 42, 2179–2187.

McBrien, N.A., Lawlor, P. and Gentle, A. (2000). Scleral remodeling in the development and recovery from axial myopia in the tree shrew. Invest. Ophthalmol. Vis. Sci. 41, 3713–3719.

Mietz, H., Kasner, L. and Green, W.R. (1997). Histopathological and electron-microscopic features of corneal and scleral collagen fibers in osteogenesis imperfecta type III. Graefes Arch. Clin. Exp. Ophthalmol. 235, 405–410.

Miyazaki, K., Hanamizu, T., Iizuka, R. and Chiba, K. (2002). Genistein and daidzein stimulates hyaluronic acid production in human keratinocyte culture and hairless mouse skin. Skin Pharmacol. Appl. Skin Physiol. 15, 175–183.

Moses, R.A., Grodzki, W.J., Jr, Starcher, B.C. and Galione, M.J. (1978). Elastin content of the scleral spur, trabecular mesh, and sclera. Invest. Ophthalmol. Vis. Sci. 17, 817–818.

Norton, T.T. and Rada, J.A. (1995). Reduced extracellular matrix in mammalian sclera with induced myopia. Vision Res. 35, 1271–1281.

Olsen, E.B., Trier, K., Eldov, K. and Ammitzbøll, T. (1988). Glycosaminoglycans in human breast cancer. Acta Obstet. Gynecol. Scand. 67, 539–542.

Pang, I., Hellberg, P.E., Fleenor, D.L., Jacobson, N. and Clark, A.F. (2003). Expression of matrix metalloproteinases and their inhibitors in human trabecular meshwork cells. Invest. Ophthalmol. Vis. Sci. 44, 3485–3493.

Pena, J.D.O., Agapova, O., Gabelt, B.T., Levin, L.A., Lucarelli, M.J., Kaufman, P.L. and Hernandez, M.R. (2001). Increased elastin expression in astrocytes of the lamina cribrosa in response to elevated intraocular pressure. Invest. Ophthalmol. Vis. Sci. 42, 2303–2314.

Poukens, V., Glascow, B.J. and Demer, J.L. (1998). Nonvascular contractile cells in sclera and choroids of humans and monkeys. Invest. Ophthalmol. Vis. Sci. 39, 1765–1774.

Prockop, D.J. and Kivirikko, K.I. (1995). Collagens: Molecular biology, diseases, and potentials for therapy. Annu. Rev. Biochem. 64, 403–434.

Purslow, P.P. and Karwatowski, W.S.S. (1996). Ocular elasticity. Ophthalmology 103, 1686–1692.

Quigley, H.A., Dorman-Pease, M.E. and Brown, A.E. (1991). Quantitative study of collagen and elastin of the optic nervehead and sclera in human and experimental monkey glaucoma. Curr. Eye Res. 10, 877–888.

Rada, J.A. (1997). Proteoglycans of the human sclera. Invest. Ophthalmol. Vis. Sci. 38, 1740–1751.

Rada, J.A., Achen, V.R., Penugonda, S., Schmidt, R.W. and Mount, B.A. (2000a). Proteoglycan composition in the human sclera during growth and aging. Invest. Ophthalmol. Vis. Sci. 41, 1639–1648.

Rada, J.A., Nickla, D.L. and Troilo, D. (2000b). Decreased proteoglycan synthesis associated with form deprivation myopia in mature primate eyes. Invest. Ophthalmol. Vis. Sci. 41, 2050–2058.

Rehnberg, M., Ammitzbøll, T. and Tengroth, B. (1987). Collagen distribution in the lamina cribrosa and the trabecular meshwork of the human eye. Br. J. Ophthalmol. 71, 886–892.

Rudnick, D.E., Noonan, J.S., Geroski, D.H., Prausnitz, M.R. and Edelhauser, H.F. (1999). The effect of intraocular pressure on human and rabbit scleral permeability. Invest. Ophthalmol. Vis. Sci. 40, 3054–3058.

Sagara, T., Gaton, D.D., Lindsey, J.D., Gabelt, B.T., Kaufmann, P.L. and Weinreb, R.N. (1999). Topical prostaglandin F2 treatment reduces collagen types I, III, and IV in the monkey uveoscleral outflow pathway. Arch. Ophthalmol. 117, 794–801.

Schachtschabel, D.O. and Binninger, E. (1990). Aging of trabecular meshwork cells of the human eye in vitro. Z. Gerontol. 23, 133–135.

Schachtschabel, D.O. and Binninger, E. (1993). Stimulatory effects of ascorbic acid on hyaluronic acid synthesis of in vitro cultured normal and glaucomatous trabecular meshwork cells of the human eye. Z. Gerontol. 26, 243–246.

Scott, J.E. (1993). Proteoglycan-fibrillar collagen interaction in tissues: Dermatan sulphate proteoglycan as a tissue organizer. In: Dermatan Sulphate Proteoglycans. (Scott, J.E., Ed.), pp. 165–181. Portland Press.

Sellheyer, K. and Spitznas, M. (1988). Development of the human sclera. A morphological study. Graefes Arch. Clin. Exp. Ophthalmol. 226, 89–100.

Shchukin, E.D., Izmailova, V.N., Krasnov, M.M., Gurov, A.N., Bessonov, A.I. and Afanaséva, G.N. (1997). Effect of the active medium on the creep of the eye sclera. Colloid J. 59, 409–411.

Spencer, W.H. (1985). Ophthalmic Pathology, p. 392. WB Saunders.

Tamm, E.R., Kock, T.A., Mayer, B., Stefani, F.H. and Lutjen-Drecoll, E. (1995). Innervation of myofibroblast-like scleral spur cells in human and monkey eyes. Invest. Ophthalmol. Vis. Sci. 36, 1633–1644.

Thale, A., Tillmann, B. and Rochels, R. (1996). Scanning electron-microscopic studies of the collagen architecture of the human sclera—normal and pathological findings. Ophthalmologica 210, 137–141.

Trier, K., Olsen, E.B. and Ammitzbøll, T. (1990). Regional glycosaminoglycan composition of the human sclera. Acta Ophthalmol. 68, 304–306.

Trier, K., Olsen, E.B. and Ammitzbøll, T. (1991). Collagen and uronic acid distribution in the human sclera. Acta Ophthalmol. 69, 99–101.

Trier, K., Olsen, E.B., Kobayashi, T. and Ribel-Madsen, S.M. (1999). Biochemical and ultrastructural changes in rabbit sclera after treatment with 7-methylxanthine, theobromine, acetazolamide, or l-ornithine. Br. J. Ophthalmol. 83, 1370–1375.

Trier, K. and Ribel-Madsen, S.M. (2004). Latanoprost eye-drops increase concentration of glycosaminoglycans in posterior rabbit sclera. J. Ocular Pharmacol. Ther. 20, 185–189.

Tripathi, B.J., Tinghui, L.I., Li, J., Tran, L. and Tripathi, R.C. (1997). Age-related changes in trabecular cells *in vitro*. Exp. Eye Res. 64, 57–66.

Umihira, J., Nagata, S., Nohara, M., Hanai, T., Usuda, N. and Segawa, K. (1994). Localization of elastin in the normal and glaucomatous human trabecular meshwork. Invest. Ophthalmol. Vis. Sci. 35, 486–494.

Uyama, M., Takahashi, K., Kozaki, J., Tagami, N., Takada, Y., Ohkuma, H., Matsunaga, H., Kimoto, T. and Mishimura, T. (2000). Uveal effusion syndrome: Clinical features, surgical treatment, histological examination of the sclera, and pathophysiology. Ophthalmology 107, 441–449.

Vogel, K.G., Ordog, A., Pogany, G. and Olah, J. (1993). Proteoglycans in the compressed region of human tibialis posterior tendon and ligaments. J. Orthop. Res. 11, 68–77.

Weinreb, R.N. (2001). Enhancement of scleral macromolecular permeability with prostaglandins. Trans. Am. Ophth. Soc. 99, 319–343.

Wollensak, G. and Spoerl, E. (2004). Collagen crosslinking of human and porcine sclera. J. Cat. Refrac. Surg. 30, 689–695.

Yamaguchi, Y., Mann, D.M. and Rouslati, E. (1990). Negative regulation of TGF-β by the proteoglycan decorin. Nature (Lond.) 346, 281–284.

Yamamoto, S., Hashizume, H., Hitomi, J., Shigeno, M., Sawaguchi, S., Abe, H. and Ushiki, T. (2000). The subfibrillar arrangement of corneal and scleral collagen fibrils as revealed by scanning electron and atomic force microscopy. Arch. Histol. Cytol. 63, 127–135.

Young, T.L., Scavello, G.S., Paluru, P.C., Choi, J.D., Rappaport, E.F. and Rada, J.A. (2004). Microarray analysis of gene expression in human donor sclera. Molecular Vis. 10, 163–176.

Yue, B.Y.J.T. (1996). The extracellular matrix and its modulation in the trabecular meshwork. Survey Ophthalmol. 40, 379–390.

INDEX

Printed and bound by CPI Group (UK) Ltd, Croydon, CR0 4YY

14/10/2024

01773629-0001